ELECTRONIC INVENTIONS
AND
DISCOVERIES

*Electronics from its earliest beginnings
to the present day*

3rd REVISED AND EXPANDED EDITION

by

G. W. A. DUMMER
M.B.E., C.Eng., F.I.E.E., F.I.E.E.E., F.I.E.R.E.

(former Supt. Applied Physics, Royal Signals and Radar Establishment)

PERGAMON PRESS
OXFORD · NEW YORK · TORONTO · SYDNEY · PARIS · FRANKFURT

U.K.	Pergamon Press Ltd., Headington Hill Hall, Oxford OX3 0BW, England
U.S.A.	Pergamon Press Inc., Maxwell House, Fairview Park, Elmsford, New York 10523, U.S.A.
CANADA	Pergamon Press Canada Ltd., Suite 104, 150 Consumers Road, Willowdale, Ontario M2J 1P9, Canada
AUSTRALIA	Pergamon Press (Aust.) Pty. Ltd., P.O. Box 544, Potts Point, N.S.W. 2011, Australia
FRANCE	Pergamon Press SARL, 24 rue des Ecoles, 75240 Paris, Cedex 05, France
FEDERAL REPUBLIC OF GERMANY	Pergamon Press GmbH, Hammerweg 6, D-6242 Kronberg-Taunus, Federal Republic of Germany

First edition 1977 published under the title
Electronic Inventions 1745-1976
Second edition 1978 (*Electronic Inventions and Discoveries*)
Third revised edition 1983

Library of Congress Cataloging in Publication Data
Dummer, G. W. A. (Geoffrey William Arnold)
Electronic inventions & discoveries.
(Pergamon international library of science, technology, engineering, and social studies)
Includes index.
1. Electronics—History. I. Title. II. Title:
Electronic inventions and discoveries. III. Series.
TK7809.D85 1983 621.381'09 83-2393

British Library Cataloguing in Publication Data
Dummer, G.W.A.
Electronic inventions & discoveries—3rd revised
and expanded ed.
1. Electronic apparatus and applications—History
2. Inventions—History
I. Title
621.381'09'03 TK7870

ISBN 0-08-029354-9 (Hardcover)
ISBN 0-08-029353-0 (Flexicover)

In order to make this volume available as economically and as rapidly as possible the typescript has been reproduced in its original form. This method unfortunately has its typographical limitations but it is hoped that they in no way distract the reader.

Printed in Great Britain by A. Wheaton & Co. Ltd., Exeter

PERGAMON INTERNATIONAL LIBRARY
of Science, Technology, Engineering and Social Studies
The 1000-volume original paperback library in aid of education,
industrial training and the enjoyment of leisure
Publisher: Robert Maxwell, M.C.

ELECTRONIC INVENTIONS
AND
DISCOVERIES

Electronics from its earliest beginnings
to the present day

3rd REVISED AND EXPANDED EDITION

TO BE
DISPOSED
BY
AUTHORITY

THE PERGAMON TEXTBOOK
INSPECTION COPY SERVICE

An inspection copy of any book published in the Pergamon International Library will gladly be sent to academic staff without obligation for their consideration for course adoption or recommendation. Copies may be retained for a period of 60 days from receipt and returned if not suitable. When a particular title is adopted or recommended for adoption for class use and the recommendation results in a sale of 12 or more copies, the inspection copy may be retained with our compliments. The Publishers will be pleased to receive suggestions for revised editions and new titles to be published in this important International Library.

Other Pergamon Titles of Interest

DEBENHAM	Microprocessors: Principles & Applications
ERA	The Engineering of Microprocessor Systems
GANDHI	Microwave Engineering & Applications
GUILE & PATERSON	Electrical Power Systems Volume 1, 2nd Edition
	Electrical Power Systems Volume 2, 2nd Edition
HINDMARSH	Electrical Machines & their Applications, 3rd Edition
	Worked Examples in Electrical Machines & Drives
MALLER & NAIDU	Advances in High Voltage Insulation & Arc Interruption in SF_6 & Vacuum
MURPHY	Power Semiconductor Control of Motors & Drives
RODDY	Introduction to Microelectronics, 2nd Edition
SMITH	Analysis of Electrical Machines
WAIT	Wave Propagation Theory
YORKE	Electric Circuit Theory

Related Journals

(Free Specimen Copy Gladly Sent on Request)

Computers & Electrical Engineering

Electric Technology USSR

Microelectronics & Reliability

Solid State Electronics

Preface

As in the first and second editions, it is not intended that this book should be a learned treatise on a particular aspect of historical electronics, but rather a summary of first dates in electronic developments over a very wide field, both for interest and for ready reference.

Because no one person can be an authority in all fields of electronics, the data given is extracted from a wide variety of published sources, i.e. books, patents, technical journals, proceedings of Societies, etc., to whom full acknowledgement is made. This present work covers inventions from Europe, U.S.A., and Japan. Obviously a survey such as this cannot be completely accurate because of — in some cases — the passage of time, and in others, conflicting claims, but gives the opinions of those knowledgeable in their fields. In addition to the summaries of well-known inventions, some little-known discoveries are included which may, one day, become important.

In this edition, for the first time, an attempt has been made to trace the development of electronics from its earliest beginnings up to the present day. An attempt has also been made to cover the whole field of electronics in a concise but comprehensive form. This section describes in six new chapters, developments in audio and sound reproduction; in radio; telecommunications; radar; television; computers; robotics; information technology; industrial; educational; automobile and medical electronics.

How does one define an electronic invention? Bearing in mind the average electronic engineer's interest in his own particular field, the first person to initiate or develop a new technique concerned with electronics has been included in this edition and the selection made on the basis of simple language and explanation.

The process of invention has changed from the individual inventor to that of the large research laboratories which have the advantage of funds and cross-fertilisation of ideas. Certainly the Bell Laboratories team in the U.S.A. made the greatest contributions to semiconductor technology, not only by inventing a working transistor, but by producing materials (Si, Ge) of a purity previously unknown. This work, basic to microelectronics, has created entirely new industries. The complexity of modern electronics has also brought together chemists, physicists, mathematicians, engineers, and others as the fields of development widen. Research, development, and production are now more closely integrated.

In the 1980s it is difficult to see more really fundamental inventions: there will be faster and faster computers, more and more advanced communications, more satellite TV, and great advances in the applications of electronics, such as in the medical and biological fields.

Looking back at the history of electronics, there seem to be three fundamental inventions on which most others depend. They are: first, Faraday's discovery of electro-magnetics, from which the dynamo was developed to generate electricity (imagine a world without electricity today!); second, Lee de Forest's thermonic tube, opening up the fields of communications and computers; and, thirdly, the Bell Laboratories transistor, because the modern "chip" in fact, consists of multiple transistors.

In the production of electronics, two inventions stand out as enabling devices to be mass produced at reasonable cost: the printed circuit with dip soldering and the planar and photo-masking techniques for microelectronics "chip" production.

In preparing this book, one major impression has emerged, instanced by Chapters 4 to 9 -how deep the penetration of electronics has become into every part of modern life, whilst the 500 inventions described in this book, together with over 1000 additional references, form a background to electronics progress which, in ever-increasing tempo, is now changing the world we live in.

G. W. A. Dummer
27 King Edwards Road, Malvern Wells,
Worcs. WR14 4AJ, England

Acknowledgements

In the section on inventions, my task in this book has been that of a compiler rather than that of an author and this has only been made possible because of the co-operation of so many authorities. In particular, the IEEE has been most helpful, as the 50th Anniversary edition of the *Proc. IRE* provided considerable information on early electronics. Thanks are also due to the IERE and the IEE for permission to quote extracts from their publications.

Many books and technical journals have provided extracts which are relevant and the author is indebted to all those detailed in the "SOURCE" following each extract. Where "SOURCE" is quoted, the words and opinions are exactly those of the authors of the extracts. The page number given in each case is that of the extract and not that of the title page. Full acknowledgement is made to all authors quoted. Thanks are due to many books and journals for their permission to quote from their publications and also to the Patent Office and to Libraries for their help. Extracts from *Science at War* are used with the permission of the Controller of Her Majesty's Stationery Office.

The author would like to record his appreciation of the help given by the Science Museum, London; in particular, Mr. W. K. E. Geddes, Dr. Denys Vaughan and Dr. B. P. Bowers, and also the following for their advice and assistance on the development of electronics in the fields described: S. W. Amos, W. Bardsley, P. J. Baxandall, C. den Brinker, T. A. Everist, J. Guest, C. Hilsum, H. G. Manfield, T. P. McLean, I. L. Powell, E. H. Putley, N. Walter and P. L. Waters.

Thanks are due to Professor Dr. Jun-ichi Nishizawa, Research Institute of Electrical Communication, Tohoku University, for data on Japanese inventions.

It is hoped that the data patiently collected for this book will be found useful, both as a review of electronics development from its earliest beginnings to the present day and as a source of reference on inventions.

Contents

Chapter 1   The Beginning of Electronics   1

Chapter 2   The Development of Components, Transistors and
             Integrated Circuits   3

Chapter 3   The Expansion of Electronics and its Effect on
             Industry   12

Chapter 4   A Concise History of Audio and Sound Reproduction   15

Chapter 5   A Concise History of Radio and Telecommunications   19

Chapter 6   A Concise History of Radar   22

Chapter 7   A Concise History of Television   27

Chapter 8   A Concise History of Computers, Robotics and
             Information Technology   29

Chapter 9   A Concise History of Industrial and Other
             Applications   35

Chapter 10   List of Inventions by Subject   38

Chapter 11   A Concise Description of Each Invention in Date
             Order.   48

Chapter 12   List of Inventors   218

Chapter 13   List of Books on Inventions and Inventors   222

             Index   227

Chapter 1
The Beginning of Electronics

For hundreds of years, two phenomena have existed - static electricity and magnetism. These remained unexplained until the early 1700s when many practical experiments commenced on both electrostatic and magnetism. By the early 1800s, work by Galvani, Oersted and Faraday on galvanism, electromagnetism and electromagnetic induction opened up a new field of experimental work which ultimately paved the way to present-day electronics.

Electrostatics

Static electricity had been known for many centuries as some substances, when rubbed together, produced static charges which could generate sparks and, in other cases, could attract small pieces of paper and other materials. The Greeks knew that friction on amber material by fur gave rise to these attractive forces and the Greek word for amber was "electron", although the word "electron" was not really used until after 1897 when J. J. Thomson discovered the electron as we know it today. "Electronics" became popular after the invention of the three-electrode valve in 1906 and has been widely used since the 1920s.

Static electricity was studied by the first of many experimenters around the end of the 1700s and one of the earliest methods of measurement of this effect was the gold-leaf electroscope, which consisted of two strips of gold leaf which moved apart when a charge was applied to it. When a rod of ebonite was rubbed with a piece of fur, the ebonite would have a negative charge and the fur a positive charge. Glass rubbed with silk exhibited a similar phenomenum. Many ingenious methods of generating static electricity were developed. Faraday, in his early experiments, showed the distribution of static charges in hollow conductors. Many attempts were made to collect the charges continuously. The Kelvin replenisher was developed as a rotary device to build up the charges, but the most important device of this time was the Wimshurst machine, built later in 1882. The problem of storing the energy was solved by the first capacitor — the Leyden jar — invented in 1745. Having produced static charges and calculated the potential voltages available (these could be quite high — Wimshurst machines were used for working x-ray tubes), measurement was now becoming important and electro-meters of various types based on the earlier gold-leaf electroscope were developed, resulting in the first electrostatic voltmeters. Electrostatics could now be generated and stored for short periods, but could not be further used and attention was focused on the other phenomena — magnetism.

Magnetism

Magnetism has also been known for centuries. It was exhibited in lodestone, found in the vicinity of Magnetia in Asia Minor and termed "magnetite". It has the property of attracting fragments of iron and when a bar of the material was suspended by its centre from a thread of silk, it aligned itself north and south. It was found that when stroked along a piece of steel, the steel also became magnetised and a knitting needle magnetised in this way became a magnet and aligned itself north and south when suspended becoming the basis of the first compass.

About 1780, Galvani of Italy began experiments on animal electricity and when performing experiments on nervous excitability in frogs, he saw that violent muscle contractions could be observed if the lumbar nerves of the frogs were touched with metal instruments carrying electrical charges.

The problem of storage was still unsolved. The Wimshurst machine could generate but not store electricity and the Leyden jar was limited in its storage capcity, but in 1800 Volta invented the electric battery. A "Volta's pile" consisted of copper and zinc discs separated by a moistened cloth electrolyte. The pile was later improved to consist of paper discs, tin one side, manganese dioxide on the other, stacked to produce 0.75 volt and between 1.0 diameter discs. This was soon followed by the first accumulator or rechargeable battery in 1803 by Ritter in Germany. The time was now ripe for the integration of electricity and magnetism and, in 1820, Oersted in Denmark reported the discovery of electromagnetism and led him to develop the Galvanometer, allowing accurate measurements of currents and voltages to be made, and from this our present range of ammeters and voltmeters was developed.

In 1831 the most important discovery was made by Faraday of electromagnetic induction. He wound an iron ring with two coils, one connected to a battery, the other to a galvanometer. On connecting and reconnecting the battery, a reading was obtained on the galvanometer, although there was no direct connection. The first application of this discovery was the dynamo. By causing a coil of wire (an armature) to rotate in a magnetic field so as to cut the lines of magnetic force, an "induced" current was produced in the coil. The current changed in direction as the coil turned through two right angles and an alternating current was produced. Direct current could be produced by using a commutator to reverse one half of the alternating current. The generation of electric power now became possible. An electric motor is similar in construction but the current is passed through the armature, the forces generated causing it to rotate.

The early 1800s was a time of great progress in invention. Infra-red and ultra-violet radiation was discovered and, in 1808, Dalton put forward his atomic theory that all chemical elements were composed of minute particles of matter called atoms. Thermoelectricity, electrolysis and the photovoltaic effect were all discovered before 1840. Work on low-pressure discharge tubes, glow discharges, new types of battery and the early microphone took place in the next 20 years.

It would be true to say that the majority of basic physical phenomena were discovered in the 75 years between 1800 and 1875, culminating in the practical applications of the telephone, phonograph, microphones and loudspeakers. Towards the end of the century, wireless telegraphy, magnetic recording and the cathode-ray oscillograph were all developed.

In 1911 Rutherford proposed the general model of the atom consisting of a nucleus of protons and neutrons, about which electrons rotated in orbits. In 1913 Bohr proposed that various stable orbits corresponded to various permissible energy levels.

The early 1900s also saw the beginnings of many present-day electronic technologies. The three-electrode valve opened the way to radio broadcasting and Campbell-Swinton put forward his theory of television.

The advent of the 1914/1918 war changed the pace of development and "electronics" now covered a wider field of applications. New radio tubes and new circuits were developed for communications and after the war, radio astronomy, xerography, early radar, and computer techniques were all ready to be further developed during the 1939/1945 war. Under the pressure of this second war, radar and computer work led to a great increase in electronics research and both governments and private industry set up large laboratories. From these came MASERS, Solar Batteries and, in particular, in the 1950s, methods of perfecting ultra-pure materials such as germanium and silicon. The stage was now set for the next major advance in electronics — the transistor, invented by the Bell Laboratories in 1948, enabling all electronics equipment to be miniaturised. The planar process invented in 1959 enabled many transistors to be manufactured simultaneously and the integrated circuit (known as the "chip") was born.

Chapter 2
The Development of Components, Transistors and Integrated Circuits

All electronic equipment are composed of components — resistors, capacitors, tubes, transistors, integrated circuits, etc. The development of such components is the story of electronic itself as, with each new invention in electronic techniques, components had to be developed and manufactured in quantity.

In the early history of components, after the invention of the three-electrode tube in 1906, radio telegraphy became commercially viable. Special tubes, transmitters and receivers were designed and built with the designer of the equipment constructing all the necessary component parts.

The 1914-1918 war gave a marked impetus to the development of radio communications and with the advent of the three-electrode tube in quantity, components such as resistors and capacitors began to assume the form roughly as we knew them up to the 1960s.

The BBC commenced programme broadcasting on 14th November 1922 and from that time, up to about 1930, many component manufacturers began to specialize in individual components, from which the home constructor used to make radio receivers. Tubes were made initially by electric-lamp manufacturers as the techniques of glass blowing and vacuum processes were similar. A typical bright-emitter three-electrode tube of the period is shown in Fig. 1. Bright emitter tubes, in rows lit many an enthusiastic amateur's living room and when these were followed by dull emitters, some of the early magic seemed to disappear. Amateur constructors may remember the pungent smell of ebonite drilled at too high a speed although, with the introduction of the screened-grid tube in 1924, a metal chassis rapidly replaced ebonite panels. The stages in construction of radio sets from the breadboard to the screened chassis are shown in Fig. 2.

Figure 1 Early bright-emitter three-electrode valve
(courtesy Mullard Radio Valve Co., Ltd).

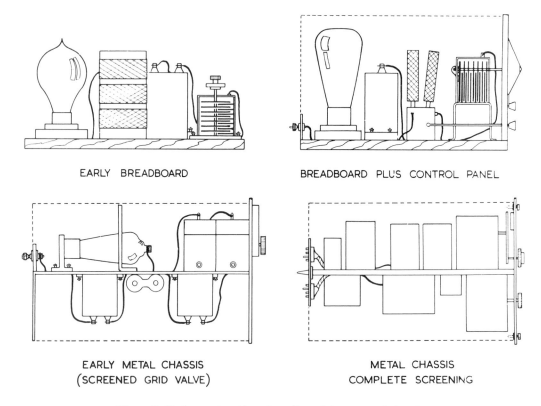

EARLY   BREADBOARD   BREADBOARD PLUS CONTROL PANEL

EARLY   METAL CHASSIS   METAL CHASSIS
(SCREENED GRID VALVE)   COMPLETE SCREENING

Figure 2 Early constructions-breadboard to screened chassis.

It might be considered that this period (the early 1920s) saw the birth of the components industry. Resistors were produced in large quantities and used as grid leaks, anode loads, etc., and consisted of carbon compositions of many kinds compressed into tubular containers and fitted with end caps. Paper-dielectric capacitors were mainly tubular types enclosed in plain bakelized cardboard tubes, with bitumen or similar material sealing the ends. Bakelite enclosed stacked-mica capacitors, fitted with screw terminals and with the bottom of the case sealed with bitumen, were also in common use. Rectangular mental-cased and plastic-cased types were also used. Electrolytic capacitors were mainly wet types in tubular mental cases. Cracked-carbon film-type resistors were introduced from Germany in about 1928 and by 1934 were being manufactured in quantity in the United Kingdom.

Figure 3 shows a front and rear view of the tuner and detector-amplifier circuits of a four-tube receiver built in 1923. Point to point wiring was used between the components. Square section wires were sometimes used with sharp right-angle bends to make all the wiring horizontal or vertical, reminiscent of the wiring patterns of some modern multi-layer printed wiring boards. This soon gave way to round wires and the home constructor grew remarkably adept at wiring up simple radio receivers.

From 1930 onwards the home construction of sets diminished and many component manufacturers and radio-set makers worked together. Techniques for component manufacture in quantities improved, and many millions of radio sets were in use throughout the world in 1939. The standard to which components were made were those of domestic radio. Pan-climatic protection was unnecessary and the self-compensating action of the radio tube made wide tolerances and poor stability generally acceptable. Apart from certain electrical engineering applications, telephone companies, a few sections of the instrument industry and the Military, no very high standard was required of the component manufacturer.

The advent of the 1939-1945 war had a tremendous effect on components because now operation of equipments in all climates of the world was essential. The spread of the war from Europe to the Far East meant that equipment had to be designed to withstand tropical climates, whereas the war in Russia required equipment to operate in arctic conditions whilst the North African desert war exposed equipment to excessive heat.

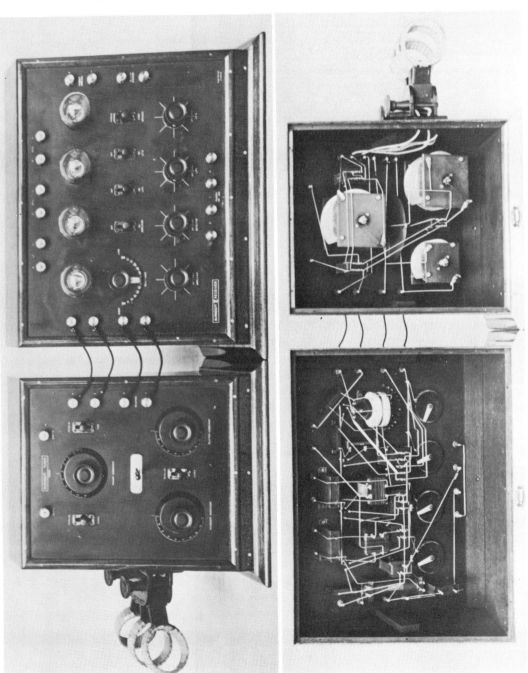

Figure 3 Four-valve receiver built in 1923 (courtesy Science Museum, London & Burndept Ltd).

Particularly damaging was the effect of tropical conditions due to the high humidity which rusted metals, lowered insulation resistance of plastics, grew fungus and swelled some moving parts, making them useless.

Directly due to war requirements many changes and developments were necessary to produce equipment to meet these arduous conditions. These changes are briefly listed below:

Standardization — increased production of few types.
Miniaturization — needed for mobile sets, for aircraft, submarines, etc.
Reliability — component failures too costly and disastrous.
Maintainability — quick replacement needed, often unskilled personnel.
Transport hazards — transport shocks and rough handling.
Mechanical shocks — impact of shells, parachute landing, etc.
Vibration — in aircraft, tanks and ships affecting components.
Storage — long periods before use, e.g. missiles.
Low temperature — use in arctic conditions.
High temperature — use in desert conditions.
Humidity — use in tropical conditions.
High altitude — high-flying aircraft (arc-over, etc.).
Combined environments — humidity/high temperature, high temperature/vibration, etc.
High powers — to increase range radio and radar.
Radiation resistance — to withstand nuclear environment.

These stringent environmental and operational requirements directly due to the 1939-1945 war resulted in major improvements to component design and manufacture. Waxed tubular paper-dielectric capacitors were replaced by metal-cased tubular types with rubber end seals. Metalized paper-dielectric capacitors were developed. Improved control of the temperature coefficient of ceramic-dielectric capacitors was introduced by manufacturers and new high-permittivity ceramic mixes developed. Many manufacturing improvements were introduced into resistor production and testing. Work on sealed variable resistors was sponsored and many types were made available. Transformers operating at higher temperatures, oil-filled and sealed in metal cans, were developed and also resin "potted" transformers. Sealed relays and indicating meters were also designed and produced to withstand the tropical conditions.

Following the war, the commercial success of component companies was concerned with the mass production of components for television receivers and radio sets. Military requirements dropped to a smaller proportion of the total components output but the lessons learned were valuable in improving the standard and reliability of commercial components.

It was this early post-war period which, in the opinion of the author, saw the beginning of the present American supremacy in technology. During the war many British scientists went to America to give details of their work so that production could commence in the U.S.A., free from bombing and war damage. Early radar experience, atomic energy research, the magnetron, advanced circuitry, Gee navigation, the proximity fused shell, radar training devices and many other technical developments were given to the U.S.A. Vigorous and enthusiastic action was taken on these and other projects by American scientists which vastly increased American technical knowledge.

Later, the Pacific war gave the U.S.A. experience in the value of electronics and built the foundations of a vast technology. In the U.K. and Europe a war-weariness prevented exploitation of electronic applications and it was probably not realized how important electronics was to become. It is significant that, up to around 1945, the majority of all electronics inventions were from Europe, whereas since 1945 the majority of inventions are from the U.S.A. The effect of the brain-drain from Europe to the U.S.A. is shown in Fig. 4, culminating in the large numbers of inventions in the 1960s.

After the war, around 1946, the printed circuit was beginning to be used in conjunction with subminiature tubes and many experimental circuits were made using both additive and subtractive techniques.

The crystal detector, invented in 1906, forerunner of the modern transistor, was used for many years until the invention of the three-electrode tube. The physics of crystal detectors was not understood until some fundamental research was done by physicists in the 1940s, where it was discovered that certain semiconductors, such as germanium and silicon, contained mobile electrons or atoms, so that a so-called

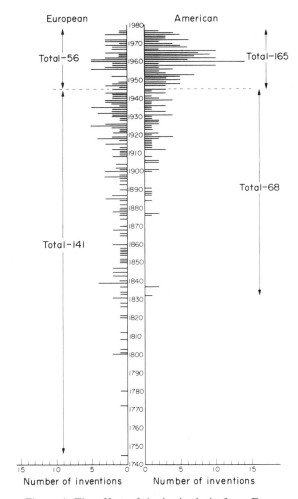

Figure 4 The effect of the brain-drain from Europe
to U.S.A. on inventions.

p–n layer in the crystal would pass current in one direction only as in the normal crystal detector, and many germanium diodes were used in the 1950s as rectifiers.

In 1948, Bell Laboratories' scientists found that amplification could be obtained by means of a third contact to the normal p–n contacts in germanium. This was termed a point-contact "transistor" or *trans*ferred res*istor* and the contact points were later replaced by an alloyed construction.

Bell Labs. then produced the diffused junction type of transistor, also in germanium, which replaced the point contact transistor. Germanium had one drawback in that temperatures higher than 75°C caused them to fail and silicon which had a much higher operating temperature began to be used and with gallium arsenide are the standard materials today.

Field effect transistors were introduced in which the metal oxide formed part of the transistor action and were known as MOSTs (Metal Oxide Silicon Transistors), having a simpler construction and were considerably smaller.

The transistor came into very wide use because of its obvious advantages of low-voltage operation (6 to 12 volts instead of 250 to 300), its low current consumption and its extremely small size. In addition, its reliability was very high compared with existing tubes, which had a limited life due to evaporation of the cathode. They were first introduced into deaf aids and then into portable radio receivers; then into computers and were widely used in all electronics until now being replaced by the integrated circuit or "chip", which is in effect a large number of transistors, all made in one operation, described later.

In the early 1950s glass-dielectric capacitors, mental-film resistors, castellated metallized paper capacitors and subminiature relays were beginning to be used more widely.

About this time, subminiature components for use in transistor circuits were being developed. For the first time, as high-tension voltages of 150 to 300 were not required, components such as capacitors could be designed to withstand only 6 to 12 V and (with the low currents at which transistors operated) resistors could be designed for very small power dissipation. This size reduction progressed to the point where handling difficulties and soldering problems arose.

However, about this time, it was realized that the actual working volume of components such as resistors and capacitors was only a very small proportion of the total. In the case of a ceramic-dielectric capacitor fitted into a ceramic case only 1/225th of the actual working volume was effective whilst in a plastic-moulded carbon film resistor only 1/280th of the total volume was effective. Attempts to fabricate simple film components led to the early thick-film and thin-film circuits now in use.

Also in the early 1950s, the concept of silicon-integrated circuits was published by the author in the U.S.A., and a model of a silicon "solid circuit" was shown at an International Components Symposium in England in September 1957. The first actual working model of an SIC was made later in the U.S.A. by Texas Instruments and a patent was filed on 6th February 1959 (Kilby). Neither of these devices was commercially viable until the invention of the planar process by Hoeni, which enabled mass production of integrated circuits to be carried out.

Other developments affecting components in the 1950s were automatic assembly techniques and "potting" or encapsulation techniques. Machines for automatic insertion in tubular components into printed-wiring boards were developed. Axial lead tubular resistors and capacitors were loaded into special feed-containers and in the machine their leads were bent over, inserted through holes in the printed wiring board and dip-soldered. Unfortunately, the capacity of these machines was so high, e.g. up to 10,000 boards a day, that production quantities were rarely sufficient to make full use of them.

This period saw the exploitation of the junction transistor and its use with subminiature components on a printed-wiring board, with the connections made by dip-soldering. Extensions to the range of transistors and diodes, both in frequency response and power output, were being rapidly made and transistorized equipment was being used for practically all electronic requirements by the mid -1960s.

During the early 1960s, the financial encouragement given by the American Air Force, Army and Navy to microelectronic manufacturers in the U.S.A. provided the necessary impetus to production techniques. The invention of the Planar process had made the process more viable and, as experience was built up, more and more applications were found. Initially, digital circuits were developed for use in computers, this being the maximum market, but linear circuits were also being developed for general purpose amplifiers, etc. A typical circuit of the 1960s is shown in Fig 5.

Fabrication processes for SICs were improved each year and, at the present time, circuit pattern delineation accuracies of the order of a few microns are being used in production. This means that many thousands of transistors can be made on one chip and packing densities of many millions of integrated circuits can be assembled in miniature equipments. A wide range of applications, including radio receivers, Hi-Fi, television sets, etc., using analogue chip devices and other applications — computers, microprocessors, communications, etc., using digital chip devices are now available.

In the manufacture of integrated circuits, the basic material is silicon, obtained by a chemical process from sand or quartz and formed into ingots or bars. The electronic properties of silicon crystals are controlled by their purity and its atomic construction is such that by adding impurities of a few atoms of phosphorus or boron, known as doping, the currents through the crystal can be changed. In order to obtain this high degree of purity, a process of "zone refining" by selectively melting along the bar of silicon moves the natural impurities to one end, which is then cut off and discarded. From this purified silicon, the crystal must be grown so that the atomic structure can be doped or altered as required to control the characteristics of the transistor circuit yet to be made. This is done by melting the silicon in a crucible and a small seed crystal is touched on the surface whilst molten and slowly drawn upwards. As it cools, the melt takes up the exact crystal structure of the seed. This process is continued until a bar of pure silicon crystal is produced, 2 or 3 inches in diameter and a foot or so long. The bar is then sliced by circular diamond saws into wafers a few thousands of an inch thick; these are then polished and cleaned and are then ready to be made into circuits.

Many hundreds of circuits, consisting of thousand of transistors, can be made simultaneously by using photo-masking techniques through which the various operations of doping, etching and metallizing are carried out. Each transistor is made very small indeed and doping the transistor with the required p-type

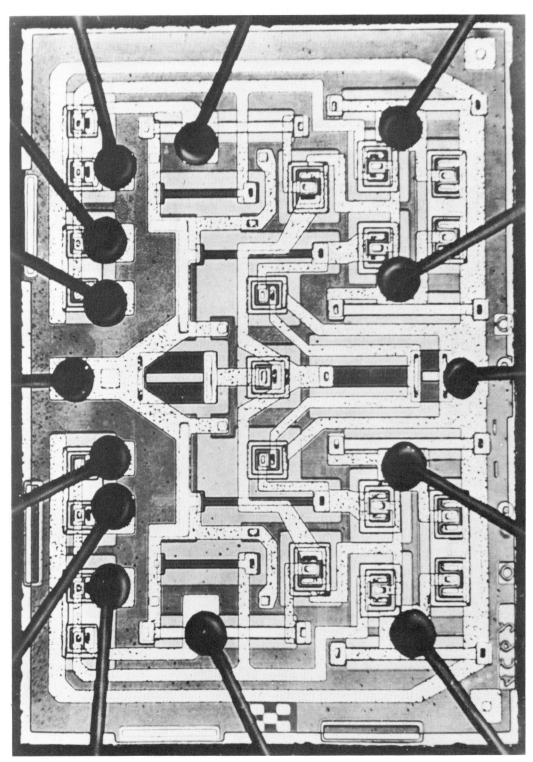

Figure 5 A typical integrated circuit of the 1960s (courtesy Mullard Ltd).

(boron) or n-type (phosphorus) impurity is done by diffusing atoms into the silicon at high temperatures of 900°C or so. The masks used in all the processes are made of silicon oxide, which is an insulator. The connections between the transistors are made by evaporating aluminium, again through a masking and etching system. Finally, the chips are mounted on a printed circuit board or ceramic plate and connected as required. The process is extremely complicated, highly precise, and absolute cleanliness is essential in their fabrication. Because they are made of such pure materials with great precision, they are very reliable.

Methods of chip assembly include plastic and ceramic sealing and the chips are mounted on printed circuit boards or hybrid assemblies, using chip carriers or TAB (Tape Automated Bonding), as well as the more common DIL (Dual-in-Line) plastic package. These closely packed assemblies use VLSI (Very Large Scale Integration) and also VHSIC (Very High Speed Integrated Circuits), whilst electron beam technologies are increasingly used to obtain the very high definition and accuracy required for these assemblies.

Over the period from about 1900 to the present day various electronics technologies have been developed, used to the full, and some have declined. Tubes were at their peak in the 1950s but, with the advent of the transistor, began to decline. Similarly, potted circuits and automatic assembly technologies had peak periods in the 1960s. We are now at the stage where TV, mini- and microcomputers and microelectronic devices of all kinds are advancing rapidly.

This history of components shows how the change from passive components (resistors, capacitors, etc.) to active components (transistors, integrated circuits) has affected electronics development. The present explosion of integrated circuits in the form of VLSI and VHSIC has been the most important development in the history of electronics.

The future of electronics, therefore, lies in the increasing use of microelectronic devices. The applications of microprocessors and minicomputers using VLSI are becoming wider and wider and this will affect both home and business life for everyone in the 1980s and 1990s.

A summary of developments in components is given in Chart 1: "History on a Page".

In the chapters, four to nine the history of electronics is traced in more detail up to the present day.

HISTORY ON A PAGE
(1) Components, Transistors and Integrated Circuits

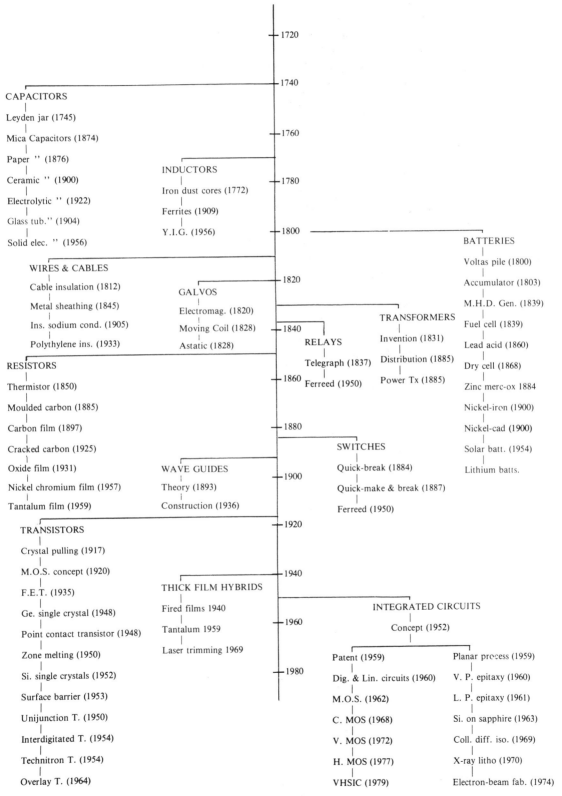

CAPACITORS
Leyden jar (1745)
Mica Capacitors (1874)
Paper " (1876)
Ceramic " (1900)
Electrolytic " (1922)
Glass tub." (1904)
Solid elec. " (1956)

INDUCTORS
Iron dust cores (1772)
Ferrites (1909)
Y.I.G. (1956)

BATTERIES
Voltas pile (1800)
Accumulator (1803)
M.H.D. Gen. (1839)
Fuel cell (1839)
Lead acid (1860)
Dry cell (1868)
Zinc merc-ox 1884
Nickel-iron (1900)
Nickel-cad (1900)
Solar batt. (1954)
Lithium batts.

WIRES & CABLES
Cable insulation (1812)
Metal sheathing (1845)
Ins. sodium cond. (1905)
Polythylene ins. (1933)

GALVOS
Electromag. (1820)
Moving Coil (1828)
Astatic (1828)

TRANSFORMERS
Invention (1831)
Distribution (1885)
Power Tx (1885)

RELAYS
Telegraph (1837)
Ferreed (1950)

RESISTORS
Thermistor (1850)
Moulded carbon (1885)
Carbon film (1897)
Cracked carbon (1925)
Oxide film (1931)
Nickel chromium film (1957)
Tantalum film (1959)

SWITCHES
Quick-break (1884)
Quick-make & break (1887)
Ferreed (1950)

WAVE GUIDES
Theory (1893)
Construction (1936)

TRANSISTORS
Crystal pulling (1917)
M.O.S. concept (1920)
F.E.T. (1935)
Ge. single crystal (1948)
Point contact transistor (1948)
Zone melting (1950)
Si. single crystals (1952)
Surface barrier (1953)
Unijunction T. (1950)
Interdigitated T. (1954)
Technitron T. (1954)
Overlay T. (1964)

THICK FILM HYBRIDS
Fired films 1940
Tantalum 1959
Laser trimming 1969

INTEGRATED CIRCUITS
Concept (1952)
Patent (1959)   Planar process (1959)
Dig. & Lin. circuits (1960) V. P. epitaxy (1960)
M.O.S. (1962)   L. P. epitaxy (1961)
C. MOS (1968)   Si. on sapphire (1963)
V. MOS (1972)   Coll. diff. iso. (1969)
H. MOS (1977)   X-ray litho (1970)
VHSIC (1979)   Electron-beam fab. (1974)

1720
1740
1760
1780
1800
1820
1840
1860
1880
1900
1920
1940
1960
1980

Chart 1

Chapter 3
The Expansion of Electronics and its Effect on Industry

In the early 1920s the main application of electronics was domestic radio and millions of radio sets were manufactured throughout the developed world. In the late 1930s, radar, black and white television and early computers were developed, each expanding the use of electronics. The 1939-1945 war created a great expansion of applications in the military field and, in 1948, the invention of the transistor saw an explosion in the commercial and home use of electronics.

The early 1960s, however, saw the greatest expansion of all time and the Integrated Circuit, together with the computer, laid the foundation for the present-day world-wide expansion of electronic applications, of which typical examples are: television, radio, gramophones, tape recorders, video games, microwave ovens, calculators, electronic watches, electronic organs, electronic alarms, telephones, test and measuring instruments, microwave and analytical equipment, data-processing systems, office equipment, automotive electronics, industrial controls, medical electronics, computers, electronic cameras, mini-computers, microprocessors, robotics, etc.

In addition, the field of telecommunications and radar, both commercial and military, used in aircraft, tanks, ships, missiles, satellites, etc., is very large. There are also the component manufacturers who supply these applications, making integrated circuits, hybrids and many specialized devices.

The list of applications is ever-widening as new applications of fibre optics, charge-control devices, bubble memories, etc., are being exploited. Certainly the field of electronics will expand into both home and business life, with teletext and television presentations combined with data-processing systems becomig more and more widely adopted.

The industrial revolution of the Victorian age created large industries such as steel, shipbuilding, cotton, railways, heavy machinery, etc. Today these industries are in decline and the changeover to light industries based on electronics has presented re-employment and re-training problems. However, the exploitation of electronic inventions can create wealth for those nations which take up the challenge, such as the U.S.A., Europe and Japan. New electronic industries are being built up employing large numbers of people. Although it is not possible to put accurate figures to the number of staff employed, an attempt has been made in the following chart to categorize inventions which have resulted in small, medium or large industries. Of the three categories, some twenty-eight inventions (Cats. II and III) can be said to have been the commencement of very large industries.

Date	Inventor	INDUSTRY (world-wide)		
		Category I Small (approx. 1-10,000)	Category II Medium (approx. 10,000- 100,000)	Category III Large (over 100,000)
1831	Faraday			Electricity
1831	Faraday	Transformers		generation
1837	Cooke *et al*	Relays		
1837	Davenport		Electric motors	
1839	Dancer	Microfilming		
1839	Grove	Fuel cells		
1843	Bain	Facsimile repro.		
1857	Wray	Mercury arc lamps		
1860	Plante	Accumulators		
1868	Léclanché		Dry batteries	
1876	Fitzgerald	Paper capacitors		
1876	Bell			Telephones
1877	Edison	Carbon microphones		
1877	Siemens	Loudspeakers		
1878	Swan/Edison *et al.*			Electric lamps
1885	Bradley		Carbon composition resistors	
1887	Berliner		Gramophones	
1895	Rontgen	X-rays		
1896	Marconi			Radio
1897	Braun	Cathode ray oscillographs		
1898	Poulsen		Magnetic recorders	
1901	Cooper- Hewitt	Flourescent lamps		
1906	de Forest		Three-electrode tubes	
1908	Cahill	Electric organs		
1910	Claude	Neon lamps		
1915	Longevin	Sonar		
1918	Northrup	Induction heating		
1919	Zworykin & Farnsworth			Television
1924	Appleton, Briet *et al.*			Radar
1926	Round	Screened grid tubes		
1927	Fox Movietone News		Sound on film recording	
1928	Tellegen- Holst	Pentode tubes		
1932	Knoll, Ruska	Transmission electron microscope		
1933	Jansky	Radio astronomy		
1934	Dreyer	Liquid crystals		
1935	Knoll *et al.*	Scanning electron microscope		
1937	Carlson		Xerography	
1938	Frisch & Meitner		Nuclear fission	
1939	Hann, Varian Bros	Klystrons		

		Category I	Category II	Category III
1939	Atiken and IBM			Computers
1940	Centralab		Thick-film circuits	
1943	Kompfner et al.	Travelling wave tubes		
1943	Eisler		Printed wiring	
1948	Gabor	Holography		
1948	Bardeen et al.			Transistors
1950's	MIT, Bell	Modems		
1953	Townes et al.	MASER		
1953	Mallina et al.	Wire wrapping connections		
1954	Chapin et al.		Solar batteries	
1958	Ampex	Video tape-recorders		
1958	Various	Explorer, Vanguard, Pioneer 1, etc. (satellites)		
1958	Schalow	LASER		
1959	Bell Labs.	Tantalum thin-film circuits		
1959	Hoeni		Planar process IC's	
1960	Bell Labs.		Electronic telephone exchanges	
1960	Various			Digital & linear integrated circuits
1960	Allen & Gibbons	LED's		
1961	Digital Equip.		Minicomputers	
1963	Bell		Electronic calculators	
1963	Rown-Sittig	Surface acoustic wave devices		
1964	Baran	Packet-switching communications		
1964	Lepselter	Beam lead conn-ections		
1964	Various	Telemedicine		
1965	Various	Intelstat Satellites		
1966	Kao and Hockham	Optical fibres		
1967	Dolby		Audio noise-reduction system	
1969	Bobeck et al	Magnetic bubbles		
1970	Boyle Smith	Charge coupled devices		
1972	Intel			Microcomputers
1972	Magnavox		Video games	
1976	Various	Marisat satellites		
1977	Sinclair	Pocket TV		
1978	Bell	Light-powered telephone		
1978	Philips	Laser optical recorder		
1978	Intel	One megabit bubble memory		
1979	Sony	CCD colour TV camera		
1980	NEC Toshiba		256 K RAM	
1980	STC	Fibre optic submarine cable		
1982	AERE	Fission track autoradiography		

Chapter 4
A Concise History of Audio and Sound Reproduction

A basic electrical sound system needs a microphone to convert the sound into an electrical waveform, a means for transmission and a means for reproducing the sound.

The earliest microphones consisted of a stretched flat membrane (Reis used a sausage skin in 1860) actuating a loose metal-to-metal contact in order to convert the sound vibrations into electric currents. Later microphones replaced the single loose metal contacts by carbon ones (Edison, 1877) and by carbon granules (Hunnings, 1878). The first intelligible human speech was transmitted over wires by Bell, however, in 1876. The microphone he developed consisted of a thin iron diaphragm which vibrated in front of a magnet with a coil wound on it, thus inducing an electric current which varied in sympathy with the voice sounds. This current was sent over a pair of wires to a similar apparatus at the other end.

Sound could now be transmitted and received by wire, and in 1877, Edison invented a method for recording it. He wrapped a sheet of tinfoil round a cylinder, set a stylus or needle attached to a diaphragm in contact with it, turned a crank to rotate the cylinder and shouted into it: "Mary had a little lamb". On cranking the cylinder again, his voice was reproduced. Ten years later, Berliner introduced the first flat-disc record and also a method for making many shellac copies.

Sound could now be transmitted, recorded and reproduced. About this time (1877), Siemens in Germany invented the first moving-coil cone loudspeaker, but it was before its time and had to wait for other developments before it could be exploited.

Following Bell's invention, the first public telephone exchange was opened in New Haven, Connecticut, in January 1878 to twenty-one subscribers. These first telephones were leased in pairs, the two telephones being permanently connected together.

In 1889, Strowger invented the first automatic switching system to allow one telephone to work with any other telephone, and in 1919 the improved crossbar selector system was developed in Sweden. In 1952, the first experimental electronic exchange was made by Bell Telephone Laboratories, and by 1958 fully automatic electronic exchanges were in use.

The spread of telephones throughout the world was rapid and there are now around 400 million telephones in service.

In 1898, another method of recording sound was invented by Poulsen in Denmark — magnetic recording. He found that by feeding the current from a microphone through an electromagnet and drawing a piano wire rapidly past it, the wire was magnetized to varying degrees, thus recording the sound. This could be played again and again and could also be wiped off and re-recorded.

Poulsen's invention of the magnetized piano wire was revived when tube amplifiers became available and spools of thinner wire were used; but thin wire was inclined to twist and was replaced by a flat metal tape developed by Blattner (the "Blattnerphone"). However, when the tape broke it was difficult to join, and I. G. Farben and AEG in Germany produced paper and plastic tapes coated with iron oxide in the form of a lacquer. Germany about 1940 introduced the "Magnetophon" tape recorder, with improved tape and a much improved quality of sound reproduction — the latter largely due to the use of radio-frequency bias as in present-day tape recorders. A plastic-based ferric-oxide tape was used, running at 56 cm/sec (22 in./sec). Nowadays open-reel tape recorders usually operate at 15 in./sec and sub-multiple (7.5, 3.75, etc.). Cassettes, which are now much more popular than open-reel tapes, because of their low cost and convenience (no tape threading), normally operate at 1.875 in./sec. Recently, cassettes of half-size have been introduced, but mostly for dictation purposes rather than music. The latest video recorder tapes can record up to 18 octaves for colour television, as against 9 octaves for normal sound recording.

Berliner's shellac gramophone discs were later replaced by plastic discs running at 78 rpm, and about 1948 the 78 rpm disc playing speed was reduced to 45 and 33½ rpm, and with microgroove plastic discs opening the era of the long-playing record.

The first sound recording to accompany films ("the talkies") was produced in 1926 by Warner Bros. A disc-recording machine called Vitaphone was synchronized to the film and the first successful talking picture was "The Jazz Singer" in October 1927, with Al Jolson. Fox Movietone News improved on this system by using a sound-on-film system in which the synchronism was automatic, using a variable-intensity light source to record the sound on the film negative. Similar systems are now used in modern cinemas.

In the early 1930s, Arthur Keller of Bell Labs in the U.S.A., and Alan Blumlein of EMI in Britain, developed stereophonic disc recording systems quite independently. These gave directional and much improved spacial effects, but the business depression of those years prevented commercial exploitation before the war. Surprisingly, the first practical demonstration of stereophonic sound reproduction was carried out by a Frenchman, Clement Ader, in 1881. He installed pairs of microphones in the Paris Opera House, connected by telephone lines to pairs of earphones in the Exhibition of Electricity held that year in Paris.

A three-channel system was used by the Bell Laboratories on 27th April 1933, when the Philadelphia Orchestra was reproduced in Washington with a most impressive degree of realism. This may be regarded as the beginning of stereophonic hi-fi.

In recent years, quadraphonic or surround sound has been introduced. Though not yet widely accepted, it aims to give the listener the illusion of being present inside an auditorium more vividly than can be achieved using ordinary two-channel stereo.

The development of video recording techniques for television, together with fast digital techniques in other fields, has begun to have a large influence on sound recording. But the first practical application of digital techniques to audio was not for recording, but was the BBC's "Pulse Code Modulation" (PCM) system for sending high-quality stereo and TV sound signals from one BBC centre to another. The instantaneous amplitude of the audio wave form is sampled at frequent intervals, and the sample amplitude is expressed as a binary number of 12 or more bits, i.e. 0's and 1's. The apparatus for doing this is called an analogue-to-digital (A/D) converter. At the other end, the stream of 0's and 1's is reconstituted into an audio waveform by a D/A converter. A large bandwidth is required to transmit this very rapid succession of 0's and 1's, but the distortion of the system is far lower than that of normal landlines. For example, programmes can be sent from London to Edinburgh and back with no perceptible degradation in quality. These techniques are now being used for sound recording, the digital bits being recorded on a video tape recorder or on video discs. There are several disc systems, one using a laser spot focused on a reflective track, and modulated by a pattern of indentations on the disc. Digital Compact Discs will soon be commercially available: a 5-in.-dia. disc will carry one hour of stereo sound, with much lower distortion and background noise level than can be achieved using ordinary analogue methods.

In Telex machines, teleprinters or computers, the output is in binary form — 0's and 1's. As this cannot be transmitted over a telephone line, it has to be converted into suitable audio tones. This is done by a MODEM (*Mo*dulator and *Dem*odulator) with a similar MODEM at the receiving end.

For transmitting speech, less bandwidth is required than for music. For good speech quality about 16 K bits/sec from an A to D converter is required. VOCODERS (or voice coders) are used to reduce this to 2.4 or 3.6 K bits/sec by making assumptions about the speech waveform. This can be converted back by a D to A converter which is aware of the assumptions made. This produces the Darlek-like speech which is just recognizable. This in turn leads to speech-analysis devices and speaking machines, i.e. synthetic speech.

A typical hi-fi loudspeaker system consists of two or three loudspeaker units of different cone size, the smallest ones being known as "tweeters" and the largest ones as "woofers", with a crossover network to direct the relevant part of the sound spectrum into each unit.

An important loudspeaker development, carried out by P. J. Walker in Britain, is the full-frequency-range electrostatic loudspeaker. This has been widely regarded, for over two decades, as setting a standard for natural, uncoloured reproduction against which moving-coil designs could be judged. It has also helped to focus attention on the very great importance of the polar radiation characteristic of loudspeakers and the relationship between the loudspeaker and the acoustics of the listening room. With improvements in other parts of the audio chain, these aspects are becoming of much greater significance.

Electrical and electronic musical instruments have been evolved over a period of many decades. The major part of this effort has probably been devoted to the electronic organ, but other instruments, such as electronic pianos, have also been produced.

A more recent development in this field of creating sound electronically is the synthesizer. This gives the user an unprecedentedly great ease and versatility in the creation and control of audio waveforms with regard to harmonic content, envelope shape, repetition rate and random noise characteristics. Musical and other sound sequences can be built up and controlled from a small keyboard. The BBC Radiophonic

Workshop uses synthesizers and other techniques, and a well-known example of radiophonic music created by them is the theme music from "Dr. Who".

Electronic organs consist of a series of electrical tone generators controlled by keyboard(s) and stops and fed to amplifiers and loudspeakers. An early and commercially successful organ was designed by Laurens Hammond and marketed in 1935. This employed a large number of synchronous-motor-driven alternators to generate the tones. Since then very many different designs have been evolved, mostly of purely electronic nature. The best of these simulate remarkably well the subtle sounds of a good traditional pipe organ, but the largest sales are of smaller instruments aimed mainly at the popular music field. As well as producing a wide range of organ-stop tonalities, often of cinema-organ character, these organs usually also provide various percussive effects, and a "rhythm box" is frequently added to give automatic drum and cymbal sounds at various selectable tempos. Techniques based on the use of microprocessors are being increasingly employed to relieve the player of the need for a high degree of performing skill.

Developments in audio and sound reproduction are summarized in Chapt 2: "History on a Page".

HISTORY ON A PAGE

(2) Audio and Sound Reproduction

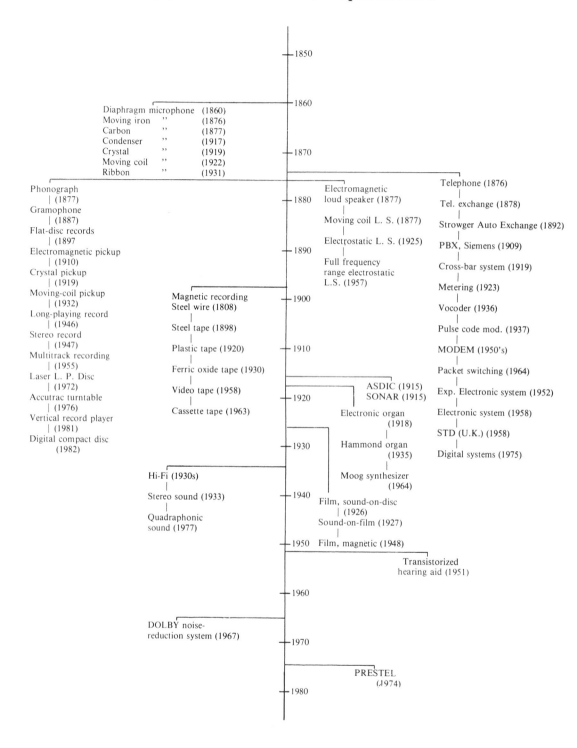

	—1850
	—1860
Diaphragm microphone (1860)	
Moving iron ,, (1876)	
Carbon ,, (1877)	
Condenser ,, (1917)	
Crystal ,, (1919)	—1870
Moving coil ,, (1922)	
Ribbon ,, (1931)	

Phonograph
| (1877)
Gramophone
| (1887)
Flat-disc records
| (1897
Electromagnetic pickup
| (1910)
Crystal pickup
| (1919)
Moving-coil pickup
| (1932)
Long-playing record
| (1946)
Stereo record
| (1947)
Multitrack recording
| (1955)
Laser L. P. Disc
| (1972)
Accutrac turntable
| (1976)
Vertical record player
| (1981)
Digital compact disc
(1982)

Electromagnetic loud speaker (1877)
|
Moving coil L. S. (1877)
|
Electrostatic L. S. (1925)
|
Full frequency range electrostatic L.S. (1957)

Telephone (1876)
|
Tel. exchange (1878)
|
Strowger Auto Exchange (1892)
|
PBX, Siemens (1909)
|
Cross-bar system (1919)
|
Metering (1923)
|
Vocoder (1936)
|
Pulse code mod. (1937)
|
MODEM (1950's)
|
Packet switching (1964)
|
Exp. Electronic system (1952)
|
Electronic system (1958)
|
STD (U.K.) (1958)
|
Digital systems (1975)

Magnetic recording
Steel wire (1808)
|
Steel tape (1898)
|
Plastic tape (1920)
|
Ferric oxide tape (1930)
|
Video tape (1958)
|
Cassette tape (1963)

ASDIC (1915)
SONAR (1915)

Electronic organ (1918)
|
Hammond organ (1935)
|
Moog synthesizer (1964)

Hi-Fi (1930s)
|
Stereo sound (1933)
|
Quadraphonic sound (1977)

Film, sound-on-disc
| (1926)
Sound-on-film (1927)
|
Film, magnetic (1948)

Transistorized hearing aid (1951)

DOLBY noise-reduction system (1967)

PRESTEL (1974)

Chart 2

Chapter 5
A Concise History of Radio and Telecommunications

At the end of the nineteenth century, the electron had been discovered and a great deal of experimental work was being done on electromagnetism. Sparks produced by discharges from a Leyden jar through an induction coil of wire fitted with a spark gap were found to be reproduced on touching nearby metal objects. This was the first instance of energy transfer over a distance without wires.

Many theories were put forward for this phenomena, but James Clerk Maxwell was the first to consider magnetic and electric fields together to be responsible and formulated his equations from which he predicted electromagnetic radiation on purely theoretical grounds. He predicted wave propagation with a finite velocity, which he showed to be the velocity of light. Heinrich Hertz succeeded in producing electromagnetic waves experimentally in 1877. He also added a loop of wire and increased the distance over which sparks could be transmitted.

A sparks detector was produced by a French physicist, Branley, who noticed that the resistance of a glass tube filled with metal filings, normally high, became low when an electric discharge occurred in the vicinity, which could be measured. It was now possible to transmit telegraphy by coding the sparks and using Branley's coherer to receive the sparks over a distance. It is interesting that Branley was the first to use the word "radio" in connection with his experiments.

Guglielmo Marconi was interested in Hertz's experiments and began experimenting in 1895 with a spark induction coil and Branley coherer, and succeeded in sending telegraph messages over a short distance. He travelled to Britain in 1896 to exploit his discoveries with W. H. Preece, Engineering Chief of the Government Telegraph Service, and Marconi demonstrated his apparatus to the Post Office officials over a distance of 100 metres.

Similar work was being done in Russia and A. S. Popov gave a demonstration on 12th March 1896, but this was not published until some thirty years later.

Marconi patented his invention in June 1896 and the first Radio Telegraph Wireless Company was formed in 1897 to acquire Marconi's patents and set up manufacturing spark transmitters and receivers. Other manufacturers such as Siemens and Halske were, by the late 1890's, also making radio equipment. By this time, aerials had become long wires raised as high as possible.

Marconi continued to improve his equipment and succeeded in linking England with France by radio in 1899, and later over longer distances. He also installed equipment on ships for communication between ship and shore.

To explain why radio communication was possible around the curvature of the earth, Heaviside in England and Kenelly in America proposed an ionized layer in the upper atmosphere which reflected the waves. This was in 1901 and was known as the Heaviside/Kenelly layer.

Spark transmitters were almost universal for radio telegraphy, but they were improved when rotary discharger sets were introduced, fed from alternators to give more transmitter powers. Atmospherics were a great enemy. Larger and larger powers were introduced, such as at Poldhu in Cornwall, operating in 1911 and at the transatlantic station at Clifden in Ireland. In 1912, the sinking of the *Titanic* focused attention on radio-telegraphy as, although over 1500 lives were lost, some 700 were saved by ships summoned by wireless.

The Great War of 1914/1918 produced a major change in technology with the introduction of the three-electrode tube in quantity, which ultimately resulted in the ending of the spark transmitter era. By 1919, high-power transmitter tubes were being made and the Marconi Company spanned the Atlantic by telephony in daylight. The superheterodyne receiver was in use and provided virtually constant gain and particularly constant selectivity over a wide tuning range. Attempts to make oscillators at much higher frequencies were made, such as the retarded field oscillator by Berkhausen and Kurtz in Germany in 1919 and the negative resistance oscillator by Gill and Morell in the United Kingdom in 1922. Crystal control of frequency was introduced and the screened grid tube and the pentode improved the performance of radio receivers, which were mostly made by home constructors in the 1920's.

Frequency standards were set up, first the quartz crystal itself, then the atomic clock, using the spectral line of ammonia and finally the caesium and rubidium beam magnetic resonance methods initially developed in 1939.

The advent of the 1939-1945 war introduced high-sensitivity communication receivers for military use, such as the R1155 for the Air Force. The first production communication receiver in wide use was the National HRO of the late 1930s. Post-war developments benefited from the research which had been done on components and high-quality FM receivers and Hi-Fi amplifiers became available. Radio astronomy, discovered in 1933 aroused considerable interest and Jodrell Bank was in operation in 1951.

The invention of the transistor revolutionized radio techniques, with its low voltage and current requirements, leading to cheap and reliable battery power supplies for portable sets. In 1954, the Regency transistor was introduced in America. Printed circuits were widely used with ferrite rod aerials in portable receivers. In 1957, the Russian Sputnik satellite was launched and carried a 1-watt transmitter, becoming the forerunner of modern satellite communication. For low-noise amplification, MASERS, parametric amplifiers and cryogenic techniques were used, particularly in satellite signals receivers. Microwave systems were now developed from the earlier oscillators and the travelling-wave tube was incorporated in high-performance microwave sets. The U.S.A. launched EXPLORER I in 1958 which discovered the Van Allen radiation belt. Also in 1958, the first communication satellite was launched by the U.S.A. — the SCORE satellite, which transmitted taped messages for 13 days. In 1960, ECHO-1 was launched as a passive communication satellite, which relayed both voice and TV signals and the first active repeater communication satellite COURIER-IB was also launched in 1960, followed by TELSTAR-1 in 1962. Since then a series of communication satellites have been launched by many countries, enabling world-wide coverage of radio and TV signals to be achieved.

Integrated circuits were now being used in radio receivers, commencing in the early 1960s, with circuit designs, which often used transistors in place of many resistors and capacitors. Improvements in circuit techniques and in packing density and complexity have resulted in modern radio receivers of high performance and great reliability. LSI (Large Scale Integration), VLSI (Very Large Scale Integration), microprocessor and ULA (Uncommitted Logic Array) devices are now used in radio systems.

Spread-spectrum communication techniques are increasingly being used, particularly for satellite communications. A spread-spectrum system is one in which the transmitted signal is spread over a wide frequency band. Various modulation techniques are used, e.g. digital code sequences (DSK), frequency hopping (which can be also "time-hopping" or "time-frequency hopping") and chirp or frequency sweeping. Because the system is based on coding, privacy is obtained by selective addressing and code-division mutiplexing. There is also freedom from interference.

High-speed converters are now being designed to translate analogue signals into digital data for microprocessor based control systems such as high-frequency video, radar signal processors, etc. The fastest method is known as parallel or flash encoding, where stray capacitances are reduced to a minimum by 1 micron lithography and other techniques.

The use of digital data systems in which machines communicate directly with other machines over radio paths is increasing rapidly, often because of the improved bandwidth/information rate that can be achieved (compared with human beings) and the trade-off flexibility that can be achieved between bandwith and data rate.

There are now strict and very detailed international CCIR (Committee Consultatif Internationale Radio) rules governing the use of radio transmissions from a few kHz to tens of GHz.

Developments in radio communication are summarised in Chart 3: "History on a Page".

HISTORY ON A PAGE
(3) Radio Communications

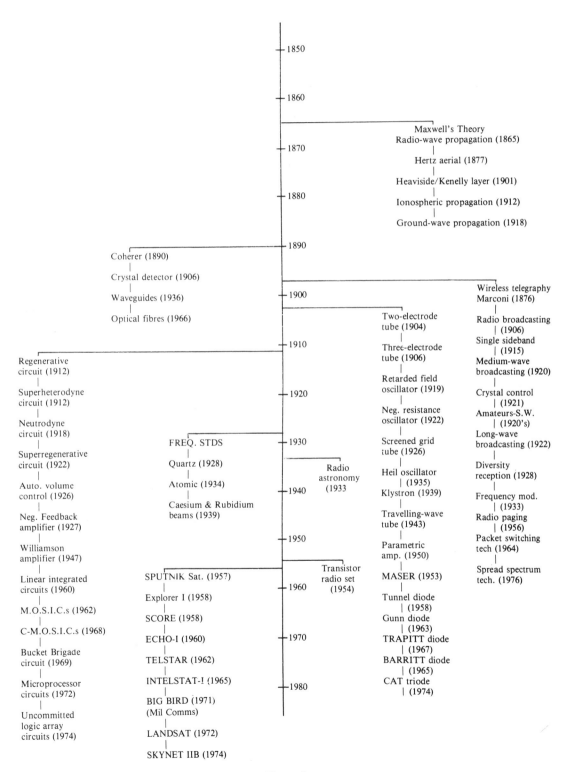

Chart 3

Chapter 6
A Concise History of Radar

Between the 1920s and the 1930s, it was noticed that signals received from high-frequency radio transmitters were affected by aircraft flying overhead. Appleton and Barnett when measuring the height of the ionosphere in the United Kingdom in 1924, and Briet and Tuve, in 1925, doing similar experiments in the U.S.A., both noticed this effect, although Briet and Tuve used a pulse technique for their measurements. During the following ten years, experimental work on the detection of aircraft and of vessels at sea by radio was done in the U.S.A., France and Germany. In Britain, in 1935, a committee was set up under Sir Henry Tizard to make a serious investigation into the air defence of the United Kingdom. Sir Robert Watson-Watt suggested lines of research and with other scientists developed the first practical radar system in the summer of 1935.

Radar operates by sending out a pulse of energy at radio frequencies, some of which is reflected back by any object in its path. By measuring the time interval between the transmitted and received pulse on a time base, the range can be accurately calculated. Three main wavelengths were used — around 10 metres for initial early warning floodlight radars; 1½ metres for more accurate positional radars and, later, around 10 cms, allowing pencil beams of high positional accuracy.

Because the United Kingdom developed and used the first practical system in the world, the following descriptions are primarily of wartime developments in British work. The success of United Kingdom radar was due to the close working liaison between the fighting Forces and the scientists. A series of CH (Chain Home) early warning radar stations was constructed, using wavelengths in the 6-15-metre band. In 1938, five stations covered the East Coast and, by 1940, the whole of the East and South Coasts were covered against aircraft flying at 15,000 feet out to a range of 120/140 miles. The CH transmitters were mounted on metal towers, 360 feet high, and the receivers on 240-feet wooden towers. A peak transmitter power of 200 kW was used with a pulse repetition frequency of 25 pulses per second. The receivers had a DF system to give approximate bearing. The complete co-operation between the scientists and the RAF, ensured that fighter aircraft were able to be alerted with time to intercept enemy bombers, enabling the "Battle of Britain" to be won in 1940.

Low-flying enemy aircraft could avoid the 10-metre CH aerial coverage so a chain of stations operating on 1½ metres was set up in 1939, known as CHL (Chain Home Low). The 1½-metre beam was directional and was swept round on its turntable one or more times per minute, searching each sector of the sky in turn. The time was now ripe for a rotating time base to be linked with the aerial rotation and the first PPI (plan position indicator) was installed in 1940. The night bomber was posing a problem so directing the fighter to link up with a bomber became a possibility with this equipment. This was known as GCI (Ground Controlled Interception) and a controller with a GCI station was able to see the echoes of both fighter and bomber on the PPI in their relative positions and advise the fighter pilot by radio. A long afterglow CRT was used to "hold" the signals.

In 1936, a start was made on airborne radar and an experimental 1½-metre ASV (Air to Surface Vessel) radar fitted in an Anson aircraft observed echoes from ships at a range of 5 miles on 4th September 1937. By 1939, a fully developed ASV system was installed in Coastal Command.

Also in 1937, the first AI (Air Interception) 1½-metre radar flew in a Battle aircraft. This was the first of a series of AI sets and the AI Mark 4 was used in the early part of the war with good success. With the invention of the magnetron by J. T. Randall and H. A. H. Boot in 1939, both AI and ASV became centimetric and the magnetron was one of the turning points in radar development. Up to the discovery of the cavity magnetron, klystron oscillators were the only source of high-frequency power. Using the same resonator principle, Randall and Boot developed the first magnetron high-power oscillator producing hundreds of kilowatts in short pulses at a wavelength of around 10 centimetres. With a small reflector, a pencil beam was now available for an AI set which could be scanned to present the pilot with a rough TV picture of the enemy aircraft. Two methods of scanning were used: a belical and a spiral scan. As an experiment, the spiral scanner was pointed downwards and a recognizable picture of towns was seen on the PPI tube. Following this, a Halifax bomber was fitted with a perspex blister underneath carrying the radar. It was flown on 27th March 1942, and the equipment was code named H_2S. It became the RAF's most successful blind navigation and bombing aid.

It was realized soon after the development of the early radar system that some form of identification was needed to avoid shooting down our own planes. A beacon system was devised (IFF) — "Identification Friend or Foe", which automatically responded to small transmitted signals and returned a stronger coded signal on the same frequency. The coding was changed to identify the aircraft as friendly. All were fitted with detonators in the event of a crash.

In 1941, the navigational problem of finding the target town in Germany for night bombers was solved by the invention of GEE (Grid navigation) by R. J. Dippy. This was a transmitted hypobolic grid pattern from three-pulse transmitters, one a master station which could activate two slave station timing pulses. The aircraft carried a cathode-ray tube display on which the arrival of the pulses was recorded. By referring the pulse positions to a map marked with the hyperbolic curve lines, the navigator could find his position and also the target. GEE could guide hundreds of bombers simultaneously to a target 350 miles away with an accuracy of about 2 miles. The first 1000-bomber raid was carried out using GEE navigation on 30th May 1942. A modified system (GEE/H) used two stations only and the aircraft transmitted to both. The signals were transmitted back with different delay times according to the aircraft's position. A still more accurate system was code named OBOE. This was so accurate that it was used to guide Pathfinders to the target for dropping marker flares, so that following bombers could unload their bombs. It used two transmitters, one enabling the aircraft to fly along a beam centred on the target. If too far one side, the navigator heard a series of dots, if too far the opposite side, a series of dashes was heard. A continuous note meant the correct course. The second transmitter measured the distance along the track and signalled the bomb aimer exactly when to release the bomb. Bombing raids from a height of 30,000 feet, at a distance of 250 miles, controlled by "Oboe" Pathfinders were accurate to within 150 yards.

Many attempts were made by Germany to jam or render ineffective the radar-guided bombing raids and, in turn, methods of confusing German radar screens were adopted by the British. One was "WINDOW" which consisted of thin tin foil strips scattered from a single aircraft giving multiple echoes to create the impression of a large bombing raid. This was first dropped in 1943 and was extremely successful. It was again used in 1945, with the invasion of France in conjunction with a further deception device called "MOONSHINE", in which an IFF set was modified to imitate the echoes from a large formation of aircraft, instead of a single one. There were also many deceptions by beam-bending towards the end of the war as radar equipment became more sophisticated.

Since the war, both in the U.S.A. and the U.K. there have been great advances in radar systems, and digital techniques have been widely introduced. Video integrators, displays and ranging equipment were the first functions to be converted to digital techniques. This digitization is necessary to handle the large amount of information collected by modern search radars. Digital computers are widely used and with an electronically scanned or phased array aerial, there are no moving parts. Array aerials have many advantages: rapid scanning, beam shaping, step scanning, so that radar aerials of this type can obtain multiple target capability and function time-sharing.

Surface-to-air guided missiles carrying radio "proximity" fuses first used with shells and described later as VT (Variable Timing) fuses, were developed such as "Bloodhound" for the RAF, and "Thunderbird" for the army and small air-to-air missiles were designed for the RAF, some carrying infra-red homing heads which homed on to the exhausts of jet aircraft. Many sophisticated devices were produced after the war, such as TV guided air-to-ground missiles, cloud-and collision-warning systems, terrain-following radars and others not yet able to be published.

So far the equipments described were used by the RAF but parallel developments were taking place in the Army and Navy.

Army Radar

Following the original optical sightings of targets, the AA guns were aimed and fuses set by information from predictors which were, in effect, analogue computers. The Army's first use of radar in 1938/1939 was to feed information to the predictor at night when the target could not be seen. The first radar set GL1 (Gun Laying Mark 1) provided range and bearing only. Two cabins were used, one containing the transmitter and one the receiver; the receiver cabin rotating to obtain the bearing. GL1 Star provided elevation by adding two more vertically displaced antennas and an extra display. GL2 was an improved system with range, bearing and elevation all built in and was later used to track the German V2s.

In 1940, the searchlight-control system, SLC (known as ELSIE) using Yagi transmit and receive antennas, (some mounted directly on to a 90-cm or 150-cm searchlight,) was used also in conjunction with Bofors AA guns. The first 10-cm AA radar GL3 was deployed in 1943, leading to auto-follow, auto-gun-aiming, auto-fuse setting and automatic firing in later models. Close co-operation between the army and the Radiation Laboratory, MIT, in the U.S.A., resulted in an auto-following anti-aircraft fire-control radar, called SCR 584, which was widely used.

In 1941, a VT (Variable Timing) fuse was developed by W.A.S. Butement in the United Kingdom, to fit into a shell. A small radio transmitter and receiver inside the shell measured reflections from the target and, when within a certain range, exploded the shell. The shells were manufactured in the U.S.A. and were fitted in the U.K., in time to meet the V1 flying-bomb attack. Nearly 100% of the German V1's approaching London were ultimately shot down by AA guns using VT fuses.

Radar was also used for field artillery fire correction. With a 3-cm pencil beam, the ground ahead could be scanned to give a rough picture of the terrain. In the same way that the Coastal Artillery sets could locate shell splashes in the sea, the field artillery could observe shell bursts on the ground, enabling fire correction to be carried out. Another type of field army radar set "Watch Dog" allowed movement to be detected in any zone in which its beam could search. "Watch Dog" used the Doppler effect to differentiate between fixed echoes and moving echoes.

After the war, target-tracking radars for AA guns were improved in accuracy. For light anti-aircraft fire, a radar was developed called "Blue Diamond", which used a "Foster" scanner, which produced a rapid azimuth scan of 60° in 20 cycles per second. The elevation scan at a lower speed was done by a nodding arrangement and the whole equipment formed part of a Fire Control System called "Yellow Fever" for the L70 AA gun in 1957.

Other search and tracking radars were developed and put into service with the Army for the tactical control and fire control of medium and heavy AA guns.

Mortar locators were developed about this time, the first being "Green Archer", which measured the range, bearing and elevation angles of two points on the trajectory of the mortar shell in order to extrapolate back to its point of origin, so that it could be attacked. This was followed by "Cymbeline" which was a lightweight and improved version of "Green Archer".

To deal with low-flying attacking aircraft or helicopters, a one-man fire-control and missile system equipment known as RAPIER was developed. Four radar-guided missiles were mounted on a lightweight trailer that could be towed by a Land Rover.

Naval Radar

Before the advent of radar, the Navy had to rely on good weather and high-power telescopes for detecting enemy planes and ships. This gave reasonable co-ordinates, but range was difficult to estimate, errors of 1000 to 2000 yards being common at ranges of 20,000 yards. At the beginning of the war, Type 79 radar operating on 7 metres and Type 81 on 3½ — 4 metres were fitted in battleships and cruisers. The GL1 Army radar was added to provide accurate range for fire-control against aircraft. For close ranges multiple pom-poms fitted with a Type 282 radar with Yagi aerials allowed ranges up to 6000 yards against dive-bomber attacks. For surface fire in capital ships and cruisers, multiple aerial systems were used with a Type 284 radar. It was fitted on the Director Control tower in HMS *King George V* and gave ranges on cruisers of 20,000 yards and on destroyers between 12,000 and 14,000 yards.

By the end of 1940, 50-cm radar range-finders were fitted in HMS *Suffolk* and Radar 281 was responsible for tracking the *Bismarck* in June 1941. About this time a 10-cm set-Type 273 had been fitted in most battleships and cruisers for surface warning. Beam-switching was adopted for blind-firing the guns at night. This technique could provide tracking in bearing of ships and aircraft with an accuracy of about ±½°. Conical scanning was used in a 10 cm high angle set against aircraft. A 3-cm radar set (Type 262) was developed to provide blind-fire with twin Bofors guns. It was designed to pick up a target at 7000 yards and track it in all three co-ordinates before it reached 5000 yards. The first "out-of-sight" fighting using radar involved the sinking of the *Scharnhorst* in 1942. Radar echoes could be obtained from shell splashes in the sea and in a low-angle gunnery set a special display was provided for the observation of "fall-of-shot" echoes. Because the Navy initially developed high-power tubes, such as the silica tubes manufactured at Portsmouth in 1939, responsibility for designing and producing transmitter and all other tubes was allocated to the Navy and the CVD (Committee for Valve Development) co-ordinated the requirements for all three services.

After the war, surface-to-air guided missiles were developed such as SEASLUG using beam-riding guidance. Other marks followed such as SEACAT and SEAWOLF. SEACAT was designed as a close-range missile to deal with low-flying aircraft or missiles and is controlled manually or by TV along a line of sight controlled by radar.

Missiles launched from submarines such as POLARIS have ranges of a thousand or more miles. POLARIS is a two-stage solid propellant missile with an inertial guidance system. It is "shot" from its launch tube by a small rocket in the submarine; its own first-stage motor firing only as it leaves the water.

Civil Maritime Systems

Civil maritime radars normally operate at a frequency in the 9-GHz range or in the 3-GHz range with a CRT display giving information on the position of other ships, buoys or coastlines. Harbour surveillance is also carried out by land-sited radars giving a picture of the movement of all ships in or approaching the harbours.

Civil Aviation

Since the war, development of radar systems has also been applied to civil aviation and used for precision approach (PAR), for airfield control (ACR), for airfield surface movement indication and for air traffic control, both long-range and terminal areas surveillance (TMA). Airborne cloud and collision warning radars were also developed.

In addition, the IFF system was adopted and known as SSR (secondary surveillance radar). Through the agency of the International Civil Aviation Organization (ICAO), there is an internationally agreed spacing and connotation for the civil system. This enables each aircraft to be identified.

Meteorological Radar

Meteorological radar echoes were first noticed in February 1941 when a rain shower was tracked a distance of 7 miles by a 10-cm radar located on the English coast. Since then, rain detection by radar of showers or heavy storms has become a useful tool for the meteorologist.

Angels

Echoes from birds and insects are also obtained at wavelengths of 20 cm or less and in the early years were known as "Angel Echoes" as no obvious aircraft or solid targets were present.

Satellite Surveillance Radars

Satellite surveillance radars operate at frequencies between 200 to 5000 MHz; below 200 MHz sky noise occurs and above 5000 MHz atmospheric absorption effects occur.

Radar Astronomy

Echoes from meteor trails were observed in the early 1940s and a radar echo was obtained from the moon in 1946. In 1961 echoes from Venus were observed at its closest approach to the earth and since then from Mercury, Mars and the Sun.

Navigational Aids

Radio aids to navigation have progressed from early direction finding (D/F) and goniometer systems to satellite navigation systems such as NAVSAT. Following D/F systems, the pre-war Lorentz system in Germany used a dots-and-dashes guidance system similar to that described in OBOE, but using CW signals instead of pulses.

Inertial navigation depends on a stable platform and high-grade gyroscopes in order to measure acceleration from a known starting point. From an accurate knowledge of acceleration in two planes and also of lapsed time, position at that time can be calculated.

The very long-range navigational system OMEGA is based on VLF CW transmitters radiating signals on a grid system throughout the world. The first positional system was the hyperbolic grid system GEE invented in 1941, followed by the American system LORAN (Long Range Navigation) in 1942. The first commercial system with high accuracy was DECCA introduced in 1946.

Some, but not all, developments in radar and navigational systems are summarized in Chart 4: "History on a Page".

HISTORY ON A PAGE
(4) Radar

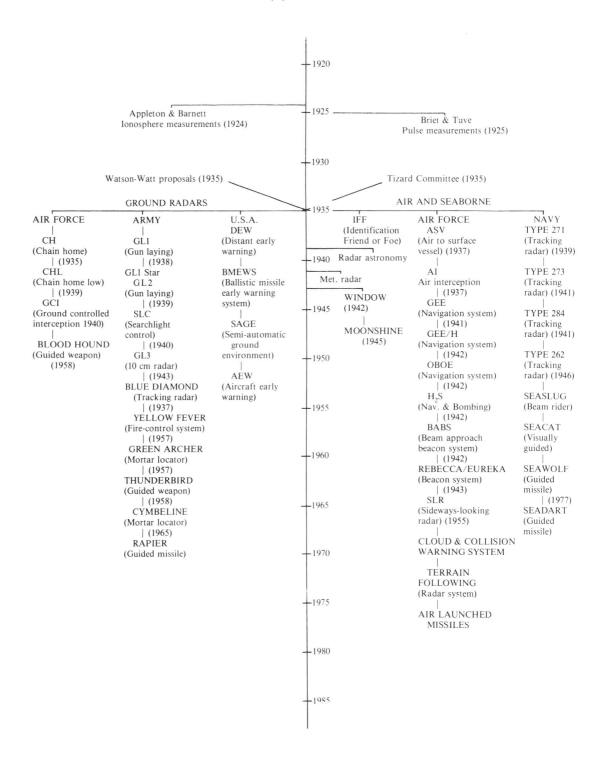

Chart 4

Chapter 7
A Concise History of Television

The heart of all modern television receivers is the cathode-ray tube, on which the transmitted moving pictures are built up in colour. It is a far cry from the initial discovery of cathode-rays by Sir William Crookes in 1878 and its development into an oscilloscope by Ferdinand Braun in 1897, to the sophisticated cathode-ray tube in present-day television receivers. The other major development was the "iconoscope" storage camera tube scanned by an electron beam, originated by Vladimir Zworykin in 1919.

Although A. A. Campbell-Swinton had proposed a system of electronic television, using cathode-ray tubes for both transmission and reception, it was Philo Farnsworth with Zworykin who first developed a practical system which was demonstrated in 1927. By 1929, television receivers were on sale in the U.S.A., under the name "Radiovisor" sets, and low-definition television broadcasting was established. The BBC opened the world's first regular high-definition (405 line) television service in 1936 using a system developed by EMI (Electrical & Musical Industries) and, although the service was shut down during the 1939/1945 war, it was resumed, in black and white, as soon as the war was over.

The principle on which television operates is that a beam of electrons scans across the screen in horizontal sweeps, whilst at the same time moving down the screen until it reaches the bottom, when one "frame" or "field" has been completed. The beam is then deflected back to the top of the screen to start the next frame. The horizontal sweeps are slower than the corresponding fly-back strokes and are synchronized with the incoming picture signals to produce light and dark shades and therefore build up a picture. The faster fly-back strokes are made invisible by blanking out the beam. The picture in contemporary British practice is built up with 625 horizontal lines and in order to save bandwidth, normally several MHz, but give acceptably low flicker, successive frames are interlaced. The television receiver itself has signal circuits to amplify and detect the incoming signals from the aerial in order to modulate the beam of electrons in the tube. The signal also provides a sound channel and synchronizing pulses for triggering the horizontal and vertical scanning at the correct timing. The deflection in both horizontal and vertical directions is provided by time bases, triggered by the synchronizing pulses, which generate currents in the picture tube yoke coils to deflect the beam of electrons.

The first public television service in colour was transmitted in 1951 in the U.S.A. and it was necessary for it to be compatible with black and white television. The colour system adopted was that specified by the NTSC (National Television System Committee). In Europe two systems were adopted in 1956; the PAL (Phase Alternation Line) and SECAM (Sequential and Memory). These three systems are now used throughout the world, digital converters being used to convert one system to another.

In colour, the visible spectrum of light starts at violet, passing through blue, green and yellow and orange to red. An approximately equal mixture of all colours gives white light. Any colour can be synthesized by mixing the three primary colours red, green and blue. The varying level of brightness of the picture elements is known as luminance, whilst in colour television when hue and colour saturation information is added, the extra signal required is known as chrominance. The television camera analyses the light from the scene to be transmitted in terms of its red, green and blue constituents by means of optical filters.

The transmitter separates the picture information into two parts, a colour signal and a brightness signal. These are decoded in the television receiver to give a colour display. Television receivers are one of the greatest technological achievements of the age with their complexity of circuitry, ingenious design, accurate alignment and great reliability. Transistorised television receivers became available in 1959 and the television set has become one of the most important items in the modern home (particularly with the introduction of TELETEXT systems such as CEEFAX in 1972 and ORACLE in 1973) and the likelihood of VIEWDATA expansion in the near future by satellite. The first satellite television pictures were transmitted across the Atlantic on 19th July 1962 by Telstar 1 and by July 1962 sixteen European countries were exchanging television programmes with the U.S.A.

Intelstat 1, the first of the internationally owned satellites, was launched in 1965, and since then a series of television and communication satellites have been in use, relaying television programmes throughout the world. Recently introduced cable TV offers an increased number of channels, less use of other space and the possibility of interactive communication.

Developments in TV since the cathode-ray tube are summarized in chart 5 as: "History on a Page".

HISTORY ON A PAGE
(5) Television

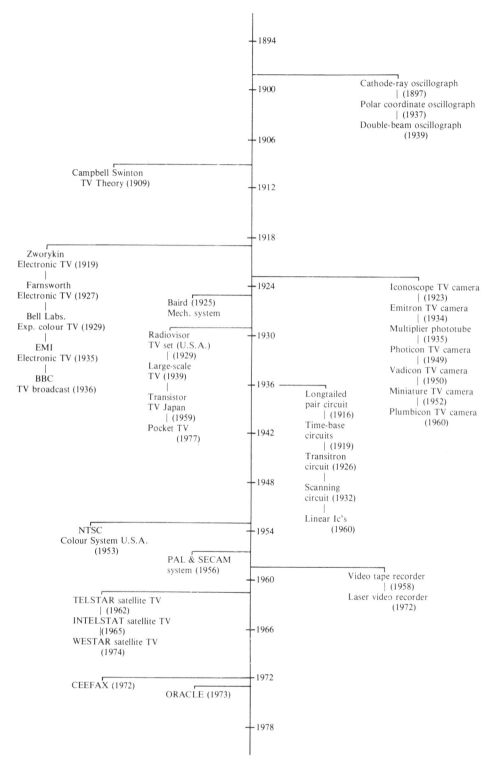

Chart 5

Chapter 8
A Concise History of Computers, Robotics and Information Technology

There is almost a 300 year gap between the invention of the first mechanical computer and the invention of the first electronic computer. In 1642, Blaise Pascal, in France, who was 19 at the time, grew tired of adding long columns of figures in his father's tax office and he designed a mechanical device consisting of a series of number wheels with gears for decimal reckoning, capable of adding and subtracting the long columns of figures. Thirty years later, a German, Gothfried Wilhelm Leibniz, invented the Leibniz wheel using similar principles which could not only do subtractions and additions, but also multiplication and division.

Almost a hundred years passed before Sir Charles Babbage conceived the first design for a universal automatic calculator. Again, a mechanical device using counting wheels, coping with 1000 words of 50 digits each, but with one vital difference, he used punched cards similar to those used in a Jacquard loom to control the programme. Punched cards were also used for input and output devices. The machine contained all the functions necessary in a modern computer — an input unit, a store or memory, an arithmetic unit, a control unit and an output unit.

Improvements were made by Pehr Georg Schuetz in Sweden and a machine similar to Babbage's was built in 1854, which was capable of printing out its own tables. Almost forty years elapsed before H. Hollerith, in America, developed a machine for tabulating population statistics for the 1890 census. Holes in punched cards were used to denote age, sex, etc., and the size of the cards was made 6 5/8 by 3¼ in. because it was the size of a dollar bill. Another forty years later, Vannevar Bush in the U.S.A. developed an early analogue computer for solving differential equations, and analogue computers were built by several universities (e.g. Manchester University in 1934).

The stage was now set for the first of a long line of electronic digital computers, although initially electromechanical rather than purely electronic. Howard H. Aitken of Harvard University designed in 1937 an electromechanical automatic sequence-controlled calculator which was built by IBM and presented to Harvard 7 years later. A relay-operated computer was built by Stibitz, of Bell Laboratories, about the same time.

The first truly electronic computer was the ENIAC (Electronic Numerator Integrator and Computer) begun in 1942 by the University of Pennsylvania and completed in 1946. It used 18,000 tubes, mostly flip-flops and pentode gates, and was 51 feet long and 8 feet high. The numbers used in this machine had 10 decimal units and the numbers could be added in 200 microseconds and multiplied in 2300 microseconds, this being the fastest calculator developed up to this time. Computer circuits progressed from tubes, mercury delay lines, CRT stores, to magnetic cores, tapes and drums. From the middle 1940s, a series of computers were built, each using later electronic techniques as they were developed, i.e. tubes to transistors, transistors to integrated circuits, each becoming smaller and smaller, until the present range of microcomputers was reached. Developments in computers are shown in chart 6: History on a Page''.

How a Computer Works
All digital computers operate by adding, subtracting , multiplying and dividing numbers at incredibly high speeds. The architecture or arrangement of a basic computer consists of an input device to feed in the data, a device to hold temporary resulted until they are required (an accumulator), an arithmetic and logic unit to perform calculations, a memory to hold data and program as required, and an output device to display data. A control unit is used to decode instructions from a predetermined schedule known as a program to control the arithmetic and memory processes.

All calculations are done by using binary arithmetic, rather than decimal arithmetic. Two numbers only are used 0 and 1 in binary instead of 0 - 9 in the decimal system. In the decimal system, a number such as 469 is made up to successive powers of 10. It could equally be written $(9 \times 10^0) \times (6 \times 10^1) + (4 \times 10^2)$ = 9 + 60 + 400 = 469. Binary numbers are made up of successive powers of 2 and can be compared with their decimal equivalents as follows:

Binary	2^0	2^1	2^2	2^3	2^4	2^5	2^6
Decimal	1	2	4	8	16	32	64

HISTORY ON A PAGE
(6) Computers

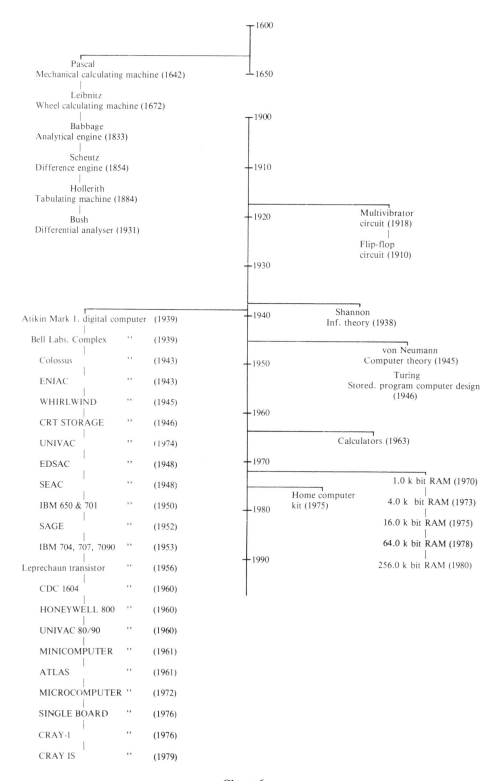

Chart 6

A single binary digit is known as a BIT (Binary Digit). Groups of 8 bits are known as "BYTES". The binary equivalent of a word or working length of the accumulator is between 1 and 8 bytes depending on the computer.

The use of the binary code greatly simplifies the design of electronic computers, for basic circuits having two states only; 0 and 1 can be used throughout.

Transistors have two properties which make them ideal for computers. They are able to store charges and release then when required, i.e. a memory, and they can also act as switches for operating binary codes.

Because the computer operates only in binary code, a programming language is necessary to convert words of instructions into binary codes. Initial low-level language used simple memories such as SUB for SUBTRACT; with these every action of the computer has to be described in detail, but with modern high-level languages, a simple instruction can generate complicated actions. To make sure that no instructions are missed out, a flow chart is used to specify the program. Programming languages such as FORTRAN (FORmula TRANslation) for scientific use, COBOL (Common Business Oriented Language) for commercial use and a BASIC (Beginners All-purpose Symbolic Instruction Code) for teaching were all developed to make the work of the programmer easier.

When a computer is switched on, a pulse generator produces a stream of millions of pulses per second known as clock pulses for timing. Pulses representing BITS of information are operated upon throughout the computer by gates which permit or inhibit the passage of the pulses being distributed through the various stages of calculations. In operation, instructions from a programmer are fed into the computer by means of magnetic tapes or floppy discs in conjunction with a keyboard. These are then stored in the form of binary numbers in rows of locations. The arithmetic unit then multiplies, adds, subtracts, etc., as dictated by the processor, stores the intermediate answers and finally after processing decodes the binary digits and feeds to a printer or VDU (Visual Display Unit) in the form of printed words and numbers.

Great advances have since been made in the storage or memory system of a computer from the earlier systems such as magnetic tapes, cores and drums. Commonly used magnetic tapes can pass through read and write heads so that pulses can be extracted at a speed of around 10 feet per second. A tape containing 9000 bits per foot travelling at 10 ft per second would result in a reading or writing speed of 90,000 bits per second. Normal tapes can hold up to 20 million bits and can be read in about 4 minutes. Magnetic drums have a number of reading heads down the side, so that rapid access can be made to the desired information. Disc storage is used to give high access speed to a large number of locations.

The integrated circuit memory chip consists of many thousands of transistors. It is able to store and extract a given pulse or bit when required. The chip is composed of a number of cells and a chip with 256 individual cells, with each cell containing a byte (a number composed of eight bits), would have eight input lines to tell which cell was being referenced and eight separate lines for the data. ROM's (Read Only Memory) and RAM's (Random Access Memory) chips have a facility on the chip to recall data from the memory. In the case of the ROM, this data is fixed, whereas the RAM chip can be instructed to remember random patterns (i.e. the ROM dictates, the RAM remembers).

The introduction of MOS (Metal Oxide Semiconductor) IC devices revolutionized memories as MOS devices are extremely small. The progress in characteristics and operating speeds over the 10 years between 1970 and 1980 is shown in the Motorola diagram (page 32) published in 1980 with acknowledgements to the Motorola Inc., IC Division, U.S.A.

Types of Computer
At the present moment there are four general categories of computer which vary widely in cost and performance.

The Super Computer
In the first and most expensive category are the "super" computers such as the CRAY, where not only is computation done at very fast speeds, but many computation processes occur in parallel — hence enabling a high throughput of work.

These computers are used for tasks where extremely large numbers of mathematical equations need to be solved numerically in a reasonably short time, often involving the simultaneous input of large volumes of data; such a task is weather forecasting.

A decade of MOS memory evolution

Product	Device count	1970–1972	1972–1974	1974–1976	1976–1978	1978–1980
IK RAM	4,000	17K $t_{Acc}=300$ ns	14K $t_{Acc}=200$ ns			
4K RAM (22 Pin)	14,000	33K $t_{Acc}=250$ ns	27K t_{Acc} 200 ns	21K $t_{Acc}=150$ ns		
4K RAM (16 Pin)	10,000			21K $t_{Acc}=250$ ns	15K $t_{Acc}=200$ ns	12K $t_{Acc}=150$ ns
16K RAM	36,000			33K	28K $t_{Acc}=200$ ns	20K $t_{Acc}=150$ ns
64K RAM	135,000		□ (sq. mils)			39K $t_{Acc}=150$ ns

The Main Frame

In the second category are the so-called "main frame" computers, examples of these are IBM 3033, Univac 90/30 and ICL 2900; in these machines computations are performed fast but with little parallelism. These machines are used for large payrolls and other accounting tasks, factory management, financial planning and for scientific and design problems where a large number of equations need to be solved.

The Minicomputer

In the third category are the minicomputers; calculations in these machines are still reasonably fast: however, the basic word length is short, typically 16 bits (2 bytes), so that the accuracy of calculation for some purposes has to be improved by using several programming steps. These machines are used in the main for tasks of moderate complexity in the accounting, scientific, computer-aided design fields. They are also used for controlling large chemical plants, steel rolling mills and other complex continuous processes. Examples of these machines are VAX 750, GEC 4080, HP 1000 and PDP11.

The Microcomputer

In the fourth category are the microcomputers, the central processing unit of which is usually on a single chip. These computers have a wide range of applications: they are often built into test equipment, central-heating systems and other places where reasonably sophisticated control is required. They are also the heart of home and small business computers, video games, and are used as the main processing element in many automated machines such as banking cash dispensers, store checkout tills and industrial robots.

Computer Systems

A modern computer installation is delivered as two basic parts, the hardware and software. The hardware is the physical manifestation of the computer components, comprising the CPU (Central Processor Unit), memory, control unit, printers, bulk storage devices and other peripheral devices. The software is the systems and utility programmes delivered with the computer to make it work. Such computer programmes include the operating system, compilers, editors and data base systems. Although software is not a physical entity the major components of this normally have to be purchased with the system installation and/or licenced on a yearly basis. The cost of this software including maintenance can be in the order of 25% of the total installation cost of the computer.

Computer Aided Design (CAD)

Computer Aided Design is using computers to aid the designer to do his job efficiently; for example in the electronics industry computers are used to analyse the function of proposed circuits, translate that design onto a printed circuit board including component layout and interconnecting tracks. Computers are used extensively in the design of integrated circuits and hence used to design themselves.

For mechanical design computers are used as drafting aids and for mathematically representing solid objects allowing such properties as mass, centre of gravity, etc., to be evaluated. They can also be used to analyse stresses in components and assemblies; for example, bridges, aircraft components and structures.

Solid modelling techniques also allow real looking computer pictures to be produced which can be used for marketing activities before a prototype has been produced.

Computers Aided Manufacture (CAM)

Computer Aided Manufacture is where computers are used to aid the manufacturing process; for example, the automatic programming of numerically controlled machines such as lathes, two- or three-dimensional milling and drilling machines. Programmes are also produced for printed circuit manufacture, mask production, drilling, assembly, etc. Computers can also produce manufacturing schedules, assembly instructions and work plans for processes to be performed manually.

Robotics

Robotics have been the subject of science fiction for many years, e.g. the book and film "R.U.R." by Karel Capek, but it is only recently that the microcomputer has enabled more sophisticated devices to be made. Servo systems were an early form of what is now termed robotics, e.g. power-assisted steering in cars, servo controls in aircraft and in industrial processes. The first robot to imitate the grasping motion of the human hand was used in a U.S. nuclear plant in 1960 (known as "Handyman"). Definitions of robots vary considerably, but a general description is "a reprogrammable multifunction manipulator designed to move materials, parts, tools or specialized devices through variably programmed motions in space for the performance of a variety of tasks". The popular conception of an industrial robot is an articulated arm with a total of 5 or 6 degrees of freedom achieved through a series of revolute and/or sliding joints. The position of these joints is controlled by a microcomputer which is capable of being variably programmed by the user of the robotic equipment.

Each robotic application requires its own "endeffector" at the end of the arm, usually in the form of gripper, spray gun or welding torch; this device also being controlled by the computer from a user-defined program. The robotics controller also has to be able to control the processes being performed by the robot, i.e. control of welding current or paint flow. The arm power source for the larger robots, intended for the heavier tasks, is frequently hydraulic, with a few notable devices being electrically powered. However, robots intended for lighter tasks, such as part sorting and assembly, are more commonly electrically powered and are rapidly becoming the preferred device.

An important part of the modern industrial robot is the software associated with the control computer. This software consists of a suite of system programs that enable the robot user to teach the required endeffector paths, operations and the control of other external equipment such as welding plant, numerically controlled machines, etc., as a series of recorded steps in the computer's memory. The computer can then move the endeffector through these recorded paths, initiating endeffector and external equipment operations at the required points of robot articulation. The method of teaching the robot varies considerably with intended application; for example, teaching the current generation of paint-spraying robots entails moving the robot arm through the required spraying motions at the proper speed, the computer recording these movements to be played back during operation. On the other hand, pick and place robots are taught the positions where operations have to occur, usually with a few intermediate points to ensure collision free operation, leaving the robot to derive the path between these points. The more sophisticated robots have software refinements that allow the arm to move from point to point along predefined straight or curved paths, possibly with a change in attitude of the endeffector; the simpler machines use joint interpolation (i.e. driving the joints at constant velocity) to move from point to point, the actual trajectory of the end-effector path depending on the robot's geometry.

The load-handling capacity, overall size and reach of these robots varies considerably. One of the largest robots is the Cincinnati Milacron: this robot can handle loads of 102 kg and position them in space with the required orientation, to within less than 1.25 mm and has a total reach of 2.5 M; this machine has a hydraulic powered source. One of the smaller machines in the Unimation PUMA 250 which can only handle 1. 5 kg, and has a reach of just 0.45 M with positional repeatability of better than 0.05 mm; this machine being electrically powered. The operational speed of the industrial robot (the maximum controllable velocity of the endeffector at the end of the arm) varies considerably with the intended task they are designed to do. For example, a fast pick and place robot can achieve speeds of up to 1.5 metres per second, whereas the welding robot will weld at a maximum of 1 meter per minute, between welds a faster speed is, of course, required.

The robot in a modern factory is an impressive mechanical arm normally performing a hot, heavy and/or dangerous task which is essentially repetitive. Such tasks are paint spraying, operations with die casting and plastic injection moulding machines, forging, spot and inert gas welding, palletising of glass sheets, deburring and fettling, dipping of wax mould forms in slurry for lost wax castings, trimming of glass reinforced plastic (GRP) components and the spraying of GRP material in confined spaces: such an example is the manufacture of boat hulls. Other applications are tasks that are essentially repetitive and where the use of the robot not only relieves a bored operator, but consequently performs a more repeatable job, hence enabling the quality of that component or product to be enhanced. Applications of this type are picking up and orientating plastic bottles, putting chocolates into boxes, assembly of small electric motors and various products in the electronic industry.

The industrial robot is in its infancy, a device made economically possible by the advent of servo technology and the microcomputer. Further advances in microelectronics and microcomputer architecture combined with advances in software will allow a large increase in sophistication for futher generations of robots. These future generations are likely to have a multiplicity of sensors to monitor the state of the work pieces, the robot, peripheral equipment and the work space. This will allow the robot control system to cope with disordered components, faulty work pieces and robot situations, by the implementation of flexible recovery strategies.

Information Technology

Allied to computers and communications technology is the increase in information now available to the public, not only television, but computer information on 'plane bookings, computer search for technical data, cable TV, satellite communication, speech recognition and synthesis, computer-aided translation, electronic publishing, electrophotographic transmission of data, computer learning aids and a whole range of developments which disseminate data of all types to users. This is increasingly becoming known as information technology (IT).

Information Technology, or as the French call it, "Télématique", is defined as follows: the acquisition, processing, storage, dissemination and use of vocal, pictorial, textual and numerical information by a micro-electronics based combination of computing and telecommunications.

The most familiar examples of information technology are the PRESTEL or VIEWDATA and the TELETEXT systems, CEEFAX and ORACLE. The PRESTEL system combines a TV set, a telephone line and a computer. A push-button control panel calls up a "page" of information required by a subscriber on to his TV screen, using a telephone line link routed into a computer data bank.

CEEFAX and ORACLE signals, building up similar pictures, are transmitted by the BBC and ITV during the vertical blanking interval, i.e. the unused part of the TV signal. The service provides around 1000 pages of constantly up-dated information, accessed by a selector panel by the owner of the TV set.

Communication systems are of two main types: Public (telephones) and Dedicated (Private Networks). The trend to digital transmission of speech and sound is linked with computer technology, so that, in effect computer can talk to computer. Initially, local networks were introduced into offices of large firms to link copiers and word processors; for example, "ETHERNET" in the U.S.A. or PROWAY (International). Other systems use telephones to link, for instance, banks to a central computer.

Communication networks now use coaxial cable or fibre optics for local use, known as "data highways", and a whole new language is developing such as BUSSING, PROTOCOL, MULTIPLEXING, SMART TERMINALS, HIERARCHIAL PROCESSING and many others, as the systems develop.

Wideband cable will provide cable TV, giving local information, sports programmes, home-security services, fire alarms, home banking and shopping, electronic mail, interactive computer-assisted learning, business communications, instant opinion polls and many other services. When linked to direct satellite broadcasting, international TV programmes become available and information technology will spread worldwide as it becomes financially sound.

Chapter 9
A Concise History of Industrial and Other Applications

Industrial Electronics

The earliest general use of electronics in industry was for control of processes, such as induction heating for the hardening of steels and setting of glues, electroplating, resistance welding for the joining of metals and ultrasonics for cleaning castings, crack testing in metals, drilling glass, etc. Photocells were used for a wide variety of devices, e.g. the automatic opening of doors, smoke-density control, sound on film ciné, sorting small objects by colour, facsimile transmission of photographs and many others.

Process control needs three functions — a transducer to convert changes in pressure, flow, temperature, colours, or levels into electrical voltages, a signal processor or amplifier and a control mechanism to open and close valves, regulate power, etc., so that the required equilibrium is obtained. A feedback system is usual to regulate the amount of control necessary.

With the introduction of the transistor and later the integrated circuit, process control expanded and with the development of computers, particularly mini-and-microcomputers, a new field of control was opened up. Large masses of data could be accommodated by the computer and assessments could be made of the effects of possible changes in production before actual decisions were made. The computers can be used to control production processes with great accuracy and there is little limit to the type of industry which can now use electronics in one form or another.

Signal processors are now all computers able to cope with process variations required in controlling production outputs. Among the industrial applications are the metal industry, the mining, glass pulp and paper, machine tool, cement, textile, rubber, plastics and chemical industry in addition to aircraft and space applications, land and marine transport applications.

Mini-or microcomputers are used in automatic machines and systems on the factory floor to carry out many different processes. For example, the industrial robots are used for paint spraying, spot welding, handling dangerous and difficult material such as hot glass and making cores for casting processes, fettling, etc. Numerically controlled machines come in many different forms, lathes, mills, drills, routers, etc. Computers are also used to control automatic test equipment which test the electronic components coming into the factory and the completed products, and are even able to isolate and report the type and position of a fault. Another important area of use is the transfer of parts between other automatic machines in the factory for automatic warehousing.

Numerically Controlled Machines

The so-called numerically controlled (NC) machines, are a special group of industrial machines controlled by electronics, usually a microcomputer, which receives commands in the form of a sequence of numbers. On receipt of these numerical commands, the machine interprets them and controls two or more axes simultaneously to achieve the required result. Examples of machines using this type of control are NC lathes which normally have two axes, NC milling machines which have three, four, or possibly five axes and automatic printed circuit drilling and routing machines which are three-axes devices.

The mode of working these machines is that the computer reads the input sequence of commands, one item at a time, then uses interpolation (the form of which depends upon the application) on the positional commands to precisely control the velocities of each of the axes to achieve the required tool path. Some of the commands in the sequence will, of course, tell the computer to change the cutting speed, the tool type, shaft speed, etc.

The control sequence or programme for these machines can be created by a variety of methods: (a) the NC machine itself can be used to create the programme sequence by moving the cutter to the required positions, recording these and the type of interpolation to be used between this and the next point; (b) the sequence can be programmed off-line by a production planner, using a human readable form of programme which is later translated by the computer into numerical sequence; (c) the sequence can be generated directly from a computer-aided design (CAD) system.

Automobile Electronics

Microprocessors are increasingly being used to control functions such as fuel conservation and also for dashboard displays of car performance. Electronically controlled automatic gear shifting, electronic ignition, alternators, voltage regulators, tachometers, warning devices, etc., are all in use in modern cars.

Engine analysers electronically measure such characteristics as r.p.m., timing, ignition, cylinder leakage, combustion efficiency, vibration, cam dwell angle and many others. A computer is now available which will instruct a four-cylinder 1400-cm^3 engine whether to operate on two cylinders or four. Sensors measure the intake manifold pressure, the water temperature, the r.p.m., the speed and choice of gears. When conditions are right the computer orders a supplementary oil pump to cut out the inlet and exhaust valves on number 1 and number 4 cylinders.

Fault diagnosis through sensors and dashboard display will tell the driver if any fault develops and automatic control of air conditioning and temperature is also being installed in modern cars. The comfort and safety for drivers is being improved by the use of more and more electronics.

Medical Electronics

Heart sound amplifiers have been used for many years, together with cardiograph displays of heartbeat waveforms. The electroencephalograph (EEG) displays brain rhythms and computers analyse their waveforms. Since the introduction of the first cardiac pacemaker in 1960, great strides have been made in their use. Defibrillators, providing electrical countershocks to the heart and electro-cardiographs observing blood circulation and functional heart phenomena, are in use. Intensive-care patient monitoring systems are standard in most hospitals and closed-circuit TV systems are also used. Other electronic aids are laser retinal photocoagulators, radio pills, ultrasonic echo analysis and hearing aids. Fibre optics are used to illuminate areas nomally inaccessible. Pressure-sensitive radio pills were used for measuring conditions in the gastro-intestinal tract. A major electronic technique is non-invasive diagnosis by computerized scanning, using nuclear, ultrasonic, fluoroscopic and X-ray equipment. Computerized axial tomography can provide "slices" of patients' anatomy and yield valuable diagnosis.

Electronic devices are also used in forensic science, using gas chromatography for drugs analysis and the scanning electron microscope linked to a semiconductor detector and data processor to enable chemical analysis of microscopic samples.

Educational Electronics

Language teaching from a recorded master tape to numbers of students in individual cubicles is well known. Learning systems are available in which step-by-step questions are set with recorded answers. These can vary from instruction in clerical procedures to TV chassis assembly and can cover many forms of industrial training. Education in physics, electronics, computers, etc., is available by simple assemblies of parts or modules to make up a complete oscilloscope, transistor receiver, microwave set, universal power supply, etc. More complex kits can demonstrate standing waves, magnetic resonance, electron spin resonance, gas lasers, etc. Electronic displays for educational purposes vary from illuminated flow lines for industrial processes and aircraft visual flight displays to complete TV educational studios. Educational TV networks can be extended to individual homes, whilst hospital TV operations can be micro-linked to other hospitals. The use of television for education with two-way interaction will certainly increase in the future.

Office and Banking Electronics

Computers can now do tasks such as cost accounting, ledger records and retrieval, updating and filing. Data collection and storage on magnetic tape with data processing can present analyses on visual displays or printout. Telex systems with data printers, facsimile document transmission and time-sharing computers allowing multiple use, all point the way to the electronic office. In the typing of letters or documents, word processors are now used. The typist starts by typing a draft document on to a VDU (Visual Display Unit). At the same time, a copy of the document is also written on a floppy disc. When the final version of the document is returned with all its corrections, modifications, additions, etc., having been made, the typist has only to type in the corrections or additions. The original copy on the floppy disc, now corrected, is used to type automatically, at 175 words per minute, a perfect copy of the document. The word processor makes logical decisions such as automatically deleting, re-spacing and repositioning words and sentences, inserting hyphens, determining line lengths and controlling format and tabulation.

In banking, bank-note counting and dispensing machines deal with both notes and cash, computers deal with chequeing transactions, mortages, loans, and dividend payments, credit-card systemization, security and customer recognition safeguards, all data being sorted on magnetic tapes. Page-reading systems use flying spot scanning of characters from a cathode-ray tube and a group of photocells pick up reflections and output them to a decision network to make the white, black or grey decision (video processing). Page readers can read 800 characters per second (half-million per hour). Electronic machines are available which can identify genuine banknotes and reject those which are counterfeit.

Consumer Electronics

The microcomputer is being built into more and more consumer products. They range from the very simplest built on a single chip to more complex units with several dozen chips.

An example of an application which uses a computer on a chip is the electronic door chime. The programme for this device is stored in a ROM which is programmed when the chip is made. Twelve different tunes can be played, the tune depending on which door the chime was initiated from and tune selected. Other applications are controllers for central-heating systems which may be programmed with different on/off cycles, home games and teaching aids — a good example is the "speak and spell" unit which even synthesizes speech! More complex micros are used in gaming machines found in arcades and pubs in which sophisticated graphics and sound effects are generated as well as complex game strategies. Micros now find their way into such mundane devices as washing machines, cookers and video tape recorders.

Security Electronics

Intruder alarms now use sound detectors, magnetic switches on door, light beams, laser beams, pressure mats and hoses, infra-red detection, microwave radar, surveillance scanners, security pass systems and TV coverage.

Miscellaneous Systems

Electronic devices are used for control of traffic lights, warehouse control systems, timing devices, mass production processes, paging systems, simulators (aircraft, etc.), railway signalling, model control, electronic cameras and many others too numerous to mention.

Chapter 10
List of inventions by subject (ie: assembly techniques, circuitry, communications, components, computers, etc.)

1. Assembly Techniques & Packaging

1940	Thick film packaged circuits	(Centralab)
1943	Printed wiring	(Eisler)
1945	Potted Circuits	(Various)
1947	Automatic circuit making equipment	(Sargrove)
1949	Dip soldering of printed circuits	(Danko & Abramson)
1950	Thermo compression bonding techniques	(Anderson, et al)
1951	"Tinkertoy" automatic assembly system	(Nat. Bur. Standards)
1953	Wire wrapping of connections	(Mallina et al)
1953	Automatic assembly systems-Autofab, Mini-Mech, etc.,	(General Mills, et al)
1956	"Flowsoldering" of printed circuits	(Fry's Metal Foundries)
1958	"Micro-module" assembly system	(U.S. Army)
1960	Printed wiring multilayer boards	(Photocircuits)
1962	"Flat-Pack" integrated circuit	(Tao)
1964	DIP or DIL (Dual in line package)	(Rogers)
1964	Etch-back plated-through hole technique	(Autonetics)
1964	Beam lead connections	(Lepselter)
1966	"Flip-chip" bonding technique	(Wiessenstern et al)
1971	Ceramic chip carrier	(3 M Co.,)
1973	Dry etching technique	(Mitsubishi)

2. Circuitry

1826	Ohm's law	(Ohm)
1843	Wheatstone bridge	(Wheatstone)
1848	Boolean algebra	(Boole)
1945	Kirchhoff's laws	(Kirchhoff
1912	Regenerative circuit	(de Forest, et al)
1912	Heterodyne & superheterodyne circuits	(Fessenden, Armstrong)
1915	Filter networks	(Campbell, Wagner)
1918	Neutrodyne circuit	(Hazeltine)
1918	Shot effect noise	(Schottky)
1918	Multivibrator circuit	(Abraham & Bloch)
1918	Dynatron circuit	(Hull)
1919	Retarded field microwave oscillator	(Barkhausen & Kurtz)
1919	Flip-flop circuit	(Eccles, Jordan)
1919	Miller time base circuit	(Miller)
1921	Crystal control of frequency	(Cady)
1922	Negative resistance oscillator	(Gill, Morrell)

1922	Super regeneration	(Armstrong)
1923	Squegger circuit	(Appleton, Herd et al)
1924	Linear saw-tooth time base circuit	(Anson)
1925	Johnson noise	(Johnson)
1926	Transitron oscillator	(van der Pol)
1926	Automatic volume control circuit	(Wheeler)
1927	Negative feedback amplifier	(Black)
1932	Energy conserving scanning circuit	(Blumlein)
1933	Hard valve time base circuit	(Puckle)
1935	Constant RC stand-off circuit	(Blumlein)
1936	Long tailed pair circuit	(Blumlein)
1939	Radio altimeter	(Bell Labs)
1942	Miller integrator circuit	(Blumlein)
1942	Phantastron circuit	(Williams & Moody)
1942	Sanatron circuit	(Williams & Moody)
1943	Magnetic amplifier	(A.S.E.A.)
1947	High quality amplifier circuit	(Williamson)
1952	Darlington pair circuit	(Darlington)
1952	Digital voltmeter	(Kay)
1952	Negative feedback tone control circuit	(Baxandall)
1960	Neuristor circuit	(Crane)
1960's	Logic circuits - microelectronics	(Various)
1960's	Linear circuits - microelectronics	(Various)
1967	Rotator circuit network	(Chua)
1968	Mutator circuit network	(Chua)
1969	Bucket-brigade delay circuit	(Sangster and Teer)
1979	Satellite echo cancelling circuit	(Bell Labs)
1980	Fibre-optic laser driven superheterodyne	(Saito et al)

3. Communications

(See also - RADIO   (History on a Page 19))

1837	Telegraphy - Morse code	(Morse)
1860	Microphone	(Reis)
1865	Radio wave propagation	(Maxwell)
1866	Transatlantic telegraph cable	(T.C. & M. Co.,)
1876	Telephone	(Bell)
1887	Aerial	(Hertz)
1890	Coherer	(Branley)
1893	Waveguide - theory	(Thomson)
1896	Wireless telegraphy	(Marconi)
1901	Radio - Heaviside/Kenelly layer	(Heaviside/Kenelly)
1906	Radio broadcasting	(Fessenden)
1906	Crystal detector	(Dunwoody)
1907	Crystal detector (Perikon)	(Pickard)
1912	Ionospheric propagation	(Eccles)
1915	Single sideband transmission	(Carson)
1918	Ground wave propagation	(Watson)
1921	Short wave radio (amateur)	(Amateurs)
1921	Crystal control of frequency	(Cady)
1925	Short wave radio (commercial)	(van Boetzelean)
1925	Ionosphere layer	(Appleton)
1926	Yagi aerial	(Yagi)
1928	Diversity reception	(Beverage, et al)
1928	Frequency standards - quartz clock	(Horton, Morrison)
1929	Microwave communication	(Clavier)
1933	Radio astronomy	(Jansky)
1933	Frequency modulation	(Armstrong)
1934	Frequency standards - atomic clock	(Cleeton, Williams)
1936	Waveguides	(Southworth, et al)
1937	Pulse code modulation	(Reeves)
1939	Frequency standards - caesium beam	(Rabi)
1945	Satellite communication theory	(Clarke)
1948	Information theory	(Shannon)
1950's	MODEM	(MIT, Bell Labs)
1954	Transistor radio set	(Regency)
1956	Transatlantic telephone cable	(Various)
1956	Radio paging	(Multitone)
1957	SPUTNIK I Satellite	(USSR)
1958	EXPLORER I Satellite	(USA)
1958	VANGUARD - I Satellite	(USA)
1958	PIONEER - I Satellite	(USA)
1958	SCORE Satellite	(USA)

1959	LUNIK - I Satellite	(USSR)
1959	DISCOVERER - I Satellite	(USA)
1960	TIROS - I Satellite	(USA)
1960	ECHO - I Satellite	(USA)
1960	COURIER - IB Satellite	(USA)
1960	TRANSIT - IB Satellite	(USA)
1961	VENUS - I Satellite	(USSR)
1961	VOSTOK - I Satellite	(USSR)
1961	MERCURY-ATLAS-4 Satellite	(USA)
1961	OSCAR-I Satellite	(USA)
1962	TELSTAR-I Satellite	(USA)
1962	MERCURY-ATLAS-6 Satellite	(USA)
1962	OSO-I Satellite	(USA)
1962	RELAY-I Satellite	(USA)
1962	MARINER-2 Satellite	(USA)
1962	ALOUETTE-I Satellite	(Canada)
1962	MARS-I Satellite	(USSR)
1962	ARIEL-I Satellite	(UK)
1963	SYNCOM-I Satellite	(USA)
1963	VELA-I Satellite	(USA)
1964	NIMBUS Satellite	(USA)
1964	VOSKHOD-I Satellite	(USSR)
1964	Packet-switching	(Baron)
1965	SNAPSHOT Satellite	(USA)
1965	GGSE-2 Satellite	(USA)
1965	LES-I Satellite	(USA)
1965	PEGASUS-I Satellite	(USA)
1965	INTELSTAT-I Satellite	(International)
1965	FR - I Satellite	(France)
1965	OVI-2 Satellite	(USA)
1965	PROTON-I Satellite	(USA)
1966	OAO - I Satellite	(USA)
1966	SURVEYOR - I Satellite	(USA)
1966	ESSA - I Satellite	(USA)
1966	BIOSATELLITE - I Satellite	(USA)
1966	LUNAR ORBITER - I Satellite	(USA)
1966	ATS-I Satellite	(USA)
1966	Optical fibre communications	(Kao, Hockham)
1966	PAGEOS - I Satellite	(USA)
1967	DIADEME - I Satellite	(France)
1967	SOYUZ - I Satellite	(USSR)
1968	APOLLO - 7 Satellite	(USA)
1968	IRIS (ESRO-I) Satellite	(Europe)
1968	SURVEYOR - 7 Satellite	(USA)
1969	TACSAT - I Satellite	(USA)
1969	SKYNET - A Satellite	(UK)
1969	AZUR Satellite	(Germany)
1970	TUNG-FANG-HUNG Satellite	(China)
1970	NATO - I Satellite	(NATO)
1970	OFO - I Satellite	(USA)
1971	SHINSEI Satellite	(Japan)
1971	OREOL - I Satellite	(France/USSR)
1971	DSCS - I Satellite	(USA)
1971	SALYUT - I Satellite	(USSR)
1972	LANDSAT - I Satellite	(USA)
1973	SKYLAB - I Satellite	(USA)
1973	SAMOS Satellite	(USA)
1974	WESTAR - I Satellite	(USA)
1974	PRESTEL system	(Fedida)
1975	VIKING - I Satellite	(USA)
1975	STARLETTE Satellite	(France)
1975	RADUGA - I Satellite	(USSR)
1976	MARISAT - I Satellite	(USA)
1978	Integrated optoelectronics	(Yariv et al)
1978	Tamed frequency modulation	(Philips)

4. Components

BATTERIES:

1800	Volta's pile	(Volta)
1803	Accumulator	(Ritter)
1839	Magnetohydrodynamic generation	(Faraday)
1839	Fuel cell	(Grove)
1860	Lead-acid cell	(Plante)

1868	Dry cell	(Léclanché)
1870	Clark standard cell	(Clark)
1884	Zinc mercuric-oxide cell	(Clark)
1891	Weston standard cell	(Weston)
1900	Nickel-iron cell	(Edison)
1900	Nickel-cadmium cell	(Junger and Berg)
1954	Solar battery	(Chapin, Fuller et al)

CAPACITORS:

1745	Leyden jar	(von Kleist et al)
1874	Mica capacitors	(Bauer et al)
1876	Rolled paper capacitors	(Fitzgerald)
1900	Ceramic capacitors	(Lombardi)
1904	Glass tubular capacitors	(Moscicki)
1956	Solid electrolyte capacitor	(McLean & Power)
1956	Semiconductor diode junction capacitor	(Giacoletto, et al)

FILTERS:

| 1915 | Filters, electromagnetic | (Campbell & Wagner) |

GALVANOMETERS:

1820	Electro-magnetism (galvanometer)	(Oersted)
1828	Moving coil	(Schweigger)
1828	Astatic	(Nobilli)

GONIOMETER:

| 1907 | Goniometer | (Artom) |

INDUCTORS:

1772	Iron dust cores	(Knight)
1909	Ferrites	(Hilpert, Snoek)
1956	Magnetic material (YIG)	(Bertaut & Forrat)
1977	Anistropic permanent magnet	(Matsuschita Elec.)

MOTORS:

1837	Electric motor	(Davenport)
1888	Induction motor	(Tesla)
1902	Synchronous induction motor	(Danielson)

RECTIFIERS:

| 1926 | Copper oxide | (Grondahl & Seiger) |

RELAYS:

| 1837 | Telegraph bell and signal | (Cooke, Davy et al) |
| 1950's | Ferreeds | (Bell Labs) |

RHEOTOME:

| 1868 | Waveform plotter | (Lenz) |

RESISTORS:

1850	Thermistor	(Faraday)
1885	Moulded carbon composition	(Bradley)
1897	Carbon film	(Gambrell et al)
1913	Sputtered metal film	(Swann)
1919	Spiralled metal film	(Kruger)
1925	Cracked carbon	(Siemens & Halske)
1926	Sprayed metal film	(Loewe)
1931	Oxide film	(Littleton)
1957	Nickel-chromium film	(Alderton et al)
1958	Field effect varistor	(Bell Labs)
1959	Tantalum film	(Bell Labs)

SWITCHES:

1884	Quick break	(Holmes)
1887	Quick make and break	(Holmes)
1950's	Ferreed switch	(Bell Labs)

TRANSFORMERS:

1831	Transformer	(Faraday)
1885	Distribution	(Deri)
1885	Power	(Zipernowski et al)

WAVEGUIDES:

1893	Theory	(Thomson)
1936	Waveguides	(Southworth et al)

WIRES AND CABLES:

1812	Cable insulation	(Sommering et al)
1845	Metallic sheathing	(Wheatstone)
1847	Submarine cable insulation	(Siemens)
1905	Insulated sodium conductor	(Betts)
1933	Polythylene insulation	(ICI)
1949	Microwire	(Ulitovsky)
1965	Wiegand wire	(Wiegand)
1965	Smooth-surfaced wire drawing	(Olsen et al)

5. Computers

(See also - History on a Page 29)

1642	Calculating machine	(Pascal)
1672	Calculating machine	(Leibniz)
1833	Calculating machine	(Babbage)
1848	Boolean algebra	(Boole)
1854	Calculating machine	(Scheutz)
1889	Tabulating machine	(Hollerith)
1931	Differential analyser	(Bush)
1938	Information theory	(Shannon)
1939	Bell Labs complex computer	(Stibitz et al)
1939	Digital computer	(Aitken)
1942	"Velodyne" analyser	(Williams & Uttley)
1943	COLOSSUS	(Newman, Turing et al)
1943	ENIAC	(Moore School)
1945	Whirlwind	(M.I.T.)
1945	Computer theory	(von Neuman)
1946	CRT storage computer	(Williams)
1946	ACE	(Turing)
1947	EDVAC	(Penn. University)
1947	UNIVAC	(Eckert, Maunchly)
1948	SEAC	(N.B.S.)
1948	EDSAC	(Wilkes)
1950	IBM 650	(I.B.M.)
1950	IBM 701	(I.B.M.)
1950	Hamming code	(Hamming)
1950's	APL language	(Iverson)
1951	Microprogramming	(Wilkes)
1952	SAGE	(I.B.M., M.I.T.)
1953	IBM 704, 709 and 7090	(I.B.M.)
1956	Transistorised computer	(Bell Labs)
1957	Plated wire memories	(Gianole)
1960	Honeywell 800	(Honeywell)
1960	UNIVAC solid state 80/90	(I.B.M.)
1960	CD 1604	(Control Data Corp.,)
1961	Minicomputer	(Digital Equip. Co.,)
1969	Semiconductor memories system	(Agusta et al)
1969	Magnetic bubble memories	(Bobeck et al)
1970	Charge coupled device memories	(Boyle, Smith)
1970	Floppy-disc recorder	(I.B.M.)
1972	Microcomputer	(Intel)
1972	1024 bit random access memory	(Intel)
1973	Logic-state analyser	(House)
1973	Logic-timing analyser	(Moore)
1974	16 bit single chip microprocessor	(National)
1975	4096 bit R.A.M.	(Fairchild)
1975	16, 384 bit R.A.M.	(Intel)
1976	One board computer with programmable 1/0	(Intel)
1976	Polysilicon resistor loaded RAM's	(Mostek)
1977	CCD analog-to-digital converter	(GE Corporation)
1978	Integrated optoelectronics	(Yariv et al)
1978	One megabit bubble memory	(Intel & Texas)
1980	256 K dynamic RAM	(NEC, Toshiba et al)

6. Industrial

1839	Microfilming	(Dancer)
1843	Facsimile reproduction	(Bain)
1908	Geiger counter	(Geiger, Rutherford)
1912	Tungar rectifier	(Langmuir)
1913	Reliability standards	(AIEE)
1914	Ultrasonics (ASDIC, SONAR)	(Langevin)
1914	Thyratrons	(Langmuir)
1916	Reliability, control system	(Bell/Western Elec.)
1918	Induction heating	(Northrup)
1920	Ultra-micrometer	(Whiddington)
1926	Copper oxide rectifier	(Grondahl & Geiger)
1931	CRO cardiograph	(Rijant)
1931	Reliability - quality control-charts	(Shewhart)
1933	Ignitron	(Westinghouse)
1937	Xerography	(Carlson)
1943	Reliability - sequential analysis	(Wald)
1943	Magnetic amplifier	(ASEA)
1944	Reliability - sampling inspection tables	(Doge & Romig)
1951	Quality control handbook	(Juran)
1961	Electronic clock	(Vogel et Cie)
1962	Electronic watch	(Vogel et Cie)
1962	Duane reliability growth theory	(Duane)
1963	Ink jet printing process	(Sweet)
1963	Electronic calculator	(Bell Punch Co.,)
1964	Telemedicine	(Various)
1964	Word processor	(IBM)
1967	Ion beam coating	(Chopra & Randlett)
1971	Electronic digital watch	(Time Computer Corp)
1972	Video games	(Magnavox)
1978	All electronic clock face	(Hosiden Elec.)
1979	Seven-colour ink-jet printer	(Siemens)

7. Microelectronics

(See also 12 - Transistors and Semiconductor Devices)

1852	Thin film sputtering process	(Grove)
1913	Sputtered metal film resistors	(Swann)
1940	Thick film circuits	(Centralab)
1949/50	Ion implantation	(Ohl & Shockley)
1952	Semiconductor integrated circuit concept	(Dummer)
1952	Zone melting technique	(Pfann)
1957	Nickel chromium thin film resistors	(Alderton, Ashworth)
1959	Semiconductor integrated circuit patent	(Kilby)
1959	Tantalum thin film circuits	(Bell Labs)
1959	Planar process	(Hoeni)
1960	Epitaxy - vapour phase	(Loor, et al)
1960	Digital and linear integrated circuits	(Various)
1961	Epitaxy - liquid phase	(Nelson)
1961	Minicomputer	(Digital Equip. Co.,)
1962	MOS integrated circuit	(Hofstein & Helman)
1963	Surface acoustic wave devices	(Rowen & Sittig)
1967	Laser trimming of thick film resistors	(Various)
1967	Ion beam coating	(Chopra and Randlett)
1968	C-MOS integrated circuit	(Various)
1968	Aluminium metallisation of I.C.'s	(Noyce)
1969	Collector diffusion isolation	(Bell Labs, Ferranti)
1970	X-ray lithography	(Feder, et al)
1971	FAMOS integrated circuit	(Frohman-Bentchowsky)
1971	Liquid crystal study of oxide defects	(Keen)
1972	Microcomputer	(Intel)
1972	1024 bit random access memory	(Intel)
1972	Nitrogen-fired copper wiring	(Grier)
1972	Two-layer resist technique	(Bell Labs)
1972	V-MOS technique	(Rodgers)
1972	Integrated injection logic	(Hart & Slob)
1973	Dry etching technique	(Mitsubishi)
1974	16 bit single chip microprocessor	(National)
1974	Electron beam lithography	(Bell Labs)
1975	Thin film direct bonded copper process	(Burgess, et al)

1975	LOCMOS integrated circuit	(Philips)
1975	Integrated optical circuits	(Reinhart, Logan)
1975	4096 bit random access memory	(Fairchild)
1975	Silicon anodisation	(Cook)
1976	Microelectronic versatile arrays	(Philips)
1976	One board with programmable 1/0 computer	(Intel)
1976	16, 384 bit random access memory	(Intel)
1977	H-MOS	(Intel)
1977	TRIMOS device	(Stanford University)
1977	CCD analog/digital converter	(G.E. Corporation)
1978	Laser annealed polysilicon	(Texas Instruments)
1978	Integrated Schottky logic	(Philips)
1981	Hydrogenated amorphous silicon films	(Grasso et al)

8. Physics

1780	Galvanic action	(Galvani)
1800	Infra-red radiation	(Herschel)
1801	Ultra-violet radiation	(Ritter)
1808	Atomic theory	(Dalton)
1820	Electro-magnetism	(Oersted)
1821	Thermoelectricity	(Seebeck)
1826	Ohm's law	(Ohm)
1831	Electromagnetic induction	(Faraday)
1832	Self induction	(Henry)
1834	Electrolysis	(Faraday)
1839	Photovoltaic effect	(Becquerel)
1840	Thermography	(Herschel)
1847	Magnetostriction	(Joule)
1851	Relation between theory of magnetism & electricity	(Kelvin)
1858	Glow discharges	(Pleucker)
1878	Cathode rays	(Crookes)
1879	Hall effect	(Hall)
1880	Piezo electricity	(Curies)
1882	Wimshurst machine	(Wimshurst)
1895	X-rays	(Rontgen)
1897	Electron	(Thomson)
1897	Cathode ray oscillograph	(Braun)
1900	Quantum theory	(Planck)
1902	Spontaneous atomic change	(Rutherford & Soddy)
1905	Theory of relativity	(Einstein)
1911	Superconductivity	(Onnes)
1911	Atomic theory	(Rutherford)
1912	Cloud chamber	(Wilson)
1913	Atomic orbit theory	(Bohr)
1918	Atomic transmutation	(Rutherford)
1921	Ferroelectricity	(Vasalek)
1929	Cyclotron	(Laurence)
1930	High field superconductivity	(de Haas & Voogd)
1930	Van de Graaf accelerator	(Van de Graaf)
1932	Neutron	(Chadwick)
1932	Transmission electron microscope	(Knoll, Ruska)
1932	Cockroft-Walton accelerator	(Cockroft & Walton)
1934	Liquid crystals	(Dreyer)
1934	Trans-uranian atoms	(Fermi)
1934	Scanning electron microscope	(Knoll, et al)
1935	Superconducting switching	(Casimir-Jonker et al)
1937	Xerography	(Carlson)
1938	Nuclear fission	(Fritsch & Meitner)
1941	Betatron	(Kerst)
1947	Molecular beam epitaxy	(Sosnowski et al)
1948	Holography	(Gabor)
1953	MASER	(Townes & Weber)
1955	Infra-red emission from GaSb	(Braunstein)
1955	Cryotron	(Buck)
1956	YIG magnetic materials	(Bertaut, Forrat)
1958	LASER	(Schalow, Townes)
1958	Mossbauer effect	(Mossbauer)
1959	Intrinsic 10μ photoconductor	(Lawson et al)
1960	Sub-millimetre photoconductive detector	(Putley)
1961	Transferred electron effect	(Ridley, Watkins)
1961	Transferred electron device	(Hilsum)
1962	Semiconductor laser	(Hall et al - also Nathan and Lasher)

1962	Josephson effect	(Josephson)
1962	LED (Gallium arsenide phosphide)	(Holonyak)
1963	Ion plating	(Mattox)
1963	Gunn diode oscillator	(Gunn)
1963	Surface acoustic wave devices	(Rowen & Sittig)
1964	"IMPATT" diode	(Johnston & de Loach)
1970	X-ray lithography	(Feder et al)
1972	X-ray scanner	(E.M.I.)
1972	Automatic crystal growth control	(Bardsley et al)
1972	Deep proton-isolated laser	(Dymeut et al)
1973	Scanning acoustic microscope	(Quate)
1975	GYROTRON	(Gapanov)
1977	FLAD display system	(Ins. App. S.S.Phy.)
1978	Light bubbles	(IBM)
1978	OMIST thyratron	(Nassibian et al)
1979	Laser enhanced plating and etching	(IBM)
1979	Amorphous silicon LCD	(RSRE & Dundee Univ)
1981	Plane-polarised light optical fibre	(Hitachi)
1982	Fission track autoradiography	(AERE)

9. Radar

(See also - History on a Page 22)

1924	Radar systems	(Appleton, Briet et al)
1937	Radar aiming anti-aircraft guns	(Pollard)
1938	"Gee" navigation	(Dippy)
1938	Klystron	(Hahn & Varian Bros.)
1939	Magnetron	(Randall & Boot)
1940	Plan position indicator	(Bowen, Dummer et al)
1940	Skiatron	(Rosenthal)
1940	"Oboe" navigation system	(Reeves et al)
1941	Radio proximity fuse	(Butement)
1941	"H_2S" navigation system	(Dee, Lovell, et al)
1942	"Velodyne" analyser	(Williams & Uttley)
1942	LORAN	(M.I.T.)
1942	Phantastron circuit	(Williams and Moody)
1942	Sanatron time base circuit	(Williams and Moody)
1942	Miller integrator circuit	(Blumlein)
1943	Ultrasonic radar navigation training device	(Dummer and Smart)
1945	DECCA navigation system	(O'Brien and Schwartz)
1947	Chirp technique	(Bell Labs)
1971	Hologram matrix radar	(Iizuka, Nguyen et al)

10. Sound reproduction

(See also History on a Page 15)

1860	Microphone, diaphragm type	(Reis)
1876	Telephone	(Bell)
1877	Phonograph	(Edison)
1877	Microphone, carbon	(Edison)
1877	Loudspeakers, moving coil	(Siemens)
1878	Carbon granule microphone	(Hunnings)
1887	Gramophone	(Berliner)
1889	Strowger auto telephone exchange	(Strowger)
1896	Telephone dial	(Keith et al)
1898	Magnetic recording (wire)	(Poulsen)
1908	Electronic organ	(Cahill)
1912	Relay auto telephone exchange	(Betulander)
1914	ASDIC	(-)
1915	Acoustic mine	(Wood)
1915	SONAR	(Langevin)
1916	Crossbar telephone exchange	(Roberts and Reynolds)
1917	Microphone, condenser	(Wente)
1919	Crystal microphone	(Nicholson)
1920	Plastic magnetic tape	(Pfleumer)
1924	Reisz microphone	(Neumann)
1925	Loudspeaker, electrostatic	(Various)
1926	Films, sound-on-disc system	(Warner Bros.)
1927	Films, sound-on-film system	(Fox Movietone News)

1930's	Radiophonic sound and music	(Grainger)
1931	Sterophonic sound reproduction	(Blumlein, Bell Labs)
1936	Vocoder	(Bell Labs)
1937	Pulse code modulation	(Reeves)
1948	Films - magnetic recording	(RCA et al)
1950's	MODEM	(MIT, Bell Labs)
1957	Full frequency range loudspeaker	(Walker)
1958	Video tape recorder	(Ampex)
1960	Telephone electronic switching	(Bell Labs)
1964	Packet switching	(Baran)
1967	Audio noise reduction system	(Dolby)
1969	PARCOR speech synthesis	(N.T.T. Japan)
1972	Video discs	(Philips)
1973	Scanning acoustic microscope	(Quate)
1974	PRESTEL system	(Fedida)
1978	Lightwave powered telephone	(Bell Labs)
1978	Laser recording system	(Philips)
1979	Satellite echo-cancelling circuit	(Bell Labs)

11. Television

(See also History on a Page 27)

1884	Nipkow television system	(Nipkow)
1897	Cathode ray oscillograph	(Braun)
1908	Electronic system - theory	(Campbell-Swinton)
1919	Electronic system	(Zworykin)
1923	Iconoscope	(Zworykin)
1925	Mechanical system	(Baird)
1929	Colour television	(Bell Labs)
1932	Scanning circuit	(Blumlein)
1933	Time base circuit	(Puckle)
1936	Long tailed pair circuit	(Blumlein)
1938	Shadow-mask TV tube	(Flechsig)
1939	Large screen TV projector	(Fischer)
1950	VIDICON TV camera tube	(RCA)
1957	PLUMBICON TV camera tube	(Philips)
1965	Integrated photodiode arrays	(Weckler)
1968	TRINITRON colour tube	(Sony)
1974	PRESTEL system	(Fedida)
1977	Pocket TV receiver	(Sinclair)
1979	CCD colour TV camera	(Sony)
1980	Large screen colour display	(Mitsubishi)

12. Transistors & Semiconductor devices

(See also 7 - Microelectronics)

1917	Crystal pulling process	(Czochralski)
1930	MOS/FET concept	(Lilienfeld)
1935	Field effect transistor	(Heil)
1948	Single crystal fabrication - germanium	(Teal and Little)
1948	Transistor	(Bardeen, et al)
1949/50	Ion implantation	(Ohl and Shockley)
1950	PIN diode	(Nishizawa)
1950's	Thermo-compression bonding	(Anderson et al)
1952	Zone melting technique	(Pfann)
1952	Single crystal fabrication - silicon	(Teal, Bueler)
1952	Alloyed transistor	(RCA)
1953	Surface barrier transistor	(Philco)
1953	Floating zone refining process	(Keck, Emeis et al)
1953	Unijunction transistor	(G.E.C.)
1954	Transistor radio set	(Regency)
1954	Silicon solar battery	(Chapin, Fuller et al)
1954	Interdigitated transistor	(Fletcher)
1955	Infra-red emission from GaSb	(Braunstein)
1956	Diffusion process	(Fuller, Reis)
1956	Semiconductor diode junction capacitor	(Giacoletto and O'Connell)

1957	Oxide masking process	(Frosch)
1958	Pedestal pulling of silicon	(Dash)
1958	Tunnel diode	(Esaki)
1958	"Technetron" FET	(Teszner)
1958	Field effect varistor	(Bell Labs)
1959	Planar process	(Hoerni)
1960	Light emitting diode (LED)	(Allen & Gibbons)
1960	Epitaxy (Vapour phase)	(Loor, et al)
1961	Epitaxy (Liquid phase)	(Nelson)
1962	LED (Gallium arsenide phosphide)	(Holonyak)
1963	Gunn diode oscillator	(Gunn)
1963	Silicon-on-sapphire technology	(Various)
1964	IMPATT diode	(Johnston, de Loach)
1964	Transistor modelling	(Gummel)
1964	Overlay transistor	(RCA)
1965	Self-scanned photodiode arrays	(Weckler)
1966	Nitride-over-oxide semiconductors	(Horn)
1967	TRAPATT diode	(Prager, Chang et al)
1968	Amorphous semiconductor switches	(Ovshinsky)
1968	BARRITT diode	(Wright)
1969	Magnetic bubbles	(Bobeck, Fischer et al)
1969	Magistor magnetic sensor	(Hudson, IBM)
1970	Charge coupled devices	(Boyle, Smith)
1970	X-ray lithography for bubble devices	(Feder et al)
1971	Carrier-domain magnetometer	(Gilbert)
1972	Auto, control of crystal growth	(Bardsley et al)
1974	CATT triode	(Tu, Cady et al)
1976	Amorphous silicon solar cell	(RCA)
1977	FLAD display system	(Ins. App. S.S. Physics)
1978	Laser cold processing semiconductors	(Quantronix)
1979	FLOTOX process	(Intel)
1980	Magnetic avalanche transistor	(IBM)
1980	MCZ silicon crystal growth	(Sony)
1981	Hydroplaning polishing of semiconductors	(M.I.T.)
1982	Amorphous photosensors	(Sony)
1982	Recrystallisation silicon process	(Texas Instruments)

13. Tubes, Lamps etc.

1855	Cold cathode discharge tube	(Gaugain)
1856	Low pressure discharge tube	(Geissler)
1857	Mercury arc lamp	(Wray)
1878	Carbon filament lamp	(Swan, Stearn et al)
1901	Fluorescent lamp	(Cooper-Hewitt)
1904	Two electrode tube	(Fleming)
1906	Three electrode tube	(de Forest)
1910	Neon lamp	(Claude)
1912	Tungar rectifier	(Langmuir)
1914	Thyratron	(Langmuir)
1919	Retarded field microwave oscillator	(Barkhausen, Kurtz)
1919	Housekeeper seal	(Housekeeper)
1922	Negative resistance oscillator	(Gill, Morrell)
1926	Screened grid tube	(Round)
1928	Pentode tube	(Tellegen, Holst)
1931	CRO cardiograph	(Rijant)
1933	Ignitron	(Westinghouse)
1935	Travelling wave microwave oscillator	(Heil)
1935	Multiplier phototube	(Zworykin, et al)
1936	Cold cathode trigger tube	(Bell Labs)
1937	Polar co-ordinate oscillograph	(von Ardenne, et al)
1938	Shadow-mask tube	(Flechsig)
1939	Double-beam oscillograph	(Fleming-Williams)
1939	Klystron	(Hahn and Varian Bros)
1939	Magnetron	(Randall and Boot)
1940	Skiatron CRO	(Rosenthal)
1943	Travelling wave tube	(Kompfner, et al)
1949	Cold cathode stepping tube	(Remington Rand)
1950	VIDICON TV Camera tube	(USA)
1956	Vapour cooling of tubes	(Beutheret)
1957	PLUMBICON TV Camera tube	(Philips)
1960	FEMITRON microwave amplifier	(Dyke)
1968	TRINITRON colour tube	(Sony)

Chapter 11

A Concise Description of Each Invention in Date Order

1642   COMPUTER (Mechanical Calculating   Blaise Pascal (France)
           Machines)

           EDITOR's NOTE:- Although non-electronic this item is included as an essential part of computer history.

           The invention of the first mechanical device capable of addition and subtraction in a digital manner has been generally credited to Pascal, who built his first machine in 1642. This claim has been contested on the basis of letters sent to Kepler in 1623 and 1624 by Wilhelm Schickhardt of Tubingen, in which the latter describes the construction of a calaculator. Pascal, who at the age of 19 had wearied of adding long columns of figures in his father's tax office in Rouen, made a number of calculators, some of which are still preserved in museums. His machines had number wheels with parallel, horizontal axes. The positions of these wheels could be observed and sums read through windows in their covers. Numbers were entered by means of horizontal telephone-dial-like wheels which were coupled to the number wheels by pin gearing. Most of the number wheels were geared for decimal reckoning but the two wheels on the extreme right had twenty and twelve divisions, respectively for sous and deniers. A carry ratchet coupled each wheel to the next higher place. The stylus-operated pocket adding machines now widely used are descendants of Pascal's machine.

           SOURCE: Serrell, Astrahan, Patterson and Pyre "The evolution of computing machines and systems" Proc. IRE May 1962. p.1041.

           SEE ALSO: "The Inventor of the First Desk Calculator" V.P.Czapla. "Computers and Automation" Vol. 10. p. 6. September 1961.

           "The computer from Pascal to von Neuman" by H H.Goldstine. Princeton Univ,. Press, 1972. p. 7.

           "The origins of digital computers" Edited by B. Randell, Springer-Verlag, Berlin 1973. p. 2.

1672   COMPUTERS (Mechanical   G.W.Leibniz (Germany)
           Calculating Machines)

           EDITOR's NOTE:- Although non-electronic, this item is included as part of computer history.

           Gottfried Wilhelm Leibniz invented the "Leibniz Wheel" which enabled him to build a calculating machine which surpassed Pascal's in that it could do, not only addition and subtraction fully automatically but also multiplication and division.

           SOURCE: "The computer from Pascal to von Neumann" by H.H.Goldstine. Princeton Univ. Press 1972. p. 7.

           SEE ALSO: "The origins of digital computers" Edited by B.Randell. Springer-Verlag. Berlin 1973. p. 2.

1745   CAPACITOR (LEYDEN JAR)   von Mushenbrock and Cunaeus (Germany)
                             and
                        von Kleist (Pomerania)

           According to the literature, the Leyden Jar was discovered almost simultaneously by Dean von Kleist of the Cathedral of Camin, Germany, in October 1745 and Peter von Muschenbrock, Professor in the University of Leyden, in January 1746, over 200 years ago. As described by them, it was a glass jar or vial with inner and outer electrodes of various things - water, mercury, metal foil etc. The modern miniature glass dielectric capacitor differs in form and structure from the 200-year-old Leyden Jar, but the principle of operation is the same.

SOURCE: "History, Present Status and Future Developments of Electronic Components" by P.S. Darnell. IRE Transaction on Component Parts. September, 1958. p. 127/8.

SEE ALSO: "Janus" C. Dorsman and C.A. Crommelin. 46, 1957. p. 274.

"Observations on the manner in which glass is charged with electric fluid" by E.W. Gray. Phil. Trans. Royal Society. London Vol. 77(1788) p. 407.

"Residual charge of the Leyden jar-dielectric properties of various glasses" by J. Hopkinson, Phil. Mag. Part 5. Vol. 4 (1877) p. 141.

1772   IRON DUST CORES   Gavin Knight(U.K.)

Iron dust cores consisting of iron filings churned in water, bound with linseed oil, moulded and fired. They were used in a Navy compass. Apparently, Knight was a secretive person and details of his process were not actually published before 1779.

SOURCE: Note from British Science Museum.

SEE ALSO: Andrade - "The early history of the permanent magnet". Endeavour. January 1958. p. 27.

Benjamin Wilson "Phil. Trans." Vol. 69.(1779) p. 51. (giving details of the process).

1780   GALVANIC ACTION   L. Galvani (Italy)

Luigi Galvani began his studies on the subject of animal electricity in 1780. When performing experiments on nervous excitability in frogs, he saw that violent muscle contractions could be observed if the lumbar nerves of the frog were touched with metal instruments in the presence of distant electrical discharges.

The word "electricity" was reserved for static electricity and the word "Galvanism" was proposed by von Humboldt for direct (continuous) current.

SOURCE: "From torpedo to telemetry" by D.W. Hill. Electronics & Power. 27th November, 1975. pp. 1110-1111.

1800   DRY BATTERY   A. Volta (1800) De Luc (1809)   (Italy)
                                        and Zamboni (1812)

Volta's invention of the electric battery was announced in a letter to Sir Joseph Banks, the President of the Royal Society and described his "Volta's Pile" - consisting of copper and zinc discs separated by a moistened cloth electrolyte. These were later improved to consist of paper discs tinned one side, manganese dioxide on the other, stacked to produce 0.75 v. between 1 in. diam. discs.

Note by Science Museum, London:-

Scyffer described experiments with dry cells carried out by Ludicke (1801) Einhof, Ritter (1802) Hachette and Desornes, Biot and many others. Scyffer regarded these as experimental and ascribes the first effective pile to Behrens and to Marechaux but considered the best performance before Zamboni to have been achieved by De Luc in 1809. Zamboni (1812) himself ascribed priority to De Luc since his paper was entitled: "Descrizione della colonna elettrica del Signore de Luc e considerazione sull analisi de lui Fatta della pile Voltiana". Work on early dry batteries was, therefore, done from about 1800 to 1812.

SOURCE: "On the electricity excited by the mere contact of conducting substances of different kinds" Phil. Trans. Vol. 90 (1800) page 403.

SEE ALSO: "A Biographical Dictionary of Scientists" T.I. Williams. p. 535 (Volta) Adam and Charles Black, London 1969.

Scyffer "Geschichtliche Dartellung der Galvanisms" 1848. p. 135-148.

1800   INFRA-RED RADIATION   W. Herschel (U.K.)

                In 1800, William Herschel, during research into the heating effects of
        the visible spectrum, discovered that the maximum heating was not within the
        visible spectrum but just beyond the red range. Herschel concluded that in
        addition to visible rays the sun emits certain invisible ones. These he called
        infra-red rays.

        SOURCE: "Electronics Engineer's Reference Book" Newnes-Butterworth
        London (1976) Chap. 4. p. 4 - 2.

1801   ULTRA-VIOLET RADIATION   J. W. Ritter   (Germany)

                In 1801, the German physicist Ritter made a further discovery. He
        took a sheet of paper freshly coated with Silver Chloride and placed it on top of
        a visible spectrum produced from sunlight falling through a prism. After a
        while he examined the paper in bright light. It was blackened, and it was black-
        ened most just beyond the violet range of the spectrum. These invisible rays
        Ritter called ultra-violet rays.

        SOURCE: "Electronics Engineer's Reference Book" Newnes-Butterworth
        London (1976) Chapter 4. p. 4 - 2.

1803   ACCUMULATOR   J. W. Ritter (Germany)

                Ritter's charging or secondary pile consists of but one metal, the discs
        of which are separated by circular pieces of cloth, flannel or cardboard, moist-
        ened in a liquid which cannot chemically affect the metal. When the extremities
        are put in communication with the poles of an ordinary voltaic pile it becomes
        electrified and can be substituted for the latter and it will retain the charge.

        SOURCE: "Biographical History of Electricity and Magnetism" Mottelay.
        Charles Griffin & Co. London 1922. p. 381.

1808   ATOMIC THEORY   J. Dalton (U.K.)

                Dalton conceived the idea that the atoms of different elements were
        distinguished by differences in their weights. In 1808, he propounded the
        theory that all chemical elements are composed of minute particles of matter
        called atoms and that these atoms, as the name implies, cannot be cut up any
        further. All atoms of one element, he said, were alike but atoms of different
        elements had different weights. The atom of hydrogen was the lightest atom
        (1.66×10^{-24} of a gram) and the weights of all other atoms were compared
        with it.

        SOURCE: "New System of Chemical Philosophy" J. Dalton 1808.

        SEE ALSO: "A Biographical Dictionary of Scientists" T. I. Williams. Adam
        and Charles Black, London 1969. p. 128.

1812   CABLE INSULATION   Sommering & Schilling (Germany)

                It was in that year (1812) that Sommering and Schilling conducted a
        series of experiments in which a soluble material, said to be indiarubber, was
        first used for insulating wire, following a suggestion made by a Spaniard,
        named Salva, in 1795 concerning the feasibility of submarine telegraphy.
        Curiously enough, this first 'cable' developed by Sommering and Schilling was
        in a sense a power cable as the objective was the detonation of mines. For at
        least another 50 years, however, practically all development was to be con-
        cerned with telegraphy.

        SOURCE: "Electric Cables" by S. E. Goodall. Proc. IEE Vol. 106 Pt. B
        No. 25. (Jan. 1959) p. 1.

1820   ELECTRO-MAGNETISM   H. C. Oersted (Denmark)
                (Galvanometer)

                In 1820, Oersted reported the discovery of electro-magnetism, and
        this led him to develop the first galvanometers. It was John Schweigger who
        constructed the first moving-coil instrument and Nobilli (1828) an Italian

physicist, developed a sensitive astatic galvanometer and compared its sensitivity with that of the most 'sensitive galvanometer' then available.

SOURCE: "From torpedo to telemetry" by D. W. Hill. Electronics & Power. 27th November, 1975. p. 111.

SEE ALSO: "Experiments on the effect of a current of electricity on the magnetic needle" by H. C. Oersted. Annals of Philosophy. Vol. 16. London (1820) p. 273.

| 1821 | THERMOELECTRICITY | T. J. Seebeck (Germany) |

The discovery of thermoelectricity is usually attributed to Professor T. J. Seebeck of Berlin in 1821, although there is some evidence that he might have been anticipated by Dessaignes in 1815. Professor Cummings of Cambridge also discovered the effect independently and published his findings in 1823.

Following Seebeck's work, J. C. A. Peltier completed Seebeck's discovery by showing that the passage of electricity through a junction of two different metals (antimony and copper) could produce a rise in temperature at the junction when passing in one direction and a drop in temperature when passing in the contrary direction.

The introduction of the first successful thermopile in the sense of an array of thermocouples (analogous to the galvanic pile) is attributed to Nobilli. Nobilli's thermopile was subsequently improved by Melloni.

SOURCE: Note from British Science Museum, London (1st and 3rd paragraphs) Author (2nd paragraph)

| 1826 | OHMS LAW | G. S. Ohm (Germany) |

Ohm was Head of the Department of Mathematics and Physics at the Polytechnic Institute of Cologne when he discovered the law bearing his name.

$$E = IR \qquad R = \frac{E}{I} \qquad I = \frac{E}{R}$$

SEE ALSO: G. S. Ohm Die Galvinische Kette Mathematisch Gearbeitet. Berlin (1827).

| 1828 | MOVING COIL GALVANOMETER | J. Schweigger (Germany) |
| 1828 | ASTATIC GALVANOMETER | C. L. Nobilli   (Italy) |

In 1820, Oersted reported the discovery of electro-magnetism, and this led him to develop the first galvanometers. It was John Schweigger who constructed the first moving-coil instruments and Nobilli (1828) an Italian physicist, developed a sensitive astatic galvanometer and compared its sensitivity with that of the most 'sensitive galvanometer' then available.

SOURCE: "From torpedo to telemetry" by D. W. Hill. Electronics & Power. 27th November, 1975. p. 111.

SEE ALSO: "Comparison entre les deux galvanometres les plus sensibles, la grenouille et le miltiplicateur a deux aiguilles, suivie de quelques resultats nouveaux" C. L. Nobilli. Chim. et Phys. RS(1828) 43, pp. 256-258.

| 1831 | ELECTROMAGNETIC INDUCTION | M. Faraday (U. K.) |
| | (see also SELF INDUCTION 1832) | |

In 1831 Faraday wound an iron ring with two coils: one connected to a voltaic battery, was to create the primary vibration - the iron ring was to concentrate the lateral vibrations from this - and another coil on the opposite side of the ring was to convert these secondary vibrations into another electric current. Thus, on 29th August 1831, Faraday discovered electro-magnetic induction.

SOURCE: "A Biographical Dictionary of Scientists" T. I. Williams. Adam & Charles Black, London 1969. p. 174.

1831   <u>TRANSFORMER</u>   Michael Faraday (U.K.)

A contrivance was used by Michael Faraday in his experiments on electromagnetic induction. This device is described in his diary under the date of
August 29th, 1831 :-

> "Expts. on the production of Electricity from Magnetism, etc. etc.
> Have had an iron ring made (soft iron) iron round and 7/8 inch thick
> and ring 6 inches in external diameter. Wound many coils of copper
> wire round one half, the coil is being separated by twine and calico -
> there were three lengths of wire each about 24 feet long and they could
> be connected as one length or used as separate lengths. By trial with
> a trough each was insulated from the other. Will call this side of the
> ring A. On the other side but separated by an interval was wound wire
> in two pieces together amounting to about 60 feet in length, the direction
> being as with the former coils: this side call B.
>
> Charges a battery of 10 pr. plates 4 inches square. Made the coil on
> B side one coil and connected its extremities by a copper wire passing
> to a distance and just over a magnetic needle (3 feet from iron ring).
> Then connected the ends of one of the pieces on A side with battery;
> immediately a sensible effort on needle. It oscillated and settled at
> last in original position. On breaking connection of A side with battery
> again a disturbance of the needle.
>
> Made all the wires on A side one coil and sent current from battery
> through the whole. Effect on needle much stronger than before".

A multiwinding transformer and a transformer experiment are described by
these words of Faraday's.

<u>SOURCE</u>: "History, present status and future developments of electronic
components" by P.S. Darnell. IRE Transactions on Component Parts.
September, 1958. p. 125.

<u>SEE ALSO</u>: "Faraday's discovery of electro-magnetic induction" by T. Martin.
Edward Arnold & Co. London. pp. 52-54. 1949.

Note by British Science Museum:-

> "The word 'transformer' was first used in the electrical sense in
> 1883 for both static transformers and rotating machines (motor generators).
> Previous to this 'induction coil' had been used. Faraday was the first to link
> two electric circuits by a magnetic circuit. The earliest application of the
> device to transfer power appears to be Jablochkoff's patent No:1996 (date of
> application 22nd May 1877).* This was really the first practical 'transformer'
> and it utilized a piece of apparatus which had existed since 1831."

> * Jablochkoff's patent covered a lighting system which includes a
>   mention of a transformer transferring power supply to a number
>   of lamps.

1832   <u>SELF INDUCTION</u> (see <u>ELECTROMAGNETIC</u>   J. Henry (U.S.A.)
            <u>INDUCTION 1831)</u>

Henry is reported to have discovered the phenomenom of self induction
in 1830 but, through his failure to publish, priority was given to Michael
Faraday to Henry's great mortification.

<u>REFERENCE</u>: "Joseph Henry" by T. Coulson. Princeton Univ. Press USA. 1950.

<u>SEE ALSO</u>: "A Biographical Dictionary of Scientists" T. I. Williams. Adam and
Charles Black, London 1969. p. 250.

1833   <u>COMPUTER</u> (Calculating Machines)   C. Babbage (U.K.)

In 1833 Babbage conceived his analytical engine, the first design for a
universal automatic calculator. He worked on it with his own money until his
death in 1871. Babbage's design had all the elements of a modern general-
purpose digital computer, namely: memory, control, arithmetic unit and

input/output. The memory was to hold 1000 words of 50 digits each, all in counting wheels. Control was to be by means of sequences of Jacquard punched cards. The very important ability to modify the course of a calculation according to the intermediate results obtained - now called conditional branching - was to be incorporated in the form of a procedure for slipping forward or backward a specified number of cards.

SOURCE: Serrell, Astrahan, Patterson and Pyne "The evolution of computing machines and systems" Proc. IRE May 1962. p.1042.

SEE ALSO: "The computer from Pascal to von Neumann" by H.H. Goldstine. Princeton Univ. Press 1972.   p.10.

"On the mathematical powers of the calculating engine" by Charles Babbage, 26th December 1837 (in Randell's book p.17).

1834   ELECTROLYSIS   M. Faraday (U.K.)

Faraday announced his two laws of electrolysis in 1834 which made explicit the amount of force required; for a given amount of electrical force, chemical substances in the ratio of their chemical equivalent were released at the electrodes of an electrochemical cell. Put another way, chemical affinity was electrical force acting on the molecular level.

SOURCE: "A Biographical Dictionary of Scientists" by T.I. Williams. Adams & Charles Black, London (1969) p.175.

1837   RELAYS   W.F. Cooke, Wheatstone and E. Davy (U.K.)

Telegraph Bell Relay
Telegraph Signal Relay

The first patent was taken out by Edward Davy in 1838 (British Patent No:7719) "I claim the mode of making telegraph signals or communications from one distant place to another by employement of relays of metallic circuits brought into operation by electric currents". In 1837 (British Patent No: 7390) Cooke and Wheatstone described an electromagnetic relay device for bringing a local battery at the distant station into action to sound an alarm bell there. However, Davy was described by Fakie as working on telegraphy as early as 1836 and entered an opposition to Cooke and Wheatstone's 1837 application for a patent, but the patent was granted. Morse in the U.S.A. is credited with a patent in 1840 (U.S. Patent No:1647) which is apparently similar to Davy's patent.

SOURCE: Note by British Science Museum, London.

SEE ALSO: J.J. Fakie "A History of Electric Telegraphy in the year 1837" London, 1884.

"A Biographical Dictionary of Scientists" T.I. Williams.   Adam and Charles Black, London 1969. (Sir W. Fothergill Cooke. p.114).

1837   TELEGRAPHY - MORSE CODE   S.B. Morse (U.S.A.)

While Cooke and Wheatstone of England had proposed electrical telegraph principles, it took the genius of Samuel B. Morse to develop the electrical hardware and also recognize the essential elements for a simple code adaptable to his on-off (binary) telegraph system. He first demonstrated his system in 1837-1838 and put it into practice in 1844, after he obtained a government grant to connect Baltimore with Washington, D.C., a distance of 37 miles. This was the first practical development of electrical telecommunications. It met a real need for fast communications and spread rapidly.

SOURCE: "Telecommunications - the resource not depleted by use. A historical and philosophical resume" by W.L. Everitt. Proc. IEEE Vol. 64. No. 9. (Sept. 1976) p.1293.

1837   ELECTRIC MOTOR   T. Davenport (U.S.A.)

The earliest known examples of a patent for an electric motor is U.S. Patent No:132 granted on 25th February, 1837 to Thomas Davenport, of Brandon,

Vermont, entitled: "Improvements in Propelling Machinery by Magnetism and Electro-magnetism."

According to the description contained in the specification, the motor, which is intended to be driven by a 'galvanic battery', is constructed on sound electro-magnetic principles.

SOURCE: "Patents for Engineers" by L.H.A. Carr and J.C. Wood. Chapman and Hall, London 1959. p.87.

1839   MICROFILMING   Dancer (U.K.)

Shortly after the publication of Daguerre's invention of making photographs in 1839, Dancer produced in England photographs of documents of strongly reduced size (1:160) having a side length of abt. 3 mm. The knowledge of the possibility to produce reduced-size photographs prompted in 1835 the English astronomer John Herschel to suggest, to store documents of general concern (e.g. reference works) in a reduced form, provided the reduction does not involve any hazard for the original documents. The same idea was advanced at the beginning of this century by members of a Belgian library, in order to make old handwritings or prints accessible to many people. Unfortunately, this idea has never been materialized.

SOURCE: "A brief historial review on microfilming" by H. Scharffenberg and R. Wendell. Jena Review. No:1 (1976) p.4.

1839   BATTERY (Magnetohydrodynamic)   M. Faraday (U.K.)

The idea of producing electricity from a moving fluid, which is the basis of an MHD generator, was proposed by Faraday in 1839. A conducting fluid is passed between the poles of a magnet and an electromotive force is produced at right angles to the field. This principle is also used in electromagnetic pumps and induction flow meters for conducting fluids. In recent proposals by Kantrowitz and Spron (1959) the working fluid is a conducting gas at high temperatures. The gas moving at a high velocity is passed through a magnetic field at right angles to the direction of flow; electrodes placed on opposite sides of the channel extract the power and are connected to the external load.

SOURCE: "The magnetohydrodynamic generation of power" by K. Phillips. AEI Engineering (Mar/Apr.1964) p.62.

SEE ALSO: "Experimental researches in electricity" by M. Faraday, London. (1839)

"Application of the MHD concept to large scale generation of electric power" by A. Kontrowitz and P. Sporn. American Electric Power Service Corporation and AVCO Research Laboratory (1959).

1839   PHOTOVOLTAIC EFFECT.   E. Becquerel (France)

The photovoltaic effect was discovered by Edmond Becquerel as early as 1839. In 1873, Willoughby Smith first observed the photoconductivity of selenium. In 1887, finally, Heinrich Hertz described the photoemissive effect of ultraviolet light on metal electrodes.

SOURCE: "Beam-deflection and photo devices" by K. Schlesinger and E.G. Ramberg. Proc. IRE (May 1962) p.991.

SEE ALSO: "On electric effects under the influence of solar radiation" by E. Becquerel. Compt. rend. acad. sci. Vol.9 (1839) p.561.

"Effect of light on selenium during the passage of an electric current" by W. Smith. American J. Sci. Vol.5. (1873) p.301.

1839   BATTERY (Fuel Cell)   W.R. Grove (U.K.)

The fuel cell principle - that is, the conversion of chemical energy to electric energy by a path that can avoid the thermodynamic limitation on efficiency imposed by the Carnot relation - has intrigued scientists and engineers for more than a century. In 1839-1842, Sir William Grove probably invented the first fuel cell. He used platinum-catalyzed electrodes to combine hydrogen and oxygen so as to produce electricity. It is interesting to note that almost 123 years later the Gemini fuel cell used the same catalyst, though probably in different physical form. One major trouble with Grove's cell was that its voltage fell off badly when an appreciable current drain was put on it. In 1889, Mond and Langer made a hydrogen-oxygen cell with perforated platinum-sheet electrodes, catalyzed by platinum black. This cell produced 1.46 watts at 0.73 volt at about 50 percent efficiency. However, it contained 1.3 grams of platinum and required pure hydrogen and oxygen. Thus, its capital cost made it a poor buy as an electric generator. In addition, to complicate things, it ran well only on pure hydrogen and oxygen.

SOURCE: "Hydrocarbon - air fuel cell systems" by C.G. Peattie, IEEE Spectrum (June 1966) p.69.

SEE ALSO: "On voltaic series in combination of gases by platinum" by W.R. Grove. Phil. Mag. Vol:14 (1839) p.127.

"On a gaseous voltaic battery" by W.R. Grove. Phil. Mag. Vol. 21. (1842) p.417.

NOTE: The principle of electrochemical fuel cells is by no means new, in fact the first cell was described in 1839 by Sir William Grove. This was a hydrogen fuel device with a sulphuric acid electrolyte and blacked platinum electrodes, generating approximately one volt at a very small current. The first power fuel cell was a 5 kilowatt unit demonstrated in 1959 by the English engineer, F.T. Bacon, and employed hydrogen-oxygen fuel with an alkaline electrolyte and sintered nickel electrodes. Rights to the development of this cell were obtained by Leesona-Moos Laboratories - a research subsidiary of Pratt & Whitney Aircraft Corp. - from N.R.D.C. and a modified version was used in the Apollo spacecraft.

SOURCE: "Fuel cells and their development in the U.K." by W.S.E. Mitchell Design Electronics (Feb.1966) p.34.

1840   THERMOGRAPHY   J. Herschel (U.K.)

Sir William Herschel's famous experiment with thermometers and a prism in 1800 showed the existence of energy beyond the red end of the visible spectrum. As long ago as 1840 his son John demonstrated thermal imaging and saw images in the dark. Considering the sophisticated techniques in use today, John Herschel's methods were both embarrassingly simple and successful. He took a blackened sheet of paper, soaked it in alcohol and focused radiation from a hot source onto the sheet. The infra-red radiation selectively heated parts of the paper, evaporating the alcohol and lightening its colour to form an image.

Variations on this theme took place over the next 100 years but not until 1940, with the pressing need for military systems with real-time infra-red tracking, was significant progress made. Since then optics, detectors, amplifiers, signal processing and displays have improved to the extent that high-performance infra-red systems can now produce television-quality pictures of a scene (at several kilometres' distance) which contains temperature differences of only a fraction of a degree. Thermal resolution, angular resolution and frame time have all been improved by a factor of ten or greater, representing an overall performance improvement of three or more decades.

SOURCE: "Infra-red imaging systems" by H. Blackburn. Systems Technology. (Plessey) No:26 (June 1977) p.15.

1843   WHEATSTONE BRIDGE   C. Wheatstone (U.K.)

Wheatstone described his bridge circuit - which he called the 'differential resistance measurer' - in a comprehensive paper on electric measurements presented to the Royal Society in 1843. That paper would have merited publication in Proceedings IEE, had there been such a Journal then, for the way in which it presented solutions to electrical-engineering problems. The

problems were the measurement of electromotive force, current strength and
resistance at a time when galvanometers were unstable instruments and there
was no sound method of calibrating them. Wheatstone found the theoretical
basis for solving these problems in an obscure German publication of 1827 in
which G.S. Ohm showed that there was a simple mathematical relationship
linking e.m.f. current and resistance.

The basis of Wheatstone's method of measurements is the use of a
calibrated variable resistance to keep the current constant and so avoid any need
for calibration of the galvanometer. Wheatstone remarked that it was easy to
make a calibrated variable resistance (for which he devised the term 'rheostat')
because Ohm's work showed that the resistance of a conductor of uniform
section was proportional to its length. He showed how to determine the value of
an unknown resistance by a simple substitution method. The unknown resistance is included in a circuit with a galvanometer whose reading is noted and then
replaced by a rheostat which is adjusted so that the galvanometer reading is the
same as before. The scale on the rheostat gives its resistance and hence the
value of the unknown resistance.

SOURCE: "Wheatstone's contribution to electrical engineering" by B.P. Bowers
Electronics & Power. May 1976. p.295.

1843   FACSIMILE REPRODUCTION   A. Bain (U.K.)

For the purpose of this review facsimile is considered to be a method
by which printed, handwritten and graphic data may be transmitted over communication channels and received in the form of a hard copy. Its origin dates
back to 1843, when the Scottish inventor Alexander Bain patented an "automatic
electrochemical recording telegraph". Next came Frederick Bakewell's
cylinder and screw arrangement on which many of the present-day facsimile
systems are based.

In Europe facsimile equipment has been commercially available since
1946. In recent years the technology advances in electronics and the drastic
fall in semiconductor prices have led to the replacement of bulky separate
facsimile transmitters and receivers by small transceivers.

SOURCE: "Facsimile - a review" by J. Malster and M.J. Bowden. The Radio
and Electronic Engineer. Vol.46. No.2. (Feb.1976) p.55.

1845   METALLIC SHEATHING OF CABLES   Wheatstone (U.K.)

The earliest attempts at metallic sheathing were made in 1845 by
Wheatstone, who folded lead strip around the cable core and then joined it with
a longitudinal soldered seam. This method bears an interesting similarity to a
modern method in which aluminium sheathing is applied using pre--formed
strip. The method was superseded by one which involved soldering 50 ft. lengths
of lead pipe end to end and subsequently sinking the pipe into contact with the
cable core by means of a die. This again has an interesting parallel in a present day method for aluminium sheathing.

In 1879 the first direct extrusion on to cable was made from a Borel
press using solid hollow billets pre-heated to 120°C. It was not then considered
good practice to recharge with molten lead because of possible thermal damage
to the insulation, so the process was limited to a one billet charge. Development of lead sheathed cable has since gone hand in hand with the development of
the lead extrusion process. Different types of press have been designed to overcome defects experienced with cable in service but extrusion in one for or
another has been universally adopted. By contrast, the present development of
aluminium sheathing processes is proceeding along three different lines and it
is hard to predict which technique will ultimately prove most successful. The
greater difficulty of extruding aluminium as compared with lead has undoubtedly
favoured this more varied approach.

SOURCE: "The metallic sheathing of cables" by A.V. Garner. AEI Engineering
(Sept/Oct 1962) p.248.

1845   CIRCUITS - KIRCHOFF'S LAWS   G.R. Kirchhoff (Germany)

Two laws that express the behaviour of an electrical network. In 1845
he gave the laws for closed circuits, extending these to general networks (1847)

and to solid conductors (1848).

    1st law - The total current arriving at any point in an
                   electric network must be zero.

    2nd law - The sum of the electromotive forces around
                   any closed circuit is equal to the sum of the
                   IR drops around the circuit.

1847   SUBMARINE CABLE INSULATION   W. Siemens (Germany)

        Telegraphic instruments had been developed, notably by Schilling, Morse and Cooke and Wheatstone, for the transmission of signals over land lines; and a suitable material for insulating the conductor had already been introduced into Europe. This was gutta percha, the gum from a Malayan tree, exhibited at the Royal Society of Arts in London by Dr. Montgomerie in 1843. The electrical and mechanical properties of gutta percha, especially when immersed in sea water, were such that it had no rival for over 70 years as an insulant in submarine cables. Moreover, the Gutta Percha Company had been formed in 1845/6 to exploit its use, and primitive extruders had been constructed for the production of rods and tubes.

        Before 1849, many attempts had been made to find a suitable insulant or protection for underground and underwater cables, including tarred rope, glass tubes, split rattan, impregnated cotton and rubber, but none of these lasted long in the sea. In 1847 Werner Siemens used gutta percha for an underground line in Germany, and some was laid in the Port of Kiel in 1848 for the purpose of detonating mines. In America in the same year, Armstrong experimented with gutta percha covered wire in the Hudson River, where Ezra Cornell had previously connected Fort Lee with New York by a rubber insulated line which worked for several months.

SOURCE: "The Story of the Submarine Cable" booklet published by Submarine Cables Limited (AEI) London (1960) p. 4.

1847   MAGNETOSTRICTION   J. Joule (U.K.)

        Magnetostriction is a well-known phenomenon in which the mechanical dimension of a magnetic material is altered as the magnetization is varied, and in which, conversely, the magnetization is altered as the dimension is changed. Thus, an alternating current applied through a coil wrapped around a specimen can induce mechanical vibrations in it; and alternately, mechanical vibrations set up in such a specimen can transform the mechanical energy into electrical energy in a coil wound around it. Electro-mechanical interactions were observed as early as 1847 with Joules discovery of magnetostriction.

SOURCE: "Solid State devices other than Semiconductors" by B. Lax and J. G. Mavroides. Proc. IRE (May 1962) p. 1014.

SEE ALSO: "On the effects of magnetism on the dimensions of iron and steel bars" J. P. Joule. Phil. Mag. Vol. 30. (April 1847) p. 226.

1848   BOOLEAN ALGEBRA   G. Boole (United Kingdom)

        Formal logic, so necessary for the workings of digital computers, could not be satisfactorily explained mathematically before George Boole. In 1848 the English logician published "The Mathematical Analysis of Logic" and in 1854 "An Investigation of the Laws of Thought", the foundation of what is now symbolic logic. With the theories expounded in these writings, it was possible to express logic in very simple algebraic systems. The equation $X^2 = X$ for every X in the system is basic to Boolean algebra and has only 0 or 1 as an answer in numerical terms. Thus modern computers can make use of this binary system, with their logic parts carrying out binary operations.

SOURCE: Electronics (April 17, 1980), p. 69.

1850   THERMISTOR   M. Faraday (U.K.)

        The temperature-sensitive non-linear resistors are known generally as thermistors, a name coined by the Bell Telephone Laboratories (of the USA). They are, however, over 100 years old, for Faraday discovered that silver sulphide possessed a high negative temperature coefficient (although in this

case the conduction is ionic and not electronic, and the material therefore
suffers from polarization effects). Uranium oxide was used in Germany, but
with this conduction is also ionic and operation is unstable. A magnesium titanate spinel was introduced in 1923, and in the U.S.A. about 1912, boron was
found to possess negative temperature characteristics. From 1930 onwards the
Bell Telephone Laboratories devoted many years of intensive research to the
problem and showed that combined oxides of manganese and nickel had valuable
properties. They also found that varying the ratio between the manganese and
nickel varied these properties. The effects of adding small amounts of copper,
cobalt and iron were also investigated.

   Today, these oxides, treated to become uniphase, are in general use,
and are made into beads, rods, blocks, etc.

SOURCE: "Fixed Resistors" 2nd Edition. G.W.A.Dummer. Pitman.
London (1967) p.147.

1851 RELATION BETWEEN THEORY OF
     MAGNETISM AND ELECTRICITY   Lord Kelvin (U.K.)

   Introduced for the first time the vectors, later termed magnetic
induction and magnetic force by Maxell.

   In a fundamental paper, he derived a result expressing the energy of a
system of permanent and temporary magnets in terms of a volume integral
throughout space.

SOURCE: "A Biographical dictionary of scientists" by T.I.Williams. Adam
and Charles Black. London (1976) p.512.

1852 THIN FILMS (Sputtering process)   W.R.Grove (U.K.)

   Although the use of cathodic sputtering as a method for the deposition
of thin films predates vacuum evaporation by many years, the latter has been
received far more widespread application because evaporation is more convenient for many materials and generally gives high deposition rates. In
recent years, however, it has been found that certain materials are more
conveniently deposited by sputtering. In some cases it is impossible to deposit
materials by any other means.

SOURCE: "Thin-film Circuit Technology" by A.E.Lessor, L.I.Maissel and
R.E.Thun. IEEE Spectrum (April 1964) p.73.

SEE ALSO: W.R.Grove Phil.Trans.Roy.Soc.London. Series B. Vol:162
(1852) p.87.

1854 COMPUTERS (Calculating Machines) P.G.Scheutz (Sweden)

   Pehr Georg Scheutz built a difference engine in Stockholm inspired by
Babbage's ideas and displayed it in London in 1854 with considerable help from
Babbge. The machine had four differences and fourteen places of figures and
was capable of printing its own Tables (Scheutz was a printer)

SOURCE: "The computer from Pascal to von Neumann" by H.H.Goldstine.
Princeton Univ.Press (1972) p.15.

1855 COLD CATHODE DISCHARGE TUBE (J.M.Gaugain (France)

   Experiments with low-pressure glow discharges started very early in
the electrical art. Most of them were in small diameter glow tubes, often in
the form called Geissler tubes. The first recognition of the fact that such a
glow tube having its two electrodes of different size was capable of rectifying
the oscillating current from an induction coil appears to have been that of
Gaugain in 1855. However, for many years, the only use for these discharges
was as light sources. The early glow lamps all required high voltage excitation.
With the availability of the rare gases neon and argon and by means of low work-

function cathodes, glow tubes were developed for low voltage applications.

SOURCE: "The development of gas discharge tubes" by J.D.Cobine. Proc.IRE (May 1962) p.972.

SEE ALSO: J.M.Gaugain, Compt.rend. Acad.Sci. Paris. Vol.40 (1855)p.640.

1856   LOW PRESSURE DISCHARGE TUBES   H.Geissler (Germany)

In 1856, Heinrich Geissler, an artist and skillful glass blower of Bonn, Germany, originated the low pressure discharge tubes that were to bear his name. The Geissler tubes were long, small-bore glass tubes, usually shorten- ed by the use of many coils and bends, which were filled with various gases at low pressures and originally excited by high-voltage alternating current. Many beautiful effects could be produced by Geissler tubes filled with different gases and they were often used for decorations. As, for example, a display used to commemorate Queen Victoria's Diamond Jubilee. However, sputtering of the electrodes together with gas clean-up resulted in a short life for the tube. The principal use was for spectral analysis and lecture demonstrations.

SOURCE: "The development of gas discharge tubes" by J.D.Cobine. Proc.IRE (May 1962) p.970.

SEE ALSO: "The electric-lamp industry" A.A.Bright Jr., MacMillan Co. New York N.Y. p.218. et seq. (1949)

W.DeLaRue, H.W.Muller and W.Spottieswoode. Proc.Roy.Soc. (London) Vol:23 p.356 (1875).

1857   MERCURY ARC LAMP   J.T.Wray   (U.K.)

The first public demonstration of a mercury arc lamp was by Prof.J.T. Wray on the Hungerford Suspension Bridge in London on 3rd September,1860. Two British patents were issued to him in 1857. The electric arc was first used com- mercially for illumination in Paris in 1863. Much later, low-pressure arc "tubes" were used for illumination. In 1879 John Rapieff described mercury arc lamps in British Patent No:211 but there appears to be no evidence that they were built. Peter Cooper Hewitt showed in public his mercury-arc lamp on April 12, 1901. Georges Claude, a French inventor, demonstrated the first Neon sign, an improvement of the Geissler tube, at the Grand Palais in Paris in 1910. Developments in luminous tube discharges were made by Moore in 1920. Since these tubes did not have a high light output, they were largely confined to sign applications.

SOURCE: "The development of gas discharge tubes" by J.D.Cobine. Proc. IRE (May 1962) p.970.

SEE ALSO: British Patents issued 1857.

1858   GLOW DISCHARGES   J.P.Pleucker (Germany)

In 1858 Pleucker investigated experimentally the luminous effects of electric discharge through gases at low pressures. He observed that the glow was deflected in a strong magnetic field.

SOURCE: "A biographical dictionary of scientists" by T.I.Williams. Adam and Charles Black. London (1976) p.422.

1860   MICROPHONE (Diaphragm type)   J.P.Reis (Germany)

Earliest among microphone diaphragms - perhaps because of its similarity to the eardrum - was a stretched flat membrane (actually a sausage skin) used by Reis to actuate a loose metal-to-metal contact. A stretched flat membrane made of metal or very thin metallized plastic is used in present-day electrostatic microphones. This diaphragm is typically clamped at its peri- phery by a ring and stretched to any desired tension by a threaded ring.

SOURCE: "A century of microphones" by B.B.Bauer . Proc.IRE (May 1962) p.720.

SEE ALSO: "Ueber Telephone durch den galvaniscen strom" Jahnesbericht d. Physikalischen. Vereins zu Frankfurt am Main. Germany (1860-61) p.57.

1860   BATTERY (Secondary)   Planté (France)

            The secondary battery business dates back to Planté's discovery of
the lead-acid system in 1860.

            Secondary or storage cells are electrochemical cells which after dis-
charge can be restored to their original chemical state by passing the current
in the reverse direction. Although they have the same set of basic components
as primary cells, the anodes and cathodes of secondary cells have a more
stringent requirement, in that the electrode reactions have to be reversible.
This requirement immediately limits the number of electrode materials
available for secondary cells. At present, lead, cadmium, iron and zinc anode
materials, and lead dioxide, nickel dioxide and silver oxide cathode materials
are the only ones used in commercial secondary cells.

SOURCE: "Batteries" by C.K. Morehouse, R. Glicksman and G.S. Lozier.
Proc. IRE (Aug. 1958) p.1474/5.

1865   RADIO WAVE PROPAGATION.   J.C. Maxwell (U.K.)

            In his first paper on electromagnetism "On Faraday's Lines of Force"
(1855-56) Maxwell set up partial analogies, between electric and magnetic lines
of force and the lines of flow of an incompressible fluid. In a series of magnifi-
cent papers in 1861-62 he gave a fully developed model of electromagnetic phenomena viewed in the light of the field concept of Michael Faraday of whose
validity Maxwell had become fully persuaded by 1858. Adopting the belief of
William Thomson (Lord Kelvin) in the rotary nature of magnetism, a magnetic
tube of induction was represented by a set of cells rotating about the axis of the
tube, interference between the rotations of neighbouring tubes being avoided by
rows of intervening cells (in the manner of idle wheels) which corresponded to
electric currents. By means of this model Maxwell was able to given an elegant
qualitative interpretation of all the known phenomena of electromagnetism. By
introducing the notion of elasticity he was then able to give a quantitative description of the propagation of a disturbance in the model. Reinterpreted in
terms of the electromagnetic field, this implied that a disturbance in the
electromagnetic field should travel with a speed equal to the ratio of the elec-
trodynamic to the electrostatic units of electric force.

SOURCE: "A Biographical Dictionary of Scientists"   T.I. Williams. Adam and
Charles Black, London (1969) p.358.

SEE ALSO: "A Dynamical Theory of the Electromagnetic Field" J.C. Maxwell.
Proc. Royal Soc. (London) Vol.13.   pp.531-536. (December, 8th, 1864)

1866   TRANSATLANTIC TELEGRAPH CABLE   T.C. & M. Co. (U.K.)

            The first successful Atlantic telegraph cables were made and laid in
1866 by The Telegraph Construction and Maintenance Company, then newly
formed by the amalgamation of the Gutta Percha Company and Glass, Elliot and
Company. The laconic telegram sent on 27th July 1866 by Mr. R.A. Glass, the
Managing Director of the Company, from Valentia on the completion of the
laying of the first of the two cables by the famous "Great Eastern", read simply
"All right", and had the more modern two-letter abbreviation been available
the message would doubtless have been even shorter.

            A few days later came the report that a message of 405 letters from the
President of the United States, replying to Queen Victoria, had been sent at a
speed of 37 letters per minute. With this performance, modest indeed compared
with the modern speed of over 2,000 letters per minute, the efforts of the int-
repid pioneers were crowned with success and the submarine cable was firmly
established as a commercial proposition for oceanic depths.

SOURCE: "The Story of the Submarine Cable" booklet published by Submarine
Cables Limited (AEI) London (1960) p.8.

1868   RHEOTOME (Waveform plotter)   H. Lenz (Germany)

            Heinrich Lenz (of Lenz's Law) developed a segmented commutator or
rheotome which could be arranged to sample a periodic waveform at known
points of the cycle and thus feed a train of pulses, each corresponding to the

amplitude of the waveform at that point, to a slowly responding galvanometer. By plotting the deflections against the time in the cycle at which they occurred, the complete waveform could be reconstructed. In 1868, Bernstein used a rheotome to chart the time course of the action potential in a nerve fibre.

By 1876, the capillary electrometer of Marey and Lippman and the string galvanometer of Einthoven were available with a sufficient sensitivity and speed of response to record bioelectric events directly. However, preceding Einthoven's studies of the electrocardiogram, Marchand in 1877 and Englemann in 1878 were able to chart the electrocardiogram using a rheotome.

SOURCE: "From torpedo to telemetry" by D. W. Hill. Electronics & Power. 27th November, 1975. p. 1111.

SEE ALSO: "Des variations electriques, des muscles et due couer an particulier etudiees au moyen de l'electrometre de M. Lippmann" E. J. Marey. Comptes. Rendues Acad. des Sci. (1876) 82, pp. 975-977.

ALSO: "Bettrage zur kenntnis der reizwelle und contractionswelle de herzmuskels" R. Marchand. Plugers Arch. f. d. ges. Physiol. (1877) 15, p. 5ll.

1868   BATTERIES (Léclanché cell)   G. Léclanché (France)

The Léclanché dry cell is perhaps the best known cell in common use today. It is widely used in flashlights and other such equipment. This type of cell was originally described by Georges Léclanché in 1868 and has undergone many improvements since that time. Basically, it consists of a nearly pure (99. 99 per cent) zinc negative terminal, a carbon positive terminal, and a mixture of ammonium chloride, manganese dioxide, acetylene black, zinc chloride, chrome inhibitor and water. The mixture acts as a depolarising agent to reduce the formation of hydrogen bubbles on the positive electrode as discharge takes place. Improvements which have been made include leak-proofing, longer shelf life, pepped-up depolarizers, improved insulation and miniaturisation.

SOURCE: "Survey of electrochemical batteries" by N. D. Wheeler. Electro-Technology (June 1963) p. 68.

SEE ALSO: Léclanché G. Les Mondes Vol. 16 (1868) p. 532. also Comptes Rendus, Vol. 83 (1876) p. 54.

1870   BATTERY (Standard Clark Cell)   L. Clark (U.K.)

In about 1870 Latimer Clark, an English engineer and electrician, introduced a new kind of voltaic cell consisting of a positive electrode of mercury covered with a paste of mercurous sulphate and a negative electrode of zinc. The electrolyte was a saturated solution of zinc sulphate. After many determinations Latimer Clark assigned to the cell a mean value of 1·457 volt at 15·5ºC. The Clark cell suffered from a very high temperature coefficient of e. m. f. (1200μV/ºC). However, in spite of this, the Chicago International Electrical Congress in 1891 adopted the Clark cell together with the silver coulometer in definitions of the ampere and the volt.

SOURCE: "Standard Cells by Muirhead" Muirhead Technique Vol. 18. No. 3. (July 1964) p. 19.

SEE ALSO: Clark Cell -- Proc. Royal Soc(1872) Vol. XX. p. 444.

Clark Cell - Philos. Trans. Vol. CIXIV (1874) p. 1

NOTE: See also Weston Cell (1891)

1874   CAPACITORS (MICA)   M. Bauer (Germany)

Mica sheet as dielectric came into commercial capacitor manufacture only about 1914-1918 - very largely because not only could it stand up to the mechanical shocks of gunfire better than glass, but it also enabled the size of the capacitors to be reduced substantially for the same effective performance. The drive of war requirements pushed this developed to the fore, although the use of mica as a capacitor dielectric had been "invented" more than 60 years earlier.

SOURCE: "Electrical Capacitors in our Everyday Life" by P.R.Coursey.
ERA Journal No.6.(Jan.1959) p.10.

SEE ALSO: "Physical properties of mica" by M.Bauer. Z.duet.geol.Ge.
Vol.26. (1874) p.137.

"Capacity of mica condensers" by A.Zeleny. Phys.Rev. Vol.22.
(1906) p.651.

1876   ROLLED PAPER CAPACITOR   D.G.Fitzgerald (U.K.)

It appears that the rolled paper capacitor was first covered by a
patent filed in 1876 by Fitzgerald, who described:-

"The construction of a condenser with layers of paper and
conductor (usually tin-foil) alternately interleaved with
each other on to a cylinder, and the impregnation of such
condenser with paraffin wax after rolling"

SOURCE: "History, Present Status and Future Developments of Electronic
Components" by P.S.Darnell. IRE Transactions on Component Parts.
September 1958. p.124.

SEE ALSO: "Improvements in Electrical Condensers or Accumulators" by
D.G.Fitzgerald. British Patent No:3466/1876. September 2, 1876.

"Paper Condensers" by X.Boucherot l'Eclairage Electrique,Feb.12th
                                                              1898
"The manufacture of paper condensers" by G.F.Mansbridge,
J.Inst. Elec.Engrs. Vol.41 (May 1908) p.535.

"The capacity of paper condensers" by A.Zeleny and Andrewes.
Phys.Rev. Vol.27. (1908) p.65.

1876   TELEPHONE   A.G.Bell (U.S.A.)

It has been a hundred years since a faint but momentous sound was
made by Thomas Watson when he plucked a reed of a rudimentary transmitter.
But that sound travelled over wire and was heard in another room by Alexander
Graham Bell who happened to be holding a similar reed-and-diaphragm
apparatus to his ear. This was the first telephone signal.

Later, on February 14th,1876, Bell filed for the now famous patent on
the apparatus that he and Watson had been working on - just three hours before
one Elisha Gray filed a caveat with the Patent Office, declaring that he was
working on a similar device but had not yet perfected it. Had the timing been
different, we might now have a "Ma" Gray instead of Ma Bell.

The first telephone patent was issued to Bell on March 7,1876, three
days before the historic moment when the first intelligible human voice was
transmitted over the new telephone. Bell, after spilling acid over his clothing,
had called out, "Mr.Watson,come here. I want you" Next year, the first commercial telephone went into service when a Boston banker leased two instruments, each consisting of a simple wooden box that contained both transmitter
and receiver. The talker had to alternately talk and listen.

According to the records, the seed idea for the Bell System was
planted when a merchant named Thomas Sanders made a verbal offer to Bell to
finance the telegraphic experiments. They reached a tentative agreement, and
shortly afterwards a lawyer named Gardiner G.Hubbard made Bell a similar
offer. The three put into writing an agreement dated Feb.27, 1875 and later
signed a deed of trust, dated July 9,1877, forming the Bell Telephone Company,
Gardiner G.Hubbard, Trustee.

SOURCE: "Electronics" December 11th, 1975. p.91.

SEE ALSO: "Alexander Graham Bell and the Invention of the Telephone"
by J.E.Flood. Electronics & Power (March 1976) p.159.
Editor's Note: This issue of Electronics & Power is devoted to the early
history of the telephone.

"The Marriage that Almost Was" by M. F. Wolff. IEEE Spectrum (Feb. 1976) p. 41.

"The Telephone, its Invention and Development" by M. Woolley. Telecommunication Journal. Vol. 43. 111/1976 p. 175.

"Bell's great invention: the first 50 years" Bell Laboratories Record. (April 1976) p. 91.

"Bell & Gray: Contrasts in Style, Politics and Etiquette" by D. A. Hounshell. Proc. IEEE Vol. 64. No. 9. (Sept. 1976) p. 1305.

1877   **PHONOGRAPH**   T. A. Edison   (U. S. A.)
            (Gramophone)

Alexander Graham Bell's invention of the telephone, in 1876, drew attention to the problems of the reproduction of speech. One of those attracted was Thomas Alva Edison, who was all the more interested in the study of sound because he was partially deaf. Sometime in the fall of 1877, Edison wrapped a sheet of tinfoil around a cylinder, set a needle in contact with it, turned a crank to rotate the cylinder, and into a mouthpiece attached to the needle he shouted the nursery rhyme that begins: "Mary had a little lamb". After making a few changes, he cranked the cylinder again, and from the horn of the instrument he heard a recognizable reproduction of his voice. Thus was born the first phonograph. The date of invention was later recollected by Edison as August 13th, 1877, but there is some question as to how precise that date is. We know, however, that the patent application was filed December 24th, 1877, and the patent was granted on February 19th, 1878.

SOURCE: "Disk Recording and Reproduction" by W. S. Bachman, B. B. Bauer and P. C. Goldmark. Proc. IRE (May 1962) p. 738.

SEE ALSO: US Patent No: 200, 521 (T. A. Edison)

1877   **MICROPHONE (Carbon)**   T. A. Edison (U. S. A.)

Among the earliest devices intended for converting vibration into electrical impulses was Reis' loose metal-contact transducer which is reported to have transmitted tones of different frequencies, but not intelligible speech. This latter event seems first to have been achieved by Bell, using a magnetic microphone, on June 3rd, 1875. However, Bell's microphone proved not to be sufficiently sensitive for telephone work and the experiments of Berliner, Edison, Hughes and others soon thereafter introduced a long era of dominance for the loose-contact carbon transducer. To Edison goes the credit of being the first to design a transducer using granules of carbonized hard coal, still used in present-day microphones.

The carbon granules are made of deep-black "anthraxylon" coal ground to pass a 60-80 mesh, treated chemically and roasted in several stages under a stream of hydrogen. This drives out volatile matter, washes out extraneous compounds and carbonizes the coal. The last step of the process is magnetic and air-stream screening to eliminate iron-bearing and flat-shaped particles.

SOURCE: "A Century of Microphones" by B. B. Bauer. Proc. IRE (May 1962) p. 721.

SEE ALSO: T. A. Edison. U. S. Patent No: 474, 230 filed April 27th, 1877. Also U. S. Patent No: 474, 231/2.

1877   **LOUDSPEAKER (Moving coil type)**   E. W. Siemens (Germany)

The motor mechanism consisting of a circular coil located in a radial magnetic field was first disclosed by Siemens. Lodge, Pridham and Jenson and others contributed to the suspension system. However, there were very few developments in loudspeakers in the twenty-seven years following Lodge's disclosure. A breakthrough in the dynamic loudspeaker was made by Rice and Kellogg in 1925. The success of the development was due to their recognition of three physical factors with relation to the action and design of a direct radiator loudspeaker. The first is that the sound-power output of a loudspeaker is the product of the mechanical resistance due to sound radiation and the square of the velocity of the diaphragm. The second is that sound radiation from a small vibrating diaphragm gives rise to a mechanical resistance which is proportional to the square of the frequency. The third is a vibrating system which is mass controlled. It follows then that, if the fundemental resonance occurs below the lowest frequency of interest, the complementary variations of the second and third factors which control the sound output as given by the first factor conspire to provide a uniform response up to the frequency region at which the assumptions begin to fail. This was the contribution of Rice and Kellogg, and it continues to be the basic precept that guides the design of all direct-radiator loud--speakers.

SOURCE: "Loudspeakers" by H.F.Olson. Proc. IRE (May 1962) p.730.

SEE ALSO: "Electroacoustics" by F.V.Hunt. John Wiley and Sons Inc. New York. N.Y. (1954)

E.W.Siemens. German Patent No: 2355 filed December 14th, 1877.

O.J.Lodge. British Patent No:9712 filed April 27th, 1898.

E.S.Pridham and P.L.Jenson. U.S.Patent No:-1,448,279 filed April 28th, 1920.

"Notes on the development of a new type of hornless loudspeaker" by C.W.Rice and E.W.Kellogg. Trans AIEE, Vol.44. pp.461-475 April, 1925.

1878   CATHODE RAYS   Sir W.Crookes (U.K.)

In his Bakerian lecture of 1878 and his British Association Lecture of 1879, he announced various striking properties of "molecular rays" including the casting of shadows, the warming of obstacles and the deflection by a magnet. The title "Radiant Matter" employed by Crookes in his British Association Lecture of 1879 referred to ordinary matter in a new state in which the mean free path was so large that collisions between molecules could be ignored.

SOURCE: "A Biographical Dictionary of Scientists" T.I.Williams. Adam and Charles Black, London 1969. p.120.

1878   CARBON FILAMENT   J.W.Swan, C.H.Stearn, F.Topham
            INCANDESCENT LAMP   and C.F.Cross (U.K.)

Swan invented a carbon filament incandescent lamp and a squirting process to make nitro-cellulose fibres for lamp filaments. Stearn was an expert in the production of high vacua and Topham the glass-blower who made the glass globes. Cross discovered the viscose process for the fibres.

SOURCE: "The Sources of Invention" by J.Jewkes, D.Sawers and R.Stellerman. MacMillan, London (1958) p.59.

NOTE: T.A.Edison

Edison's main interest was to perfect the incandescent electric lamp, of which the construction was now being rapidly advanced by Swan. After researches involving the examination of thousands of alternatives, he devised the cotton thread filament which, when joined with bulb patents, made the "Edison" lamp a commercial success.

SOURCE: "A biographical dictionary of scientists" by T.I.Williams (2nd Ed.) Adam & Charles Black, London (1976) p.159.

1878   CARBON GRANULE MICROPHONE   H. Hunnings   (United Kingdom)

To Hunnings goes the credit for inventing a carbon transmitter using a multiple loose contact, which soon took the form of a disk electrode projecting into a contact cup containing granulated carbon particles. From that time on, one improvement followed another in seemingly endless parade.

SOURCE: "Electroacoustics"'by F.V. Hunt, Harvard University Press/ John Wiley (1954), p.37.

SEE ALSO: Henry Hunnings, (granulated-carbon microphone) British Pat. No. 3647 dated 16th September 1878; U.S. Pats. No. 246,512 (filed 14 May 1881) issued 30 August 1881, and No. 250,250 (filed 30 September 1881) issued 29 November 1881; both assigned to American Bell Telephone Co.,

1879   HALL EFFECT   E.H.Hall (U.K.)

Hall voltage, linear function of magnetic flux.

If a current of particles bearing charges of a single sign and constrained to move in a given direction is subjected to a transverse magnetic field, a potential gradient will exist in a direction perpendicular to both the current and the magnetic field.

SOURCE: The Encyclopaedia of Physics. 2nd Ed. Editor R.M.Besencon. Van Nostrand, New York 1974. p.400.

1880   PIEZO ELECTRICITY   J.Curie and P.Curie (France)

The relation between voltage generated and mechanical pressure on crystallographic materials.

SOURCE: "Developpement, par pression, de l'electricite polaire dans les cristaux hemiedres a faces inclinees" J.Curie and P.Curie. Comp. Rend. Vol.91. pp.294-295. July-December 1880.

1882   WIMSHURST MACHINE   J.Wimshurst (U.K.)

The Wimshurst machine generates electrostatic electricity by friction. It consists of two discs which revolve in opposite directions. The discs are made of an insulating material such as glass and near the peripheries of the discs are mounted small sections of sheet conductor. The charges are generated on the conductor sections by the action of brushes which graze the sectors as they rotate and are picked up by combs. These charges are stored in capacitors and are used to produce a spark between a pair of ball conductors.

REFERENCE: "Text Book of Physics" by J.Duncan and S.C.Starling. Part 5 - Magnetism and Electricity. MacMillan & Company.London (1926) p.960.

REFERENCE: "Encyclopeadic Dictionary of Electronics and Nuclear Engineering" by R.I.Sarbacher. Pitman, London (1959) p.1402.

1884   NIPKOW TELEVISION SYSTEM   P. Nipkow

The first television invention that had practical consequences was the "electrical telescope", patented by Paul Nipkow in 1884. At the heart of his camera was the now famous Nipkow disk. It had 24 holes equally spaced along a spiral near the periphery of the disk. The image to be transmitted was focused on a small region at the disk's periphery, and the disk was made to spin at 600 revolutions per minute. As the disk rotated, the sequence of holes scanned the image in a straight line. A lens behind the image region collected the sequential light samples and focused them on a single selenium cell. The cell would then produce a succession of currents, each proportional to the intensity of the light on a different element of the image.

At the receiving end, Nipkow proposed using a magneto-optic (Faraday-effect) light modulator to vary the intensity of the reconstructed image. To form the image, a second disk, identical to and rotating synchronously with the one at the transmitter, would be needed.

Nipkow built no hardware - which is probably just as well, because the technology of the time would not have permitted him to build his system; the light modulator alone would have required some 10 watts of control power. His disk, however, was a model for several later television systems that were built, most notably those of British inventor John Logie Baird.

SOURCE: Electronics (April 17, 1980), p.70 and 75.

1884   <u>SWITCH, QUICK BREAK</u>   J. H. Holmes (U.K.)

            The loose-handle quick-break switch was invented by J. H. Holmes in
1884. For this device he was granted British Patent No:3256 of 1884, under the
title: "Improvements in or applicable to switches or circuit closers for
electrical conducting apparatus".

<u>SOURCE</u>:   "Patents for Engineers" by L. H. A. Carr and J. C. Wood.
Chapman and Hall, London (1959) p. 95.

1884   <u>BATTERY (Zinc-mercuric oxide cell)</u>   C. L. Clarke (U.S.A.)

            The alkaline zinc-mercuric oxide system was first suggested by Clarke
in 1884. Although there were a number of additional attempts made over the
years to design a practical cell using this system, it was not until early in
World War II that a commercially usable mercuric oxide dry cell was invented
by Ruben.

<u>SOURCE</u>:   "Batteries" by C. K. Morehouse, R. Glicksman & G. S. Lozier. Proc.
IRE (Aug. 1958) p. 1467.

<u>SEE ALSO</u>: C. L. Clarke. U. S. Patent No:298, 175 (May 6, 1884)

            "Balanced alkaline dry cells" by S. Ruben. Trans. Electrochem. Soc.
Vol. 92 (1947) p. 183.

1885   <u>TRANSFORMER (Power)</u>   C. Zipernowski, M. Deri and O. T. Blathy
                                                          (Hungary)

            The earliest patent covering the construction of the transformer
appears to be that of Carl Zipernowski, Max Deri and Otto Titus Blathy, all of
Budapest, and since this patent was applied for and granted in this country, the
British version can be quoted here.

            The date of application was 27th April, 1885, the patent being numbered
5201 in that year under the title:"Improvements in Induction Apparatus for
transforming Electric Currents".

<u>SOURCE</u>:   "Patents for Engineers" by L. H. A. Carr and J. C. Wood. Chapman
and Hall. London (1959) pp. 89/90.

1885   <u>RESISTOR (Moulded carbon composition type)</u>   C. S. Bradley (U.K.)

            It is of passing interest to note that the earliest moulded-rod
composition-type resistor of which the author has been able to trace any record,
dates back before the days of radio. In 1885, a moulded-composition resistor
was patented, comprising a mixture of carbon and rubber heated and moulded to
shape and subsequently vulcanized to a hard body.

<u>SOURCE</u>:   "Fixed resistors for use in communication equipment" by P. R.
Coursey. Proc. IEE Vol. 96 Pt. III (1949) p. 169.

<u>SEE ALSO</u>: C. S. Bradley. British Patent No: 8076/1885.

            M. Slattery U. S. Patent No:354, 275 (1885)

            D. C. Voss U. S. Patent No:573, 558 (1896)

1885   <u>TRANSFORMER (Distribution)</u>   M. Deri (Austria)

            For the earliest patent covering the use of the transformer in a dis-
tribution system it is necessary to refer to German Patent No:33951 of 1885,
since no corresponding patent was applied for in the United Kingdom.

            This application was made on 18th February 1885 by Max Deri (in this
case described as being of Vienna) under the title (translated) of "Improvements

in the Distribution of Electricity".

SOURCE: "Patents for Engineers" by L.H.A.Carr and J.C.Wood. Chapman and Hall. London(1959) p.91.

1887   GRAMOPHONE   E.Berliner (U.S.A.)
            (Phonograph)

On May 4th, 1887, Emile Berliner applied for a patent on what he called a "Gramophone" to distinguish it from Edison's phonograph of ten years earlier and from Bell's and Tainter's Graphophone of one year earlier. The first figure of the drawings in Berliner's patent shows a record wound on a cylindrical support but by 1888, when he introduced his first model, he had changed to a flat-disk record. The groove in this record had a lateral side-to-side movement, as against the vertical "hill and dale" system which had been employed by others. This lateral recording process was reminiscent of Leon Scott's phonoantograph, which used a diaphragm and hog bristle to trace a record of sound vibrations on lamp-blacked paper some thirty years earlier.

Berliner also used lamp-black as the recording medium, and combined this method with an etching process which permitted transfer of the original engraving to copper or nickel. Thus Berliner achieved a permanent master recording, and for the first time mass duplication of records was possible. No longer did artists have to repeat each number endless times.

By 1895, Berliner had developed a system utilizing many ideas of his own and others: Scott's lateral groove, his own flat disc, and a coating of Bell's and Tainter's wax. The system stood up as the industry standard for half a century, thus Berliner deserves a mantle as the father of disk recording and reproduction.

SOURCE: "Disk Recording and Reproduction" by W.S.Bachman, B.B.Bauer and P.C.Goldmark. Proc.IRE (May 1962) p.738-739.

SEE ALSO: "The gramophone and the mechanical recording and reproduction of musical sounds" by L.N.Reddie. J.Roy.Soc.Arts. Vol:LVI, pp.633-649 May 8th, 1908.

1887   AERIALS (Radio Wave Propagation)   H.R.Hertz   (Germany)

It remained for Heinrich Hertz to prove the existence of electric waves in space as predicted by Maxwell. The first true antenna appears to have been used by Hertz in his classical experiments at Karlsruhe in 1887. His antenna consisted of two flat metallic plates, 40 cm square, each attached to a rod 30 cm long. The two rods were placed in the same straight line, and were provided at their nearer ends with balls separated by a spark gap about 7-mm long. The spark gap was energized by a Ruhmkorff coil. In order to detect the radiated waves, Hertz employed a receiving circuit consisting of a circular loop of wire broken by a microscopic gap. The radius of the loop was 35 cm. which was found by experiment to be the proper size to be in resonance with the oscillator.

SOURCE: "Early history of the antennas and propagation field until the end of World War I, Part I - Antennas" by P.S.Carter and H.H.Beverage. Proc. IRE (May 1962) p.680.

SEE ALSO: "Ueber sehr schnelle elektrische Schwingungen" H.Hertz. Ann. Physik und Chemie (Wiedeman) N.F.Vol.31. pp.421-448.(May 15,1887)

1887   SWITCH,QUICK MAKE & BREAK   J.H.Holmes (U.K.)

Following Holmes invention of the quick-break switch (see 1884) a later variant provided for a loose-handle operating a spring over a dead centre by means of a toggle action, so that both quick-make and quick-break were obtained. (British Patent No:5648 of 1887)

SOURCE: "Patents for Engineers" by L.H.A.Carr and J.C.Wood. Chapman and Hall. London (1959) p.95.

1888   INDUCTION MOTOR   N. Tesla (U.S.A.)

Tesla's "master" patent covering the polyphase induction motor was taken out, inter alia, in Great Britain; the British Patent, No:6481 of 1888, may therefore be used as a reference.

Its title was "Improvements relating to the electrical transmission of power and to apparatus therefor", and it was granted in the name of Nikola Tesla of the City and State of New York, U.S.A.

The specification is long and detailed with 18 diagrammatic figures.

SOURCE: "Patents for Engineers" by L.H.A. Carr and J.C. Wood. Chapman and Hall, London (1959) pp. 96/7

1889   COMPUTERS (Tabulating Machines)   H. Hollerith (U.S.A.)

Hollerith worked on a machine for tabulating population statistics for the 1890 Census in the U.S.A. which he patented in 1889. He used a system of holes in a punch card to represent various characteristics: such as male or female, black or white, age, etc. The cards were 6 5/8 by $3\frac{1}{4}$ inches in size, which he chose because it was the size of a dollar bill. Each card contained 288 locations at which holes could be made.

SOURCE: "The computer from Pascal to von Neumann" by H.H. Goldstine. Princeton Univ. Press (1972) p. 65. (Reprinted by permission)

SEE ALSO: "An electric tabulating system" by H. Hollerith, reprinted in "The Origins of Digital Computers" edited by B. Randell, Springer-Verlag, Berlin (1973) p. 129.

1889   STROWGER AUTOMATIC TELEPHONE EXCHANGE   A.B. Strowger (U.S.A.)

The man who is generally given the credit of inventing the first practical system of automatic telephony to be used commercially is Almon Brown Strowger, an undertaker from Kansas City, U.S.A. It is said that Strowger's business was losing money because, when people called him, the operators at the telephone exchange would connect the call to other undertakers instead, and he was determined to invent a system that would work without operators. He patented his ideas in 1889, and three years later his equipment was installed in the first public automatic telephone exchange, at La Porte, Indiana, U.S.A.

SOURCE: "The Telephone and the Exchange" by P.J. Povey. Post Office Publication, 1974, p. 35.

SEE ALSO: "The Story of Telephone Switching". TELONDE, No. 2/1977, p. 16.

1890   COHERER   E. Branly   (France)

In Paris, Edouard Branly, physics professor at the Catholic University observed in 1890 that metal filings, when subjected to "Hertzian waves" behaved very strangely. Normally, filings do not transmit an electric current because there are air spaces between them - but when placed with the range of electro-magnetic waves, the filings fuse a little together, enough to offer a conducting path to an electric current. The filings remain a conductor until they are disturbed by shaking or tapping.

Branly called the little glass tube in which he placed his filings "coherer"; it was the first form of a "detector" for electro-magnetic waves.

SOURCE: "A History of Invention" by E. Larsen. J.M. Dent & Sons, London and Roy Publishers, New York (1971) p. 278..

1891   BATTERY (Standard Weston Cell)   E. Weston (U.S.A.)

       In 1891, Dr. Edward Weston, an Anglo-American from New Jersey, filed a patent which was granted in 1893 disclosing a cell in which the electrolyte of the Clark cell was replaced by a saturated solution of cadmium sulphate and the zinc negative electrode was replaced by cadmium amalgam, the depolariser being mercurous sulphate as before. This cell was a definite improvement on the Clark cell since the temperature coefficient of e.m.f. was only about $-40\mu V/^{\circ}C$. Considerable improvements were made by various workers from 1893 onwards and finally the London Conference held in 1908 authorised the appointment of a special international committee. As a result of the work of this committee, Weston's cadmium cell was assigned the value of $1\cdot 01830$ International volts at $20^{\circ}C$.

SOURCE: "Standard cells by Muirhead" Muirhead Technique. Vol. 18. No. 3. (July 1964) p. 19.

SEE ALSO: Weston cell - U.S.A. Patent No:494, 827 (1893) also British Patent 640, 812797, 381

1893   WAVEGUIDES   J.J. Thomson (U.K.)

       Perhaps the first analysis suggesting the possibility of waves in hollow pipes appeared in 1893 in the book "Recent Researches in Electricity and Magnetism" by J.J. Thomson. This book, which was written as a sequel to Maxwell's "Treatise on Electricity and Magnetism" examined mathematically the hypothetical question of what might result if an electric charge should be released on the interior wall of a closed metal cylinder. Even now, this problem is of considerable interest in connection with resonance in hollow metal chambers. A much more significant analysis, relating particularly to propagation through dielectrically filled pipes, both of circular and rectangular cross section, was published in 1897 by Lord Rayleigh.

SOURCE: "Survey and History of the Progress of the Microwave Arts" by G.C. Southworth. Proc. IRE (May 1962) p. 1199.

SEE ALSO: "Recent researches in electricity and magnetism" by J.J. Thomson (1893) p. 344.

       "On the passage of electric waves through tubes or the vibrations of dielectric cylinders" Lord Rayleigh. Phil Mag. Vol. 43 (Feb. 1897) p. 125.

1895   X-RAYS   W.K. Rontgen (Germany)

       On 8th November, 1895, while experimenting with a Crookes's tube (see W. Crookes) covered with an opaque shield of black cardboard, Rontgen noticed that, when a current passed through the tube, a nearby piece of paper painted with barium platinocyanide fluoresced. In a series of classical papers (1895-7) he described the properties of the new, so-called X-rays, but his attempts to detect their interference by crystals were unsuccessful.

SOURCE: "A Biographical Dictionary of Scientists" by T.I. Williams. Adam and Charles Black, London (1969) p. 448.

1896   TELEPHONE DIAL   E.A. Keith   (United
                                                    C.J. Erickson   Kingdom)
                                                    J. Erickson

       The early automatic telephone systems of Connolly, Connolly, and McTighe, of Sinclair, and of Strowger, all used push-buttons which the caller was required to press a number of times depending on the number of the telephone he wanted. When automatic exchanges beame larger, and telephone numbers became longer, this type of signalling was no longer practical, and in 1896, E.A. Keith, C.J. Erickson, and John Erickson invented the telephone dial.

SOURCE: "The Telephone and the Exchange" by P.J. Povey. Post Office Publication, 1974, p. 60.

1896   WIRELESS TELEGRAPHY   G. Marconi   (Italy)

            On 2nd June 1896, Marconi took out in the United Kingdom the first
patent for wireless telegraphy based on Hertz's discoveries, though exploiting
radiations of a much longer wavelength. His apparatus consisted of a tube-like
receiver or "coherer" connected to an earth and an elevated aerial; its signals
were at first transmitted over one hundred yards, a satisfactory demonstration
being arranged from the roof of the London General Post Office. Ship to shore
communication was established in the following year, when Marconi formed a
Wireless Telegraph Company in London for the exploitation of his patents in all
countries except Italy; this later developed world-wide affiliations. His first
transatlantic signals were made on 12th December, 1901 from Poldhu in Cornwall to St. John's, Newfoundland, where they were received through an aerial
suspended from a kite.

SOURCE: "A Biographical Dictionary of Scientists"   by T.I. Williams. Adam
and Charles Black, London (1969) p. 352.

1897   ELECTRON   J.J. Thomson   (U.K.)

            By improving the vacuum of Hertz's cathode ray experiments, Thomson
got deflections which, combined with the long-known deflections by a magnet
determined the ratio of the charge to the mass of the supposed particles. This
ratio e/m was over 1,000 times larger than the ratio for hydrogen, the lightest
atom known. Thomson considered that this was due to the smallness of the mass
and that particles with this small mass were universal constituents of matter,
since they were the same whatever the chemical nature of the gas carrying the
discharge and the electrodes through which it entered and left. He examined
two other cases of the discharge of electricity, namely, those from a hot wire
negatively charged and from a negatively charged zinc plate illuminated by
ultra-violet light. In both he found charged particles with the same e/m ratio
as the cathode rays, and in the second was able to measure the actual charge
by condensing drops of water on the particles to form a mist, finding the size
of the drops from their rate of fall. The results agreed with the supposed value
of the charge on a hydrogen atom as far as this was then known. Thomson
called these new light particles "corpuscles"but later adopted the word "electron"
invented a few years before by J.G. Stoney for the charge on a hydrogen atom
regarded as a natural unit of charge.

SOURCE: "A Biographical Dictionary of Scientists"   T.I. Williams. Adam and
Charles Black, London(1969) p. 511.

SEE ALSO: "The discovery of the electron" by D.L. Anderson. Van Nostrand
Reinhold, New York (1964).

            "The first subatomic particle" by Otto Frisch. New Scientist.
(17th Nov.1977) p. 408.

1897   CATHODE RAY OSCILLOGRAPH   F. Braun (Germany)

            The cathode-ray oscilloscope for the study of the time variation of
electron currents was developed by Ferdinand Braun in 1897, the same year in
which J.J. Thomson measured the specific charge of the electron by its
deflection in electric and magnetic fields. Ferdinand Braun constructed the
first cathode-ray oscillograph at the University of Strassbourg in 1897.

            Just like the early X-ray tube, the "Braun tube" used gas discharge
phenomena for the emission and formation of an electron beam. Even after the
introduction of a thermionic cathode into cathode-ray tubes by Wehnelt in 1905
an argon atmosphere of about 10^{-3} mm Hg was still retained in commercial
oscilloscopes for another 25 years. The effect of ion focusing facilitated the
formation of long, filamentary, electron beams.

SOURCE: "Beam-deflection and photo devices" by K. Schlesinger and E.G.
Ramberg. Proc.IRE (May 1962) p. 991 and 995.

SEE ALSO: "On a method for the demonstration and study of currents varying
with time". by F. Braun. Wiedemann's Ann. Vol. 60 (1897) p. 552.

"Cathode Rays" by J.J. Thomson.   Phil. Mag. Vol. 4. (1897), p. 293.

"Ferdinand Braun and the cathode ray tube" by G. Shiers. Scientific American,
Volume 230 (March 1974), p. 92.

1897   RESISTOR (Carbon film type)   T.E.Gambrell
                                                     and   (U.K.)
                                                 A.F.Harris

         The earliest type of carbon-film resistor was also used many years
before broadcasting and it was apparently forgotten when the greater demand
arose. Some of these carbon-film resistors, notably those formed by spraying
or otherwise applying the conducting coating and then baking on to a glass
filament, were enclosed inside a glass tube with metal end-caps sealed on for
terminal connections and with terminals connected to the filament coating by
casting into a metal such as type-metal.

         SOURCE: "Fixed resistors for use in communication equipment" by P.R.
Coursey. Proc.IEE Vol.96 Pt.III (1949) p.170.

         SEE ALSO: T.E.Gambrell and A.F.Harris. British Patent No:25412/1897.

1898   MAGNETIC RECORDING   Valdemar Poulsen   (Denmark)

         No one really knowns how the idea of magnetic recording occurred
to Poulsen. During his experiments he found that:

         "it would be possible to magnetize a wire to different degrees so
         close together that sound could be recorded on it by running the
         current from a microphone through an electromagnet and by
         either drawing the wire rapidly past the electromagnet or drawing
         the electromagnet rapidly past the wire."

This invention had several things to commend it: the wire or tape could be
used over and over again by de-magnetizing it and recordings could be played
thousands of times without destroying the quality. Poulsen invented this
Telegraphone in 1898; with it he won the Grand Prix at the Paris Exposition
in 1900. He filed an application for a Danish patent in 1898 and within two
years he had filed additional patent applications in the United States and most
European countries. These early patents suggested, as recording media,
steel wires and tapes and discs of material coated with magnetisable metallic
dust, though he himself used only steel wire and tape in his machinery. The
basic principles enunciated by Poulsen are still applied in all types of modern
magnetic recorders.

         SOURCE: "The Source of Invention" by J.Jewkes, D.Sawers and
R.Stillerman. MacMillan & Co.London (1958) p.326.

         SEE ALSO: "The Development of the Magnetic Tape Recorder" Engineer.
(March 18th,1949)

1900   CAPACITORS (Ceramic)   L.Lombardi (Italy)

         Ceramic materials have been used for many years as electrical
insulators. They are able to withstand severe working conditions because they
are vitrified by firing at temperatures of the order of 1,200°C. Being completely inert they will withstand their rated working voltage indefinitely and
retain their shape and physical characteristics under normal conditions.

         SOURCE: "Fixed Capacitors" (2nd Edition) by G.W.A.Dummer, Pitman,
London (1956) p.115.

         SEE ALSO: "An improved process for manufacturing thin homogenous plates,
more particularly applicable for use in electrical condensers" L.Lombardi,
British Patent No:9133 filed May 17,1900.

         "Permittivity of titania" by W.Schmidt. Ann.Phys.Lpz.4, 9(1902) p.959.
"Dielectric losses and breakdown strength of porcelain" by F.Beldi. Brown
Boveri Rev.18 (May 1931) p.172.
"Insulating materials of the steatite group" by E.Schonberg. Elektrotech.Z.54.
(June 1933) p.545.
Tubular metallized ceramic capacitors. Porzellanfabrikkahla, Germany.
Brit.Pat. No:440951/1934.

1900   BATTERY (Nickel-iron cell)   T.A. Edison (U.S.A.)

            The nickel-iron-alkaline batteries as they exist today are essentially
the same as discovered by Edison around 1900 and marketed in 1908. The
negative electrode consists of pockets of active material grouped together to
form a plate, while the positive plate is an assembly of perforated nickel tubes
filled with nickel hydroxide and nickel.

SOURCE: "Batteries" by C.K. Morehouse, R. Glicksman and G.S. Lozier.
Proc. IRE (August 1958) p. 1478.

SEE ALSO: "The Edison nickel-iron-alkaline cell" by F.C. Anderson, J.
Electrochem. Soc. Vol. 99. (1952) p. 244C.

1900   QUANTUM THEORY   M. Planck (Germany)

            Max Planck had propounded in 1900 the theory that energy is not emitted
in a continuous flow, but always in a collection of packets of finite size. He was
driven to accept this view in order to explain the distribution of the energy found
in the light from hot bodies such as the sun. His conception struck at the
principle of continuity as the comprehensive basis of nature.

SOURCE: "Science at War" by J.G. Crowther and R. Whiddington H.M.S.O.
London (1947) p. 124.

1900   BATTERY (Nickel-cadmium cell)   Junger & Berg (Sweden)

            The nickel-cadmium battery discovered by Junger and Berg about 1900
is closely related to the nickel-iron Edison battery. Both are mechanically
rugged and will withstand electrochemical abuse in that they can be overcharged,
overdischarged, or stand idle in a discharged condition. The nickel-cadmium
battery differs from the Edison battery in the use of cadmium anodes in place of
iron.

SOURCE: "Batteries" by C.K. Morehouse, R. Glicksman and G.S. Lozier.
Proc. IRE (August 1958) p. 1478.

SEE ALSO: "Storage Batteries" by G.W. Vinal. J. Wiley - New York 4th Ed.
(1955).

1901   RADIO- HEAVISIDE/KENNELLY LAYER   O. Heaviside   (U.K.)
                                                     A. Kennelly   (U.S.A.)

            When Marconi first succeeded in establishing radio communication
around the curvature of the earth in 1901, Oliver Heaviside in England and
Arthur Kennelly in the U.S.A. proposed that the phenomena might be due to
the existance of an ionized layer in the upper atmostphere surrounding the
earth which could reflect radio waves. However, another explanation was that
the waves were detectable because of diffraction around the earth's curvature,
a phenomenon which would be more observable, the longer the wavelength used.
Another explanation could be that of the bending of the wavefront which would
occur if the dielectric constant of the atmosphere should progressively change
with altitude due to the effects of temperature, density, and moisture.

            The observation of the waves of high frequency drew renewed attention
to the proposals of Heaviside and Kennelly. Their hypothetical layer was
dubbed the Heaviside-Kennelly layer. Breit and Tuve in 1926, definitely con-
firmed the existence of several such layers by sounding the atmosphere with
short pulses sent from a transmitter on earth which were reflected from the
layers back to a receiver adjacent to the transmitter. The time delay identified
the height of each reflection and determined physical characteristics of the
nature of the reflection and the degree of penetration as a function of frequency.
The Breit-Tuve experiments were a predecessor of later radar techniques.
The vagaries observed by the earlier experimenters began to be explainable
and even predictable. The region of the several layers was renamed the
"Ionosphere" and the individual layers designated the D, E, F_1 and F_2 layers,
the D layer being the lowest.

SOURCE: "Telecommunications - The resource not depleted by use. A Historical and Philosophical Resume" by W. L. Levitt. Proc. IEEE. Vol. 64 No. 9. (Sept. 1976) p. 1297.

SEE ALSO: "A test of the existence of the conducting layers" by G. Breit and M. Tuve. Phys. Rev. Vol. 28 (Sept. 1926) p. 554.

1901   FLUORESCENT LAMP   P. Cooper-Hewitt   (U.S.A.)

The first low-pressure mercury discharge lamp was introduced at the beginning of the twentieth century by Peter Cooper-Hewitt, the American individual inventor. Sir Humphrey Davy had discovered the effect of a discharge through mercury early in the nineteenth century. Cooper-Hewitt's lamp was inefficient by modern standards, though better than contemporary incandescent lamps; it also produced the characteristic blue light of the mercury discharge lamp. It differed from the modern fluorescent lamp in being designed to produce visible radiation, not ultra-violet. In 1901 he used rhodamine dye, which fluoresces red, to improve the light's colour, but the rhodamine deteriorated too rapidly for this to be a success. The Moore and neon discharge lamps, introduced at much the same time as the Cooper-Hewitt, also contributed to the development of the fluorescent lamp: D. McFarlan Moore, an American individual inventor, was the first to apply to hot cathode used on it, and he also anticipated Wehnelt in constructing lasting electrodes. Georges Claude's introduction of the neon tube, and the desire to modify its colour, stimulated interest in fluorescent powders and in means of employing them with a lamp.

SOURCE: "The Sources of Invention" by J. Jewkes, D. Sawers and R. Stellerman. MacMillan, London(1958) pp. 298/9.

SEE ALSO: "Lighting by Luminescence" by A. Claude. Light and Lighting. (June 3, 1939).

1902   INDUCTION MOTOR   E. Danielson (Sweden)
            (SYNCHRONOUS)

Danielson's synchronous induction motor is an excellent example of the sudden appearance of an invention which was complete at its first inception.

It was not patented in Great Britain, but the full information is available in United States Patent No:694092 of 25th February, 1902 granted to Ernst Danielson of Westeras, Sweden.

After relating the advantages of the over-magnetised synchronous motors for eliminating lag (of current) and referring to the difficulty of starting such machines, the specification proceeds:

> "The invention consists, briefly, in combination with an ordinary induction-motor and a suitable resistance for connecting to the secondary element of said motor, a source of continuous electric currents (and) a switch arrangement so connected that the secondary part of the motor may by means of said switch arrangement either be connected to the said resistance or to the said source of continuous currents".

With reference to the self-synchronizing action, the specification states:-

> "When the exciting current is supplied, the motor is changed from an asynchronous to a synchronous one, provided, however, that the exciting-current is strong enough to pull the motor in step."

The drawing and diagram of connections (Showing a three-phase-secondary winding) which are attached to the specification are completely up-to-date, and were it not for the somewhat archaic outlines of the machines, might have been taken from a present-day text book or manufacturer's pamphlet.

SOURCE: "Patents for Engineers" by L. H. A. Carr and J. C. Wood. Chapman and Hall, London (1959) pp. 97/8.

1902   SPONTANEOUS ATOMIC CHANGE   E. Rutherford and F. Soddy (U. K.)

            Very swift and brilliant analysis of the phenomena led Rutherford and
Soddy to announce in 1902, only six years after the original discovery of radio-
activity, their theory of the spontaneous disintegration of atoms. They asserted
that atoms, the very foundation of matter and nature, were exploding, and not
according to any rule, but merely by chance. Einstein has said that he "could
not believe that the Almighty had organised the world according to the throwing
of dice." It needed a bold spirit to adopt chance as the first principle in the
explanation of the transmutation of the fundamental atoms of matter.

            Rutherford forthwith bent his full genius to the task. Within nine years
he had made the rays reveal the general structure of the atom. He proved that
atoms are not little hard balls, but very spacious structures, consisting of a
few distant electrons circulating round a relatively heavy nucleus, like the
planets round the sun of a miniature solar system. Nearly all the mass of the
atom was concentrated in the nucleus, which carried a positive electric charge
exactly balancing the sum of the negative electric charges carried by the
circulating electrons.

SOURCE: "Science at War" by J. G. Crowther and R. Whiddington. H. M. S. O.
London (1947) p. 123.

1904   TWO ELECTRODE TUBE   J. A. Fleming (U. K.)

            Fleming received his early education in London, where he attended
London University; later he spent four years at Cambridge, and he was
appointed "electrician" to the Edison Electric Light Company in 1882. During
a visit to the United States in 1884 he visited Edison to discuss electric
lighting problems, and it is of particular moment to the story of the valve that,
during the visit, Edison demonstrated a discovery he had made a year before -
the Edison effect. Using a carbon-filament lamp in which a metal plate had
been sealed, Edison found that when the plate was connected through a
galvanometer to the positive terminal of the filament a current flowed, but no
current flowed when connection was made to the negative terminal. He used
the device to regulate the supply voltage in power stations. Fleming was very
interested in the phenomenon, and, on his return home, carried out researches
in which he showed that "the space between the filament and the metal plate is
a one-way street for electricity".

            It was on the 16th November, 1904, that Professor J. A. Fleming
(1849-1945) described in British Patent Specification No:24850 a two-electrode
valve for the rectification of high-frequency alternating currents.

SOURCE: "Fleming and de Forest - an appreciation" by Captain C. F. Booth.
IEE Pub. Thermionic Valves 1904-1954. IEE London (1955) p.1.

1904   CAPACITORS. GLASS (TUBULAR)   I. Moscicki (U. K.)

            The glass dielectric capacitor was later manufactured in tubular form
known as the Moscicki tube - which provided the only form of capacitor available for Marconi's early experiments in practical wireless communication; and
in a modified form, using flat glass plates interleaved with zinc sheets and all
immersed in oil, continued to provide the condensers for the spark wireless
transmitting apparatus up to and including part at least of the 1914-18 war
period.

SOURCE: "Electrical Capacitors in our Everyday Life" by P. R. Coursey.
ERA Journal. No. 6. (Jan. 1959) p. 10.

SEE ALSO: I. Moscicki "Improvements in electric condensers" British Patent
No:1307, filed January 18th, 1904.

1905   INSULATED SODIUM CONDUCTOR   A. G. Betts (U. S. A.)

            The concept of using sodium as an electrical conductor is hardly a new
one. Back in 1901, a basic Swiss patent was issued for this purpose. In 1905
and 1906, French and American patents, respectively, were issued to Anson G.
Betts, who recognised the favourable economics of the metal. His patent

specified that "the sodium be enclosed - prefereably hermetically - by a sheathing of substantially nonoxidizable reinforcing material."

In 1927, H. H. Dow and R. H. Boundy of the Dow Chemical Company began to experiment with sodium in steel pipes, and in 1930, Dow constructed a line, 10 cm in diameter by 260 meters long, by joining together 6-meter lengths of sodium-filled steel pipe. This uninsulated conductor operated at Dow's plant in Midland, Mich., for about ten years at currents ranging from 500 to 4000 amperes direct current. The results of this experimental work have been published in considerable detail.

SOURCE: "Insulated sodium conductors - a future trend" by L. E. Humphrey, R. C. Hess, G. I. Addis, A. E. Ruprecht, P. H. Ware, E. J. Steeve, J. A. Schneider, I. F. Matthysse and E. M. Scoran. IEEE Spectrum (Nov. 1966) p. 73.

SEE ALSO: "Sodium in pipe successful as electrical conductor" by R. H. Boundy. Elec. World. Vol. 100. No. 26. p. 852. (Dec. 24, 1932)

"A 4000-ampere sodium conductor" by R. H. Boundy. Trans. Electrochemical Soc. pp. 151-160 (Sept. 1932)

T. De Koning. Cooling Electrical Machines and Cables. The Hague, Netherlands: Groothertoginne, (1955). pp. 202-232.

1905   THEORY OF RELATIVITY   A. Einstein (U. S. A.)

The brilliant theory of relativity was proposed by Einstein in 1905 to explain the observed fact that the speed of light is constant under all conditions. The theory of relativity was extended to all the motions of nature, and it began to be evident that many apparently different things were the same fundamental thing seen from different points of view. Space and time, which according to the old ideas seemed utterly different, were in the light of relativity seen to be two different aspects of one underlying unity.

From this line of development, Einstein showed that mass and energy were one of these pairs of interchangeable aspects. Mass could be conceived as congealed energy; and he even calculated how much energy there would be in a unit of mass. He found that it would be enormous. When energy was released, mass disappeared. The annihilation of mass would be accompanied by a vast and proporti onate output of energy. If a mass of one ounce of matter could be transformed into energy, it would produce enough to turn nearly one million tons of water into steam.

SOURCE: "Science at War" by J. G. Crowther and R. Whiddington. H. M. S. O. London (1947) p. 125.

1906   CRYSTAL DETECTOR (CARBORUNDUM)   H. C. Dunwoody (U. S. A.)

Silicon shares with carborundum the distinction of being used in one of the first crystal rectifiers. The use of carborundum was discovered by an associate of de Forest, Gen. Henry H. C. Dunwoody, who applied for a patent on March 23, 1906. It saved the day for de Forest's company, which had just been enjoined from infringing on the electrolytic detector of Reginald Fessenden after a lengthy legal battle. De Forest's Audion, only recently invented, was not yet commercially practicable. Ironically, it was Pickard who helped de Forest's company use its newly invented carborundum, showing the company the proper form of holder and battery bias for good results. He was never compensated for this consulting work.

SOURCE: "The crystal detector" by A. Douglas. IEEE Spectrum (April 1981), p. 66.

1906   RADIO BROADCASTING   R. Fessenden (U. S. A.)

The first documented successful broadcasting of speech and music was conducted by Dr. Reginald Fessenden at Brant Rock, Mass., on Christmas eve, 1906, utilizing a 50 kc radio-frequency alternator which produced about 1 kw of power and which was built by the General Electric Co., under the direction of Dr. E. F. W. Alexanderson. Modulation was accomplished by means of a microphone which is belived to have been water-cooled and which was connected in the antenna circuit. Clear reception was obtained at many locations including ships at sea.

Subsequently, Dr. Lee de Forest conducted experimental broadcasting in 1907 from his laboratory in New York City, in 1908 from the Eiffel Tower in Paris, and in 1910 from the Metropolitan Opera House in New York City. These experiments, which were conducted with arc transmitters of about 500 watts power, modulated by microphones in the antenna-ground system, while successful, were handicapped by the high noise level inherent in arc transmitters.

SOURCE: "AM and FM Broadcasting" by R.F.Guy. Proc. IRE (May 1962) p. 811.

SEE ALSO: "History of radio to 1926" by G.L.Archer. The American Historical Society. New York. N.Y.1938.

1906   THREE ELECTRODE TUBE   L. de Forest (U.S.A.)

While Fleming was developing his two-electrode valve, Dr. Lee de Forest was working in the United States on somewhat similar lines, and on the 25th October, 1906, de Forest applied for a patent for a three-electrode valve - a triode - as a device for amplifying feeble electric currents, the amplification being achieved by using a voltage on the intermediate electrode (grid) to control the plate current. A few months later de Forest extended the patent to cover the use of the valve as a detector. The introduction of the third electrode to provide an amplifier as compared with the two-electrode rectifier very greatly extended the potential applications of the thermionic valve, and much credit is due to de Forest for his achievement.

Unfortunately, the invention of the triode led to considerable bitterness and litigation involving Fleming and de Forest, the former insisting to the end of his long life that de Forest's work was dependent on his own two-electrode valve. On the other hand, de Forest has always maintained that he was not aware of Fleming's patent before taking out his own. Initially the American courts held that de Forest's addition of the grid was dependent on Fleming's work. The story of the patent litigation did not end until 1943, when the United States Supreme Court decided that the original Fleming patent had always been invalid.

SOURCE: "Fleming and de Forest - an appreciation" by Captain C.F.Booth IEE Pub. Thermionic Valves 1904-1954. IEE London (1955) p. 2.

1907   GONIOMETER   A. Artom   (Italy)

The direction of aircraft on long-distance routes can be plotted more accurately than ever before, thanks to the invention of the radio-goniometer by Italian Alessandro Artom. Now navigation by radio waves is possible, making the old, cumbersome slide-rule calculations obsolete.

SOURCE :   "The Timetable of Technology", published by Michael Joseph, London, and Marshall Editions, London (November 1982), p. 26.

1907   CRYSTAL DETECTOR (PERIKON)   G. W. Pickard (U.S.A.)

In addition to the silicon detectors, Wireless Specialty sold detectors using other minerals discovered by Pickard in November 1907: Pyron (iron pyrite) and Perikon (zincite in contact with chalcopyrite). The name Perikon was coined from PERfect pIcKard cONtact. Each mineral had its own field: Perikon was most sensitive, but had to be readjusted often, whereas silicon was very stable and able to withstand heavy static discharges.

SOURCE: "The crystal detector" by A. Douglas. IEEE Spectrum (April 1981), p. 66.

1908   ELECTRONIC ORGAN   T. Cahill (U.S.A.)

Beginning with the 1900s there were many attempts to offer electric or electronic substitutes for organs, but none enjoyed any degree of commercial success. They were based on photo-optics, magnetic or electrostatic pre-recordings, vacuum tube or neon lamp oscillators, or amplified blown reeds, etc. One of these, called the Telharmonium, was invented by Thadius Cahill and demonstrated in 1908. The size of a small power-generating station, it consisted of almost a hundred alternator generators for all the frequencies of the scale. Then through a console of switches, synthesised musical signals were transmitted over telephone lines without benefit of amplifiers.

In 1935 Mr. Laurens Hammond, based on his synchronous electric clock, invented the first commercially successful mass-produced electric organ that started an industry. Since then, many manufacturers all over the world have joined this industry, offering instruments in a variety of sizes and prices that have transformed modern music.

SOURCE: "Electronics Engineer's Reference Book" Newnes-Butterworth London (1976) Chap. 17. p. 17 - 2.

SEE ALSO: T. Cahill   U.S. Patent No: 1295691 (25th Feb. 1919)

L. Hammond U.S. Patent No: 1956350 (24th April, 1934)

1908   GEIGER COUNTER   E. Rutherford and H. Geiger (U.K.)

About this time, experiments were being conducted with the collection of current by a positive wire and negative cylinder arrangement that were to have a profound effect on science. These experiments were reported in a paper by Rutherford and Geiger, which showed that the number of charges of an ionizing event could be multiplied several thousand times by the ionizing action of electrons in the high field region near the wire. This was the start of what for a time were called Gieger-Muller counter tubes and now simply Geiger counter tubes. The technique of the proportional type of counter was established in 1928, and the following year, schemes for determining the coincidence of ionizing events were presented. Thus direction, scattering, absorption, etc. types of experiments were possible and the modern era of cosmic-ray and nuclear research developed.

SOURCE: "The development of gas discharge tubes" by J. D. Cobine. Proc. IRE (May 1962) p. 971.

SEE ALSO: E. Rutherford and H. Geiger. Proc. Royal Soc. London A. Vol. 81. (1908) p. 612.

1908   TELEVISION (Electronic)   A. A. Campbell-Swinton (U.K.)

Boris Rosing, of the St. Petersburg Technological Institute, seems to have been the first physicist who thought of using Braun's tube for the reception of images. As early as 1907 he suggested a system of remote electric vision, with a Nipkow disc for scanning the scene to be transmitted and a cathode-ray tube as the receiver. At about the same time the English inventor, A. A. Campbell-Swinton, also proposed a system of electronic television, but with cathode-ray tubes for transmission as well as for reception. He published his ideas in the scientific magazine NATURE in 1908, and elaborated them again in 1911 and 1920, explaining that the image transmitted in this way could be split up into, and reassembled from, about 400,000 points of different light value within $\frac{1}{25}$ of a second.

SOURCE: "A History of Invention" by E. Larsen. J. M. Dent & Sons, London, and Roy Publishers, New York (1971) p. 323.

1909   FERRITES (H.F.)   G. Hilpert   (Germany)
                                      J. L. Snoek   (Holland)

The first proposal for high-frequency application was made in Germany in 1909 by Hilpert, who first synthesized such ferrites. However, the practical development and exploration of these materials did not come about until Snoek from the Philips Laboratories in Holland carried out extensive investigations on the high-frequency properties of such materials as manganese and nickel ferrite well into the UHF region.

SOURCE: "Solid-State devices other than semiconductors" by B. Lax and J. G. Mavroides. Proc. IRE (May 1962) p. 1012.

SEE ALSO: "Genetische und konstitutive Zusammenhange in den magnetischen Eigenschaften bei Ferriten und Eisenoxyden" G. Hilpert. Berichte deutsch chemisch, Gesell. Vol. 42. p. 2248-2261. (1909)

"New developments in ferromagnetic materials" J. L. Snoek. Elsevier Pub. Co. New York (1947).

1910   NEON LAMP   G. Claude   (France)

          The majority of cold cathode tubes use a gas filling of which neon is the
principal constituent. Development of the modern cold cathode tube may thus
be said to date from 1898, the year in which Sir William Ramsay discovered
neon. Ten years later Georges Claude began to isolate substantial quantities
of a helium-neon mixture and in 1910 Claude exhibited two 38-ft. neon tubes.
These were the precursors of the familiar neon sign and decorative lighting
tube of today, in which the light   comes principally from the long positive
column.

          Filament lamps and electric power were both expensive at that time.
There was accordingly a strong incentive to develop a cheap and robust lamp
of low power consumption and suitable for use on domestic supply voltages.
Professor H.E. Watson has described the work which led to the appearance of
the domestic neon lamp, first in Germany in 1918 and later in Holland. Once
sputtering had been substantially reduced, the design of a 'beehive' neon for
220 V supplies presented few problems. For 110 V supplies, however, an alloy
or activated cathode was required in order sufficiently to reduce the cathode
fall. The successful development of such a tube by Philips in Holland stimulated renewed activity in the U.S.A. and eventually, in 1929, the General
Electric Co., produced a miniature neon indicator.

SOURCE: "A survey of cold cathode discharge tybes" by D.M. Neale
The Radio and Electronic Engineering. (February 1964) p.87.

SEE ALSO: "The development of the neon glow lamp (1911-61)" by H.E. Watson
Nature 191, No. 4793. p.1040-1 (9th September, 1961).

1911   ATOMIC THEORY   Lord Rutherford   (United Kingdom)

          Atomic Theory.   The general model of the atom was first proposed by
Rutherford in 1911 and consisted of a nucleus of protons and neutrons, about
which rotated electrons in orbits.   Such a system appears not unlike our
solar system in a scale relative to the size of the components.

SOURCE :   "Semiconductor Electronics - 1. Solid-State Physics".
Electro-Technology (October 1960), p.95.

1911   SUPERCONDUCTIVITY   K. Onnes (Netherlands)

          The discovery of superconductivity dates back to 1911 when Kamerlingh
Onnes was carrying out a systematic investigation at Leiden of the electrical
resistivity of metals in the range of low temperatures opened up by his recent
liquefaction of helium in 1908. In the course of these measurements he observed
a sudden drop in the resistance of mercury at $4.15^{\circ}K$, and was unable to detect
any remaining resistance bekow that temperature; in short, the metal had
become superconducting.

SOURCE: "Materials for conductive and resistive functions " by G. W. A.
Dummer. Hayden·Book Co. N.Y.   p.121

SEE ALSO: 1911   Onnes, H.K. Comm. Phys. Lab., University of Leiden,
                         No.119b ("Further experiments with liquid helium")
                         No.120b ("The resistance of pure mercury at helium
                         temperatures. Further experiments with liquid helium")
                         No.122b ("Disappearance of the electrical resistance
                         of mercury at helium temperatures"; No.124c.

             1913   Onnes, H.K. Comm. Phys. Lab. University of Leiden.
                         No.133b, Suppl. No.34.

             1914   Onnes H.K. Comm. Phys. Lab., University of Leiden.
                         No.139f ("Appearance of galvanic resistance in supra
                         conductors when brought into a magnetic field").

1912   TUNGAR RECTIFIER   I. Langmuir (U.S.A.)

          During the spring and summer of 1912, Dr. Irving Langmuir of General
Electric became interested and active in the development of the gas-filled,

tungsten-filament, incandescent lamp. These early laboratory lamps were constructed with heavy-coiled tungsten-spiral filaments that operated at low voltage. The bulb space around the filament was relatively small so that it would operate at high temperature and a few drops of mercury were placed in the bulb. There was usually another portion of the bulb where the mercury vapour condensed and ran back into the filament portion.

These lamps were frequently tested by operation from a 110 v. d.c. line through a series resistance. As one of the objects of these tests was to study the life and characteristics of the lamp at very high temperatures, there were frequent burnouts. It was noticed that, when the filament sometimes burned out during operations, an arc formed at the break.

In all of this early work the primary interest was in the arc as a source of light and apparently no tests on the arc as a rectifier were made at that time. However, one of the men sketched a tube with a hot filament and a separate anode plus liquid mercury.

In the early part of 1915 the possible need for a garage battery charger was considered and during 1916 it became a going concern.

SOURCE: "Early History of Industrial Electronics" W.C. White. Proc. IRE (May 1962) p. 1130.

1912   CLOUD CHAMBER   C.T.R. Wilson (U.K.)
(For revealing the ionisation tracks
of radio-active particles)

His intention was to produce an artificial cloud by the adiabatic expansion of moist air. For condensation to occur it was then believed that each droplet required a nucleus of dust. However, Wilson showed that even in the complete absence of dust particles some condensation was possible, and that it was greatly facilitated by exposure to X-rays. He was led to conclude that charged atoms (or ions) were the necessary nuclei (1896-7). After much labour he produced, in 1911, his Cloud Chamber, in which the paths of single charged particles showed up as trails of minute water droplets.

The Wilson Cloud Chamber, sometimes modified or refined, has been an indispensable tool of modern physics ever since.

SOURCE: "A Biographical Dictionary of Scientists" by T.I. Williams. Adam and Charles Black, London (1969) p. 564.

1912   RELAY AUTOMATIC TELEPHONE EXCHANGE   G.A. Betulander   (Sweden)
                                                 N.G. Palmgren

The idea of using relays as switching circuits in a telephone exchange was by no means new; it was first conceived in 1912 by G.A. Betulander and N.G. Palmgren of Sweden who also developed the crossbar selector. Although his name is less well known, Betulander is to the relay exchange what Strowger is to the selector exchange.

SOURCE: "The Telephone and the Exchange" by P.J. Povey. Post Office Publication, 1974, p. 84.

1912   RADIO   (Ionospheric propagation) W.H. Eccles (U.K.)

Following Marconi's success in 1901 in transmitting signals across the Atlantic, Kennelly and Heaviside postulated the existence of a conducting (ionized) layer in the earth's upper atmosphere and suggested that such a layer might cause the waves to follow the curvature of the earth. After it became clear that diffraction could not explain the substantial field strengths actually received at great distance, increased attention was directed to this proposal of an ionized region.

The theory of radio-wave propagation through the ionosphere is based on work by Eccles in 1912, on the ionizing effect of solar radiation, and on the effective refractive index of an ionized medium. Larmor in 1924 re-examined the work of Eccles and others and ascribed the major part of the refractive effect to the presence of free electrons in large numbers. The Eccles-Larmor theory, as later extended and developed by Appleton, Hartree and others to

include the effect of anisotropy due to the earth's magnetic field, is now considered the basic theory of radio-wave propagation in the ionosphere. This work
was later extended by Booker and others to cover oblique propagation in a non-
homogeneous ionosphere.

SOURCE: "Radio-wave propagation between World Wars I and II" S.S.Attwood
Proc.IRE (May 1962) p.689.

SEE ALSO: "On the diurnal variations of the electric waves occurring in
nature and on the propagation of electric waves round the bend of the earth"
by W.H.Eccles. Proc.Roy.Soc.(London) A. Vol.87.pp.79-99. (June 1912)

1912   CIRCUITRY (REGENERATIVE)   de Forest, E.H.Armstrong
                                                  and I.Langmuir   (U.S.A.)
                                                        and
                                                  Meissner   (Germany)

        The regenerative circuit was invented in 1912 by de Forest, Armstrong
and Langmuir in the United States and by Meissner in Germany. After 20 years
of litigation, the United States Supreme Court finally decided in de Forest's
favour.
        In the regenerative detector circuit RF energy is fed back from the
anode circuit to the grid circuit to give positive feedback at the carrier
frequency, thereby increasing the sensitivity of the circuit.

        Regenerative receivers marked a big step forward in providing greatly
increased sensitivity. Inherently they provided large amplification of small
signals and small amplification of large signals. By 1922 they had reached the
high point in their development and had almost entirely superseded crystal sets.

SOURCE: "The development of the art of radio receiving from the early 1920's
to the present" by W.O.Swinyard. Proc.IRE (May 1962) p.794.

SEE ALSO: "Some recent developments of regenerative circuits" by E.H.
Armstrong. Proc.IRE Vol.10. pp.244-260. August 1922.

        "The regenerative circuit" by E.H.Armstrong. Proc.Radio Club of
America. (April 1915).

1912   CIRCUITRY - Heterodyne and
            Superheterodyne   H.M.Fessenden
                                                            (U.S.A.)
                                               E.H.Armstrong

        Professor Fessenden, in his search for an improved receiver, inven-
ted the heterodyne system in 1912. Previous receivers had merely acted as
valves, detecting by turning a direct current on and off in amounts proportional
to the received signal. In contrast, the heterodyne system operated through
the joint action of the received signal and a local wave generated at the receiving station. Combination of these two alternating currents resulted in an audio
beat-note, the difference frequency between the two waves. Although Fessenden's local oscillator was an arc source, very bulky and troublesome, it was
nevertheless the forerunner of superheterodyne and single-banded reception.

        The next advance in double-detection technique involved amplification
of the beat-note or intermediate frequency. Several parallel developments
took place in the United States and in Europe. It is difficult to name an
inventor since the superheterodyne system as a basic idea seemed to appear
from several sources at about the same time. The works of J.H.Hammon,
A.Meissner, Lucian Levy, E.F.Alexanderson, and E.H.Armstrong stand out.
Armstrong fully appreciated the problem and obtained a patent in 1920 that was
of major importance in the practical application of the superheterodyne system.

SOURCE: "Radio Receivers - past and present" by C.Buff. Proc.IRE (May
1962) p.887.

SEE ALSO: "The superheterodyne - its origin, development and some recent
improvements" by E.H.Armstrong. Proc.IRE Vol.12 (Oct.1924) p.549.

1913   RELIABILITY (Standards)   AIEE (U.S.A.)

            At the risk of some controversy, it can be stated that the first step in
the new field of reliability can be traced to discussion generated by the AIEE
Standard No.1 published in 1913. The title of this document was "General
Principles upon which Temperature Limits are Based in the Rating of Electric
Machines and Apparatus". This Standard started the first cycle of appreciation for the chemical composition of electrical materials. During the subsequent development of the electrical and electronic field, there has been a
cyclic recurrence of this emphasis which can be recognised today as "analysis
for the basic mechanism of failure in materials used in highly reliable electronic parts". However, in all fairness, it should be stated that the element of
time-dependent degradation was not a part of the referenced first document.
This element did not become a formally recognized factor until the great up-
surge period for reliability in the 1950's.

SOURCE: "The reliability and quality control field from its inception to the
present" by C.M.Ryerson. Proc.IRE (May 1962) p.1323.

1913   ATOMIC ORBIT THEORY   N. Bohr   (Denmark)

            In 1913 Bohr proposed certain changes in the earlier atomic concept.
These were : (1) there are stable orbits in which the electrons do not
radiate; (2) in these orbits the angular momentum is an integral number
times some constant; (3) if an electron changes from one orbit to another,
energy is emitted or absorbed corresponding to the difference in energy
between those orbits.   Thus, various stable orbits correspond to various
permissible energy levels.

SOURCE :   "Semiconductor Electronics - 1. Solid-State Physics".
Electro-Technology (October 1960), p.95.

1913   RESISTORS - (Thin metal film)   W.F.G.Swann (U.K.)

            Since this is a review article it is necessary to refer to some of the
earlier papers on the subject of thin films. A typical example is the paper that
Dr.W.F.G.Swann presented to the British Association meeting over sixty years
ago in 1913, which illustrates how little our ideas have changed in this period.
Swann measured the resistivity of sputtered platinum films as a function of the
sputtering time and hence of the film thickness; he also recorded the temperature coefficient of resistance of these films between liquid nitrogen temperature and 100°C and observed the abnormally high resistivity associated with
very thin films and also a negative temperature coefficient of resistance in the
thinnest films.

SOURCE: "Resistive thin films and thin-film resistors - History, Science and
Technology" by J.A.Bennett. Electronic Components (Sept.1964) p.737.

SEE ALSO: W.F.G.Swann, Phil Mag. Vol.28 (1914) p.467.

1914   THYRATRONS   I. Langmuir (U.S.A.)

            In 1914 Dr. Langmuir first suggested a method of controlling the arc in
a mercury-pool tube by means of a grid. He showed how a grid voltage could be
used to control the starting of the main arc in each rectifying cycle. Thus, the
average arc current through the tube was controlled when an a.c. anode voltage
was used.

            In 1922 Toulon, a French scientist, improved on this method of control
by varying the phase of the grid voltage with respect to the anode voltage rather
than to its amplitude. Thus, the arc could be made to start at any point in the
anode voltage cycle and this resulted in a very practical and convenient method
of controlling the average value of a rectified anode current.

            In 1936 Dr.A.W.Hull developed the idea of operating a hot-cathode
diode in a low pressure of an inert gas or vapour. As a result, the space charge
effect was eliminated and the voltage drop of the discharge dropped to a low and
more or less constant value of 5 to 10 volts. At this low arc drop voltage, the
positive -ion bombardment of the cathode is not destructive.

SOURCE: "Early history of industrial electronics" by W. C. White. Proc. IRE (May 1962) p. 1132.

1914   ASDIC   (United Kingdom)

The British traced their sonar system to World War I, when a piezo-electric oscillator was used to emit the sound waves in a system called the ASDIC, named for the Allied Submarine Detection Investigation Committee), set up in 1917. At the outbreak of World War II, Britain outfitted its fleet with ASDICs that had the quartz-steel transducer. In the U.S., meanwhile, work on the magnetostriction-tube transducer was being done under contract by the Submarine Signal Company of Newport, R.I.

SOURCE: Electronics (April 17, 1980), p. 174.

1915   ACOUSTIC MINE   Wood (U.K.)

The acoustic mine was invented by Wood, later Deputy Superintendent of the Admiralty Research Laboratory, in the war of 1914-1918.

The first encounter with a German acoustic mine in the recent war occurred in the Firth of Forth on August 31st, 1940, when one was exploded by a motor boat. Then the cruiser 'Galatea' reported that twice in a week mines had exploded well ahead of her. Destroyers began to explode mines half a mile away. One suggestion was that the phenomenon was due to unexploded bombs. Even at the end of September, some experts doubted the existence of acoustic mines.

The mine contained a reed tuned to vibrate on a frequency of 240 per second. The noise was picked up and communicated by the reed to a carbon microphone. The later forms of mine contained counters which operate like a telephone exchange. The mine will not go off until it is called up, so to speak, for, say, the seventh time. The first six ships will pass over it safely and it will explode under the seventh. The mines contained clocks which could keep them disarmed for many days, until a fixed date.

SOURCE: "Science at War" by J. G. Crowther and D. Whiddington. H. M. S. O. London (1947) p. 172.

1915   SONAR (Submerged submarine detection   P. Langevin (France)
                    by ultra-sonics)

The detection of submerged submarines by sound has been particularly highly developed by British scientists. The method has been evolved from a suggestion made in 1912, after the sinking of the liner Titanic by collision with an iceberg. A British engineer named Richardson suggested that icebergs might be detected by the echo of a pulse of sound waves emitted from the approaching ship. The use of short sound waves which can be readily "beamed" like a searchlight is necessary in order to secure precise definition of the direction of the berg. In 1914 the only known practical way of obtaining such short or supersonic waves was from oscillators made of mica and caused to vibrate by electrical stresses.

An Allied Committee was at this time formed to develop anti-submarine technique. It included Dr. R. W. Boyle of Canada, Professor W. H. (later Sir William) Bragg, Professor P. Langevin of France and Rutherford. Langevin succeeded in 1915 in producing ultra-sonic waves by applying the piezo-electric effect discovered by Jacques and Pierre Curie. When quartz crystals are cut in the appropriate way electrification expands or contracts them; and conversely when mechanically stretched or compressed they produce an electric charge. Thus an alternating electric potential applied to a quartz crystal causes it to vibrate and these vibrations are of the type that produces sound waves. Hence by suspending a quartz crystal in water, and applying to it a pulse of alternating current of appropriate frequency, the crystal will be made to vibrate and communicate to the water a pulse of sound waves of the desired length. This will be transmitted through the water and be reflected by any obstacle. If such an obstacle is of sufficient size an echo can be received on the quartz which sent out the transmission. The echo creates a weak pulse of alternating current which can be detected by electrical instruments. In the spring of 1918, scientists at the Admiralty Experimental Station at Harwich succeeded with an apparatus of this kind in securing super-sonic echoes from a British submarine at a range of a few hundred yards.

SOURCE: "Science at War" by J.G. Crowther and R. Whiddington. H.M.S.O. London (1947) p.153.

SEE ALSO: "Uses of Ultrasonics in Radio, Radar and Sonar Systems" by W.P. Mason. Proc. IRE (May 1962) p. 1374.

"Electroacoustics" by F.V. Hunt. Harvard Univ. Press. Cambridge. Mass. (1954) Chapter 1.

1915   FILTERS (ELECTROMAGNETIC)   G. Campbell & K.W. Wagner (U.K.)

In 1831 Michael Faraday formulated the law of electromagnetic induction and self-induction. Some 84 years later, in 1915, G. Campbell and K.W. Wagner utilized Faraday's law in their invention of the first electromagnetic or LC wave filter. Significant advances in filter theory and technology then followed rapidly, until today, filters have so permeated electronic technology that it is hard to conceive of a modern world without them. Consumer, industrial, and military electronic systems all require some kind of signal filter, and, in the past, LC filter networks provided one of the most efficient and economical methods of implementing them.

SOURCE: "Inductorless filters: a survey " by G.S. Moschytz. IEEE Spectrum. (August 1970) p. 30.

SEE ALSO: "The golden anniversary of electric wave filters" by A.I. Zverev. IEEE Spectrum. Vol. 3. pp. 129-131 (March 1966)

"Introduction to filters" by A.I. Zverev. Electro-Technol. (New York) pp. 63-90 (June 1964)

1915   RADIO (Single sideband communication)   J.R. Carson (U.S.A.)

The military and international companies with a need to communicate abroad leaned heavily on radio, and by the 1950s designers were hunting for ways to ease spectrum crowding. Single-sideband transmission appeared to hold great promise.

Like so many other radio techniques, it had been developed originally for use in telephone systems. In 1915, J.R. Carson of AT & T's Development and Research Dept. had proved that only one sideband was needed for transmitting intelligence. Three years later the first commercial application of single-sideband showed that it was possible to use this technique to increase channel capacity. By the mid-1930s single-sideband transmission at high radio frequencies had proved successful. In the late 50s, after its first major application in the Strategic Air Command became known, single-sideband really caught on.

SOURCE: "Communications" by J.H. Gilder. "Electronic Design" 24. (November 22, 1972) p. 96.

1916   RELIABILITY (Control)   Bell / Western Electric (U.S.A.)

Perhaps one of the first clear-cut examples of what would now be called a reliability-control programme occurred in 1916. The Western Electric Company and the Bell Telephone Laboratories co-operated in a planned programme to produce good performing and trouble-free telephone equipment for public use. The following elements of a good reliability programme were involved:-

    (1)   An ably planned, forward-looking R & D programme considering the system needs.

    (2)   Development programmes leading to mature designs verified by design reviews and engineering model tests.

    (3)   Part improvement projects using test-to-failure methods, consideration of tolerances and performance under stress loading.

    (4)   Standardization and simplification in design.

(5)   Product evaluation under field application conditions for prototypes and pilot production models.

(6)   Quality control during manufacture to assure achieving the
reliability inherent in the design.

(7)   Feedback from the field to provide information for the designers.

SOURCE: "The reliability and quality control field from its inception to the
present" by C.M.Ryerson. Proc. IRE(May 1962) p.1323.

1916   "CROSSBAR" TELEPHONE EXCHANGE   J.G. Roberts
                                                J.N. Reynolds (U.S.A.)

The first "crossbar", based on an idea conceived as long ago as 1901,
by Homer J. Roberts, of the U.S.A., but the idea needed many years of
development work before it could be transformed into a practical system.
This was done by John G. Roberts, and John N. Reynolds of the U.S.A., who
patented the system in 1916, and by Gotthelf A. Betulander and Nils G. Palmgren
of Sweden who designed the first satisfactory mechanism. The first fully
operational crossbar exchange was installed in Sweden in 1926. By the late
1930's, crossbar had been developed to the point where it could offer several
advantages over Strowger, although it was more expensive. At that time,
crossbar, like Strowger, was a step-by-step system but it used a totally
different kind of selector.

SOURCE: "The Telephone and the Exchange" by P.J. Povey.   Post Office
Publication, 1974, p.75.

1917   CONDENSER MICROPHONE   E.C.Wente (U.S.A.)

One electrostatic device that has earned a secure place for itself in
the transducer art is the condenser microphone. Wente's "uniformly sensitive
instrument" of 1917 probably represents the first transducer design in which
sensitivity was deliberately traded for uniformity of response, and it was almost certainly the first in which electronic amplification was relied on to gain
back the ground lost by eschewing resonance.

SOURCE: "Electroacoustics" by F.V.Hunt. Harvard Univ.Press (1954) p.169.

SEE ALSO: "The sensitivity and precision of the electrostatic transmitter for
measuring sound intensities " by E.C.Wente, Phys.Res.Vol.19 (May 1922)
p.498.

1917   CRYSTAL PULLING TECHNIQUE   J.Czochralski

Most of the single crystal silicon used for p-n junction devices is produced by one of three techniques. The oldest of these is the pulling process
commonly called the Czochralski technique whereby the crystal is grown from
a charge of Si melted in a crucible (usually quartz). Historically, the next development was the float-zone technique of crystal growth, involving the movement
of a narrow molten zone up or down a vertical rod of initially polycrystalline
silicon. This technique has been widely used because it eliminates the problem
of crucible contamination. The third crystal growth technique,which has only
relatively recently been commercially exploited, is a crucibleless pulling
technique whereby the molten region is formed on the end of a thick silicon rod
and the crystal is seeded and pulled from this molten pool in a manner combining, to some extent, features of Czochralski and float-zone growth.

SOURCE: "Microinhomoheneity problems in silicon" by H.F.John, J.W.Faust
and R.Stickler. IEEE Trans. on Parts, Materials and Packaging. Vol.PMP-2
No.3. (September 1966) p.51.

SEE ALSO: "Measuring the velocity of crystallisation of metals" by
J.Czochralski. Z.Phys.Chem. Vol.92 (April 1917) p.219.

1918   INDUCTION HEATING (High Frequency)   E.F.Northrup (U.S.A.)

The production of heat by induced currents was recognised as early as
1880. The heating of transformer cores due to eddy currents was, of course,

understood at an early date. Probably the real engineering and developmental pioneer in induction heating with frequencies above the power range was Professor E.F. Northrup of Princeton University. In 1918 he built practical "furnaces" for frequencies above 10,000 cyrcles and powers as high as 60 kw using a spark gas oscillator.

SOURCE: "Early history of industrial electronics" by W. C. White. Proc.IRE (May 1962) p.1134.

1918   RADIO (Ground wave propagation)   G. N. Watson (U.K.)

At all frequencies, electromagnetic energy is diffracted into the geometric shadow of an obstacle, the amount increasing with decreasing frequency. In the radio-frequency range, diffraction yields significant and usable signal strengths in the shadow zone beyond the geometric field strength in the shadow zone of the earth was attacked by many mathematicians over 15 years with resultant divergent answers. The first significant break-through was produced by Watson in 1918, who showed that waves radiated by an antenna on the surface of a perfectly conducting sphere would be attenuated exponentially at great distances. The numerical values of field strength predicted by Watson under these conditions proved to be far lower than known experimental values. This discrepancy promoted increased interest in the ionosphere as a mechanism that might explain the wide divergence in the mathematical and experimental values of field strength, particularly in the kilocycle range of frequencies.

SOURCE: "Radio-wave propagation between World Wars I and II" by S. S. Attwood. Proc.IRE (May 1962) p.688.

SEE ALSO: "The diffraction of electric waves by the earth" G. N. Watson. Proc.Roy.Soc.(London) A, Vol.95. pp.83-89. (October 1918) and Vol.96 pp.546-563 (July 1919).

1918   THE DYNATRON   A. W. Hull (U.S.A.)

The dynatron possesses substantial advantages over other types of oscillator that are available for the same purposes and, in addition, has applications for which other types are not available. For example, a two-terminal oscillatory circuit, with one terminal "earthy" can be maintained in oscillation, there being no necessity for tappings, or auxiliary circuit elements. The oscillators are in general more stable in frequency than those maintained by triodes and are under more precise control.

SOURCE: "Applications of the Dynatron" by M. G. Scroggie.   "The Wireless Engineer", No. 121, Volume X (October 1933), p.527.

SEE ALSO: "The Dynatron" by A. W. Hull. Proc., IRE, February 1918, Volume 6, p. 5 - 35.

"The Dynatron Detector" by Hull, Kennelly and Elder. Proc., IRE (October 1922), p.320.

1918   ATOMIC TRANSMUTATION   Lord Rutherford (U.K.)

When certain radioactive atoms explode, they release some of their energy by communicating it to the atomic fragments flung out in the explosion. Some of these fragments consist of nuclei of atoms of helium. They are flung out with immense speed and hence high energy of movement. Why not direct these against the citadel of an atom, against its nucleus. Rutherford was pre-occupied with this idea when the war of 1914 began. He was called by the British Government to assist in the scientific struggle against the German submarines and was very busy with this work. But he kept the attacks on the atom going. While the Germans were shelling Verdun, Rutherford was bombarding the nucleus in his laboratory at Manchester, when he could find the time.

In 1918 he failed to appear at the meeting of an important war science committee. When one of his colleagues asked him why he had failed to appear, he said that he had just got definite evidence that it might be possible to disintegrate an atom at will, and that if this proved to be true "it was far more important than the war."

Immediately after the conclusion of the war, while still in Manchester, he completed this investigation and in 1919 published a conclusive proof that

atoms of nitrogen could be transmuted by bombardment with atomic projectiles
from natural radioactive substances; and later, in Cambridge, he extended his
investigations to the study of the artificial disintegration of the atoms of many
light elements. He was assisted in this especially by Chadwick, who had been
one of his most brilliant pupils and collaborators at Manchester.

SOURCE: "Science at War" by J.G. Crowther and R. Whiddington H.M.S.O.
London (1947) p.126

1918   MULTIVIBRATOR CIRCUIT   H. Abraham & E. Bloch (France)

        The multivibrator, described by Abraham and Bloch in 1918, is a very
important circuit, and is in use for many purposes. In its original form,
however, it is more commonly used for the generation of pulses than as a time
base or discharger circuit.

        The multivibrator consists of two valves, each having the anode
coupled to the grid of the other via a condenser, and with a leak resistance to
earth. The method of operation of the circuit is as follows: when the anode
current in one valve increases, it passes a negative signal to the grid of the
other, due to the increased voltage drop across the anode load. The anode
current of the second valve is thus reduced, which results in the grid of the first
valve becoming more positive and increases the anode current of this valve
still further. It will thus be apparent that the effect is a cumulative one and
that the anode current in the first valve rapidly reaches a maximum while the
anode current in the second is cut off. This circuit condition then remains
until the charge in the condenser C_1 has leaked away through R_2 sufficiently to
permit anode current to flow once more in the second valve. When this occurs,
the cumulative effect again takes place, but in the reverse direction. The
multivibrator, therefore, has two unstable limiting conditions which occur when
either of the valves is at cut-off and the other at zero grid potential.

SOURCE: "Time Bases" by O.S. Puckle. Chapman & Hall, London (1944) p.25.

SEE ALSO: "Notice sur les Lampes-valves a 3 Electrodes et leurs applications" Publication No:27 of the French Ministere de la Geurre (April 1918)

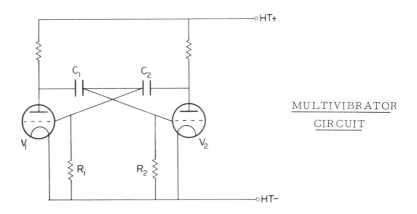

MULTIVIBRATOR

CIRCUIT

1918   CIRCUITRY (Neutrodyne)   L.A. Hazeltine (U.S.A.)

        The neutrodyne circuit was invented in 1918 by Hazeltine. Basically this
was a tuned radio-frequency (TRF) amplifier which employed a specific type of
neutralization. A current obtained from the plate circuit was fed back into the
grid circuit in the proper magnitude and phase to balance out, or neutralize, the
effect of grid-to-plate capacitance inside the tube, and thus it achieved stability
and prevented oscillation.

SOURCE: "The development of the art of radio receiving from the early 1920's
to the present" by W.O. Swinyard. Proc. IRE (May 1962) p.794.

SEE ALSO: "Tuned radio frequency amplification with neutralization of capacity
coupling" by L.A. Hazeltine. Proc. Radio Club of America. Vol. 2. (March 1923).

1918   <u>NOISE</u> (Shot effect)   W. Schottky (Germany)

The first realisation that unwanted random noise was a factor to
contend with in the field of communications came during World War I when
attempts were being made to design high-gain vacuum-tube amplifiers. It was
soon found that there was a limit to the number of stages which could be cas-
caded in the quest for high gain due to an unacceptably high background noise
which masked the weak signals being amplified. In his classic paper Schottky
first explained one of these effects and formulated the random component in the
plate current of a vacuum tube.

Schottky ascribed the random fluctuations in the plate current to the
fact that this current is composed not of a continuous but rather of a sequence
of discrete increments of charge carried by each electron arriving at the plate
at random times. The average rate of charge arrival constitutes the dc component of the plate current on which is superimposed a flucuation component as
each discrete charge arrives. He referred to this phenomenon as "schroteffekt"
or "shot-effect".

SOURCE: "Noise and Random Processes" by J. R. Ragazzini and S. S. L. Chang
Proc. IRE (May 1962) p. 1146.

SEE ALSO: "Theory of shot effect" by W. Schottky. Ann. Phys. Vol. 57.
(Dec. 1918) p. 541.

1919   <u>TELEVISION (Electronic System)</u>   Vladimir Zworykin   (U.S.A.)
1927   <u>TELEVISION (Electronic System)</u>   Philo Farnsworth   (U.S.A.)

It was already recognised that a complete cathode-ray system
provided the answer, but no experimental work was started and the idea lay
dormant until, after going to America in 1919, Zworykin joined the Westinghouse
Company. His major difficulty centred on the electronic camera tube. He, like
Campbell-Swinton, had conceived the idea of charge storage by 1919. Zworykin
lacked the necessary funds to carry his ideas forward into practical form at
the time he conceived them, and it was several years before he could concentrate on their elaboration. Westinghouse took him on to their research staff
but their laboratory devoted itself mainly to radio research, and, since
Zworykin was given no freedom to pursue his ideas on television, he resigned
to join a development company in Kansas. Returning to Westinghouse in 1923,
he drew up an agreement whereby he retained the rights to the television
inventions he had disclosed in 1919, while Westinghouse acquired the exclusive
option to purchase the patents at a later date.

Philo Farnsworth was essentially an individual inventor who, though
fortunate enough to find substantial financial backing, always retained his
autonomy in research. Largely self-taught, he appears at an early age to have
conceived of a completely electronic system. He was of the type which prefers
to work on a small scale with relatively simple equipment. Working in laboratories in Los Angeles and later in San Francisco he was able to demonstrate a
complete electronic system in 1927, when he filed his first patent application,
including his image dissector tube, which constitutes his most important inventive contribution. After long-drawn-out patent interference proceedings,
Farnsworth and Zworykin each received basic patents on their different systems
of television transmission.

SOURCE: "The Sources of Invention" by J. Jewkes, D. Sawers and
R. Stillerman. MacMillan & Co. London (1958) p. 385/6

SEE ALSO: "The History of TV" by G. R. M. Garratt and A. H. Mumford
Proc. of Institution of Electrical Engineers. Part III A, Television, Vol. 99.
(1952)

1919   <u>RETARDED FIELD MICRO-WAVE</u>   H. Barkhausen & K. Kurtz (Germany)
            <u>OSCILLATOR</u>

In 1919 it was noted by Barkhausen and Kurtz, during tests for the
presence of gas in transmitting valves in which the grid was held at a high
positive potential and the anode at a negative one, that oscillation could be
maintained in a circuit connected between grid and anode, or between other
pairs of electrodes.

The explanation was given that electrons are accelerated from the cathode to the positive grid through which some of them pass. These are then retarded in the grid anode space and turn back to the grid, some of them again passing through, and being reflected at the cathode to repeat the behaviour.

SOURCE: "Microwave Valves: A survey of evolution, principles of operation and basic characteristics" by C.H.Dix and W.E.Willshaw. Journal Brit. IRE (August 1960) p.578.

SEE ALSO: H.Barkhausen and K.Kurtz "The shortest waves obtainable with vacuum tubes" Z.Phys. Vol.21. p.1 - 6 (1920).

1919   CIRCUITRY (Flip-Flop Circuits)   Eccles and Jordan (U.S.A.)

In digital circuitry probably the most important single device is the bistaple circuit or "flip-flop". Used today in innumerable modifications the original form of this circuit was developed in 1919 by Eccles and Jordan. During the late 1930's flip-flops in tandem connection began to be used by nuclear physicists to scale down the counting rate from Geiger-Muller counters by factors of 2^n. These devices were all essentially binary counters in which the last "carry" pulse was used as the output to a mechanical register. In 1944, Potter described a scale-of-ten counter in which feedback caused a binary counter to skip six of its possible 16 states. A neon lamp indicating the state of each flip-flop provided a binary-coded readout for each decade. In 1946, Grosdoff described another binary counter with feedback to provide a scale-of-ten circuit. Grosdoff's circuit featured a direct readout obtained by a matrix of resistors arranged to light one of ten neon lamps for each stablestate of the counter. By connecting such decades in tandem direct reading decimal counting registers of arbitrarily large capacity were achieved and the way was cleared for the widespread application of high speed electronic counting.

SOURCE: "Digital display of measurements in instrumentation" by B.M.Oliver Proc.IRE (May 1962) p.1170.

SEE ALSO "A highly accurate continuously variable frequency control system" by C.H.Vincent. Instruments and Methods. Vol 7 (1960)p325

1919   VALVES - HOUSKEEPER SEAL   W.G.Houskeeper (U.S.A.)

The breakthrough came in 1919 when W.G.Houskeeper patented a method of sealing base metals through glass. Before this the leads brought through the glass had to be wires or ribbons. With this invention it was possible to make large diameter seals using metals such as high purity copper which has high conductivity, and is easily worked into any desired shape. The expansion of copper is $165 \times 10^{-7}/^\circ C$ and that of a common glass is $52 \times 10^{-7}/^\circ C$ so a matched seal is not possible. The technique developed by Houskeeper was to taper the metal to a feather edge and then seal the glass to the thin edge of the taper. The copper is cleaned, oxidised and then sealed to the glass. The thin copper is sufficiently flexible to equalise the difference in expansion and contraction between the glass and metal.

With this development it became possible to make the anode part of the vacuum envelope and cool its outer surface directly. By 1925 valves with water cooled anodes were in commercial production and they mark the beginning of the era of high-power broadcast transmitters.

SOURCE: "Electronics Engineer's Reference Book" Newnes-Butterworth London (1976) Chap.7. p.7 - 31.

SEE ALSO: "The art of sealing base metals through glass" by W.G. Houskeeper. Jour.Amer.Inst. EE Vol.41 (1923) p.870.

1919   CRYSTAL MICROPHONE   Nicholson (U.S.A.)

The crystal microphone, destined to be widely used in home tape recorders and public address systems, is made in the USA by Nicholson. The mike works on the piezo-electric principle, by which small voltages are produced on the surface of a solid such as a crystal. Sound quality is good, and costs low, but the mike reacts adversely to heat, humidity and rough handling.

SOURCE : "The Timetable of Technology", published by Michael Joseph, London, and Marshall Editions, London (November 1982), p.61.

1919   "MILLER" TIME BASE CIRCUIT   J.M.Miller (U.S.A.)

There is a family of time bases which utilize the device known as a
Miller-Capacitance Valve, i.e. a valve circuit with a condenser C connected
directly between the grid and the anode. The insertion of this condenser results
in feed-back from the anode of the valve to its grid. When such a valve circuit
is provided with a resistive anode load the effect on the grid circuit is substantially the same as if a capacitance of C (A + 1) in series with a resistance equal
to I/g had been connected between the grid and the cathode, where A is the gain
of the valve stage and g is the mutual conductance in amperes per volt. This is
usually known as the "Miller Effect".

SOURCE: "Time Bases" by O.S.Puckle (2nd Ed.)Chapman & Hall London(1951)
          Page 158.
SEE ALSO: "Dependence of the input impedance of a three-electrode vacuum
tube upon the load in the plate circuit" by J.M.Miller. Sci.papers of the Bureau
of Standards. Vol.15 (1919) p.367.

N.B. See also 1942 (Blumlein)

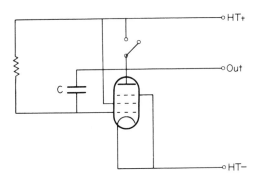

"MILLER TIME BASE CIRCUIT"

1919   RESISTOR (Metal-film type)   F.Kruger (Germany)

Proposals for the production of metallic-film resistors that will
possess a high degree of stability date back over 25 years, but mostly they
could be prepared only with comparatively low resistance values. By spirall-
ing the coating in a manner similar to that used for high-stability resistors, it
has been found possible to increase the resistance of resistors made with gold
films up to 200 000 ohms.

SOURCE: "Fixed resistors for use in communication equipment" by P.R.
Coursey. Proc.IEE Vol.96. Pt.III (1949) p.173.

SEE ALSO: F.Kruger. British Patent No: 157909/1919.

1920   PLASTIC MAGNETIC TAPE   Dr. Pfleumer (Germany)

Magnetic tape recording becomes possible outside the laboratory.   The
reason is the introduction, by Austrian researcher Dr. Pfleumer, of plastic
tapes in place of steel wires or tapes.   The magnetophones marketed by the
Germans in the thirties will incorporate these plastic tapes.

SOURCE: "The Timetable of Technology", published by Michael Joseph,
London, and Marshall Editions, London (November 1982), p.65.

1920   <u>ULTRA-MICROMETER</u>   R. Whiddington   (U.K.)

Few would suspect that one of the first authentic disclosures of practical electronic instrumentation was published in the November 1920 issue of
The Philosophical Magazine by R. Whiddington M.A., D.Sc., Cavendish Professor of Physics in the University of Leeds.

The title of Professor Whiddington's communication is "The Ultra-
Micrometer; an Application of the Thermionic Valve to the Measurement of
Very Small Distances". The summary of the paper reads: "If a circuit consisting of a parallel-plate condenser and inductance be maintained in oscillation by
means of a thermionic valve, a small change in distance apart of the plates
produces a change in the frequency of the oscillations which can be accurately
determined by methods described. It is shown that changes so small as 1/200
millionth of an inch can easily be detected. The name "ultra-micrometer" is
tentatively suggested for the apparatus".

SOURCE: Letter: "A pioneer of electronics" by F.G.Diver. The Radio and
Electronic Engineer". Vol.45. No.11 (Nov.1975) p.687.

SEE ALSO: "The ultra-micrometer - an application of the thermionic valve to
the measurement of very small distances" by R. Whiddington. Phil.Mag.
Series 6, Vol.40 (Nov.1920) p.634.

1921   <u>CRYSTAL CONTROL OF FREQUENCY</u>   W.G.Cady (U.S.A.)

While working with a quartz crystal connected in the circuit of a self-
excited vacuum-tube oscillator, Cady discovered that the frequency of self-
oscillation could be stabilized over a small range by the vibration of the quartz
crystal. In retrospect it appeared that a good many other experiments had made
use of crystals associated with vacuum-tube circuits with which they might have
observed this stabilizing action; but most of them sought to avoid the "anomalous"
effects that occurred in the neighbourhood of resonance in the crystal, and it
remained for Cady to discover this remarkable stabilizing action of a high-Q
resonator. Cady extended these results by applying two pairs of terminals to
the crystal and connecting these as a feedback path for a three-tube amplifier
in such a way that the circuit would oscillate but only at the resonance frequency
of the crystal.

Cady described and demonstrated his crystal circuits (January 1923)
to Professor G.W.Pierce of Harvard University. Pierce took up the study of
such crystal-oscillator circuits immediately, and within a few months his
experiments had led him to the invention of several improved forms of crystal
oscillator in which a two-terminal crystal could be made to control uniquely
the frequency of oscillation in a single-tube circuit.

SOURCE: "Electroacoustics" by F.V.Hunt. Harvard Univ.Press (1954)
p.53 & 55.

SEE ALSO: "The Piezo-electric resonator" by W.G.Cady.Phys.Rev.17, 531
(A) (April 1921) Proc.Inst.Radio Engrs. 10, 83-114 (April 1922); also
Piezoelectricity. New York and London, McGraw-Hill Book Co.,Inc.,(1946).

Walter G.Cady (crystal resonator) U.S.Patent No:1,450,246 (filed
28 January 1920) issued 3 April 1923; also (crystal stabilization and a 3-tube
oscillator controlled by a 4-terminal crystal) U.S.Patent No:1,472,583 (filed
28 May 1921) issued 30 October 1923.

1921   <u>FERROELECTRICITY</u>   J.Valasek

Ferroelectricity was first discovered by Valasek in 1921 in Rochelle
salt, which is piezoelectric above its Curie point; however, the most important
ferroelectric material today is barium titanate, which was discovered in 1942
and developed shortly thereafter at M.I.T. for ferroelectric applications. Since
then, other materials of the Perovskite structure and of other structures have
been found to be ferroelectric.

SOURCE: "Solid State devices other than Semiconductors" by B.Lax and
J.G.Mavroides. Proc.IRE (May 1962) p.1014.

SEE ALSO: "Piezoelectric and allied phenomena in Rochelle salt "by
J. Valasek. Phys. Rev. Vol. 17. (April 1921) p. 475.

ALSO: "High dielectric constant ceramics" by A. von Hippel, R. E. Breckenridge
F. E. Chesley and L. Tisya. Ind. Engrg. Chem. Vol. 38 (Nov. 1946) p. 1097.

1921   SHORT WAVE RADIO   Amateurs   (U.S.A.) and (Europe)

          One of the main developments in radio after 1918 was the discovery of
the usefulness of the shorter wave-bands. It was generally considered that
wavelengths below 200 metres were useless except for short distance transmission, though cases were known of long ranges being obtained on short waves.
These were regarded as freaks, however, and wave-lengths below 200 metres
were, after 1918, allocated to amateurs who encouraged by these 'freak' results,
arranged trial broadcasts from America to England in December 1921 on 200
metres. Their success showed that short-wave low-power broadcasts could be
heard over long distances.

SOURCE: "Sources of Invention" by J. Jewkes, D. Sawers and R. Stellerman.
MacMillan. London (1958) p. 353.

1922   CIRCUITRY (Super regeneration)   E. H. Armstrong (U.S.A.)

          Superregeneration was discovered by E. H. Armstrong during the
defense of his patent case for the regenerative receiver. In a superregenerative
receiver, sustained oscillations are squelched by periodic variation of the
effective resistance of the input resonant circuit. Oscillations periodically
build up in a circuit resonant at the signal frequency. Sustained oscillations
are prevented by periodic application to the grid of the superregenerative tube
of a signal that damps the oscillations. The quenching frequency is usually
between 20,000 and 100,000 cps. The superregenerative detector, because of
its broad tuning and considerable sensitivity, was practical and popular earlier
when unstable modulated oscillators were used as transmitters at frequencies
above 30 Mc.

SOURCE: "Radio receivers - past and present" by C. Buff. Proc. IRE (May
1962) p. 887.

SEE ALSO: "Super-regenerative receivers" by J. R. Whitehead. Cambridge
Univ. Press (1950).

1922   NEGATIVE RESISTANCE   E. W. B. Gill and J. H. Morell (U.K.)
          OSCILLATOR

          Following the work of Barkhausen and Kurtz (1919) which for the first
time reported the generation of electrical oscillations depending primarily on
the oscillatory motion of electrons in a vacuum, and not on the excitation of
oscillatory currents in a tuned circuit, and this discovery might be said to
represent the starting point of the whole field of modern microwave valves.

          Subsequent work on this device called generally the "retarding field"
generator, by Gill and Morrell, reported in 1922 and subsequently by many
others, showed that this simple type of operation had many variants including
one in which with adequate emission, a negative resistance could be provided
by the tube over a frequency band.

SOURCE: "Microwave Valves: A survey of evolution, principles of operation
and basic characteristics" by C. H. Dix and W. E. Willshaw. Journal Brit. IRE
(August 1960) p. 578.

1923   "SQUEGGER" CIRCUIT   E. V. Appleton, J. F. Herd
                                      and R. A. Watson Watt   (U.K.)

          The first hard valve time base was of the transformer-coupled type and
was developed about 1923 by Appleton, Herd and Watson Watt. The circuit was
known as a squegging oscillator" or "squegger".

The circuit consists of a transformer-coupled valve V_1, oscillating fairly violently at a radio frequency and having a condenser C_1 in the cathode-grid circuit. At each positive peak of potential at the grid, current flows from the transformer secondary winding through the valve V_1 from grid to cathode and into the condenser C_1 which thus accumulates a charge such that the mean grid potential becomes increasingly negative. When the negative potential reaches the cut-off bias potential of the valve, the anode current and, hence, also the alternating potential at the grid, is cut off and remains so until the charge on the condenser leaks away sufficiently through the diode to permit the resumption of oscillation. While the anode current is cut off, the condenser C_1 loses its charge via V_2 and the grid makes a potential excursion towards zero volts. Upon this excursion there is superimposed a damped oscillatory motion which is the second half-cycle of the oscillation.

SOURCE: "Time Bases" by O.S. Puckle (2nd Ed.)Chapman & Hall, London (1951) page 89.

SEE ALSO: E.V. Appleton, J.F. Herd and R.A. Watson Watt. British Patent No:235254.

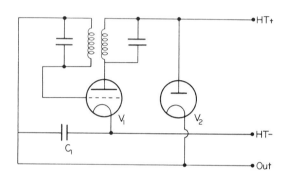

SQUEGGING OSCILLATOR CIRCUIT

1923   ICONOSCOPE   V.K. Zworykin (U.S.A.)

The combination of electron beam scanning and storage was first proposed and carried into practice by V.K. Zworykin in 1923 - at a time preceding any other electronic pickup system. The target was an aluminium film oxidised on one side, which was photosensitized with cesium vapour and faced a metal grill serving as collector for photoelectronics. The metal side, which served as signal plate, was scanned by a high-speed electron beam, which penetrated through the oxide layer, forming a temporary conducting path permitting the locally stored charge to flow off through the signal plate.

While the tube as described was capable of transmitting only rudimentary patterns, it became the ancestor of the extended line of storage camera tubes which dominate all phases of television picture transmission. Their common features are electron beam scanning and a picture target with small transverse conductivity capable of storing charge released in response to light.

The first practical storage camera tube was the iconoscope. Here, the picture was projected on a mosaic of photosensitive elements capacitively coupled to a signal plate. The mosaic was then scanned by a high-velocity beam, restoring the mosaic to a uniform potential and releasing photoelectrically stored charge for forming the picture signal. The secondary-emission ratio was greater than unity, so that the equilibrium potential under the beam was close to that of the electrodes facing the mosaic.

SOURCE: "Beam-deflection and photo devices" by K. Schlesinger and
E. G. Ramberg. Proc. IRE (May 1962) p. 993.

SEE ALSO: V. K. Zworykin. U.S. Patent No: 2, 141, 059 (Dec. 20th, 1938) filed
December 29th 1923.

"The Iconoscope - a modern version of the electric eye" by V. K.
Zworykin. Proc. IRE Vol. 22. (Jan. 1934) p. 16.

1924   REISZ TRANSVERSE-CURRENT CARBON MICROPHONE   G. Neumann

                                                          (Germany)

A very significant invention in the microphone field was the Reisz
transverse-current microphone, which became the standard BBC microphone
from 1926 to 1935 - the familiar white marble octagonal microphone. It was
invented by Georg Neumann, who worked for the Reisz Company in Germany
at the time, about 1924.

A block of marble or other insulating material had two deep recesses
cut in it, these being joined by a shallow surface trough. Each recess had a
carbon electrode in it, connected to a terminal at the rear. The block was
covered by a thin diaphragm of mica or paper and the space between this and
the trough-plus-recesses was filled with fine granulated carbon.

The diaphragm was very light, and well damped by the carbon granules,
giving the microphone a performance very much superior to that of other
carbon microphones, such as those used in telephones.

SOURCE: Communication from P. J. Baxandall, Malvern (22nd July 1982).

SEE ALSO: "BBC Engineering 1922 - 1972", p. 42.   BBC Publications 1972.

"Georg Neumann - In Memoriam", J. Audio Eng. Soc., (October 1976), p. 708.

1924   LINEAR SAW-TOOTH TIME
        BASE CIRCUIT   R. Anson (U.K.)

The earliest attempt at the construction of a linear saw-tooth time base
is believed to be due to Ansom, and was developed about the year 1924. The
author believes that this instrument appeared as a result of the development of
the Anson relay in which a neon tube was used for signal-shaping purposes in
conjunction with telegraph receivers. The neon tube is a two-electrode valve,
filled with neon gas at a low pressure, in which neither of the electrodes is
heated. When a potential, which is dependent upon the gas pressure, the
proximity of the electrodes, the material of which the electrodes are made and
their surface condition, is applied between the electrodes the gas becomes
ionized. This potential is normally about 130 volts. When the potential falls to
about 100 volts the gas deionizes.

The neon time base consists of a condenser charged through a high
resistance and discharged by the neon tube when the charge upon the condenser
reaches the striking voltage of the tube.

SOURCE: "Time Bases" by O. S. Puckle. Chapman & Hall, London (1944) p. 13

SEE ALSO: R. St. G. Anson, British Patent No: 214754.

1924/5   RADAR   E. Appleton, G. Briet, R. A. Watson Watt
                       S. F. R., G. E. M. A. et al (primarily U.K.)

The first use found for the reflecting properties of radio waves was in
measuring the height of the Heaviside layer. This was done in Britain in 1924
by Sir. Edward Appleton and M. F. Barnett, and in the U.S.A. in 1925 by Dr.
Gregory Briet and Dr. Merle A. Tuve of the Carnegie Institute. Breit and Tuve
were the first to apply the pulse principle.

By 1939, Germany, Great Britain, Holland and the USA all possessed
military radar apparatus, while the first peaceful application had been made in

France in 1935. In France and Germany the work was done by the scientists of radio companies: in Britain and the U.S.A. by scientists in government research stations.

Scientists of the Societe Francaise Radioelectrique began to study the use of metric and decimetric radio waves to detect obstacles in 1934, first with a view to saving life at sea and then for military uses. An 'obstacle detector' working on decimetric waves and employing magnetrons and the pulse principle was fitted to the liner Normandie in 1935; it appears to have been successful and equipment was installed to detect ships entering and leaving the harbour at Le Havre in 1936. Work was meanwhile done on equipment for detecting aircraft, although no radar warning system had been installed by 1939.

German work, begun before 1935, was carried on under contracts from the Navy, by a new firm G.E.M.A., and later by other firms. As a result, by 1939 a large number of radar sets were in operation for detecting aircraft. Though the German work began at least as early as that in other countries and had reached a very similar level of development by 1939, it lagged behind after the outbreak of war. German policy was based on the assumption that the war would be short, and consequently less effort was put into such basic work as radar than in Britain or the U.S.A.

Interest in the possibility of using radio waves to detect aircraft arose in Britain about 1934. The reflection of radio waves from aircraft had been observed in 1931 and 1933; on the latter occasion the possibilities for aircraft detection were carefully analysed. H.E.Wimperis, Director of Scientific Research at the Air Ministry and his assistant, A.P.Rowe suggested that the country should increase its efforts to develop a method of detecting aircraft at a distance and an investigating committee of three scientists, Sir Henry Tizard, Professor A.V.Hill and Professor Patrick Blackett, was set up. It was soon realised that radio beam would be the ideal alternative to the existing inadequate acoustic warning equipment.

Robert A. Watson-Watt (now Sir Robert) who was a lecturer in physics at University College, Dundee, before he began a research career in the government laboratories, played the major role in developing practical radar equipment in Britain. He was superintendent of the Radio Division of the National Physical Laboratories at the time the pressure for improved air defence reached its peak. He felt confident that radio waves could be employed to detect aircraft. His two memoranda of February 1935 described his suggested means for so using them; after a demonstration of the echoes produced by aircraft from the B.B.C. Daventry short-wave station, the Tizard Committee recommended that work on the lines suggested by Watson-Watt should be started. Working with six assistants, of whom A.F.Wilkins was the principal, Watson-Watt developed the first practical radar equipment for the detection of aircraft on the Suffolk coast in the summer of 1935. The main problems he solved were the construction of a high-power transmitter, the modulation of it with short pulses, the development of receivers to handle the pulses and of suitable transmitting and receiving aerials. The performance of the first equipment was considered promising enough for the Air Ministry to build a chain of five radar stations.

The development of radar had meanwhile been proceeding independently in the United States. Military interest began after L.A.Hyland, an associate of A.H.Taylor, discovered accidentally in 1930 that aircraft cause interference in radio waves and Leo Young successfully applied the pulse apparatus to this. Despite the fact that radar looked so promising, the Navy was reluctant to spend any significant amount on it, but, through the persistent efforts of Harold G. Bowen, chief of the Naval Laboratory, $100,000 was allocated for radar research. Robert M.Page, head of the research section of the Naval Laboratory's radio division, developed some of the first modern radar equipment. In 1938, two years after successful laboratory demonstrations of the equipment, the American Navy finally fitted radar devices to some of its ships.

After 1940 Great Britain and the United States co-operated in radar development.

SOURCE: "The Sources of Invention" by J.Jewkes, D.Sawers and R.Stillerman. MacMillan & Co. London (1958) p.346, 347 and 348.

1925   R̲E̲S̲I̲S̲T̲O̲R̲ ̲ ̲(̲C̲r̲a̲c̲k̲e̲d̲-̲c̲a̲r̲b̲o̲n̲ ̲t̲y̲p̲e̲)̲   Siemens and Halske   (Germany)

It was undoubtedly in Germany that the first practical use was made of
a cracked-carbon film in lieu of metal to form a highly stable resistance coating
and resistors of this general type were manufactured by several firms in that
country for a number of years before the war. Amongst these, the Siemens and
Halske organisation seems to have produced the largest quantities, so that this
type of resistor commonly came to be known as the "Seimens resistor".

One of the earliest disclosures of the cracking of hydrocarbon vapour
to produce a hard carbon layer is contained in the Seibt patent of 1930 and in the
Stemag patent of the same year. The Siemens and Halske patent relating to
these resistors is dated March 1932, bu the fundamental method is already
there referred to as "well known".

S̲O̲U̲R̲C̲E̲:̲ "Fixed resistors for use in communication equipment" by P.R.
Coursey. Proc. IEE Vol. 96. Pt. III (1949) pp. 174-175.

S̲E̲E̲ ̲A̲L̲S̲O̲:̲ Siemens and Halske Akt. Ges. German Patent No:438429/1925.

Siemens and Halske Akt. Ges. British Patent No: 387150 (1932)

C.A.Hartman. German Patent No: 438, 429 (1925)

1925   E̲L̲E̲C̲T̲R̲O̲S̲T̲A̲T̲I̲C̲ ̲L̲O̲U̲D̲S̲P̲E̲A̲K̲E̲R̲   -   (Various)

The electrostatic loudspeaker failed to gain wide commercial acceptance, in spite of extensive development activity devoted to it during 1925-1935,
for the very sound reason that several serious shortcomings still adhered to its
design. Either the diaphragm or the air gap itself had usually been relied on to
provide the protective insulation against electrical breakdown, but this protection was often inadequate and limits were thereby imposed on the voltages that
could be used and on the specific power output. Close spacings, a film of trap-
ped air, stiff diaphragm materials, and vulnerability to harmonic distortion
combined to restrict to very small amplitudes both the allowable and the attainable diaphragm motion: and as a consequence, large active areas had to be
employed in order to radiate useful amounts of sound power, especially at low
frequencies. But when large areas were employed, the sound radiation was
much too highly directional at high frequencies. Several of the patents cited
below bear on one or another of these features, and it is now apparent that an
integration of such improvements would have made it possible to overcome
almost - but not quite - every one of these performance handicaps. Occurring
singly as they did, however, no one of these good ideas was able by itself to
rescue the electrostatic units from the burden of their other shortcomings.
Taken together, however, with the newly available diaphragm materials and
with the important addition of one or two new ideas, the modern form of electrostatic loud-speaker can so completely surmount these former handicaps that
it merits careful consideration as a potential competitor for the moving-coil
loudspeaker in many applications.

S̲O̲U̲R̲C̲E̲:̲ "Electroacoustics" by F.V.Hunt. Harvard Univ. Press (1954)
p. 173 & 174.

S̲E̲E̲ ̲A̲L̲S̲O̲:̲ For example, Colin Kyle, U.S. Pats. No:1, 644, 387 (filed 4 October
1926) issued 4 October 1927, and No:1, 746, 540 (filed 25 May 1927) issued 11
February 1930: Ernst Klar (Berlin) German Pat. No:611, 783(filed 22 May 1926)
issued 5 April 1935, and U.S. Pat. No:1, 813, 555(filed 21 May 1927, renewed 14
November 1930) issued 7 July 1931 (insulating spacers, perforated plate coated
with a dielectric);Hans Vogt (Berlin), more than a score of contemporary and
relevant German patents, for example, German Pats. No:583, 769 (filed 25
December 1926)issued 9 September 1933 and No:601, 117 (filed 17 May 1928)
issued 8 August 1934, and U.S. Pat. No:1, 881, 107 (filed 15 September 1928)
issued 4 October 1932(tightly stretched diaphragm between perforated rigid
electrodes): Edward W.Kellogg (G.E. Co) U.S. Pat. No:1, 983, 377 (filed 27
September 1929) issued 4 December 1934 (sectionalized diaphragm with inductances for impedance correction):William Colvin, Jr., U.S. Pat. No:2000, 437
(filed 19 February 1931) issued 7 May 1935 (woven-wire electrodes): D.E.L.

Shorter, British Pat. No:537,931 (filed 21 February 1940, complete spec. 23 January 1941, accepted 14 July 1941) (diaphragm segmentation with external dividing networks for improving directivity and impedance).

"Wide range electrostatic loudspeakers" by P.J.Walker. "Wireless World" Pt.1. (May 1955 p. 208, Pt. 2 (June 1955) p. 265, Pt. 3 (August 1955) p. 381.

1925   SHORT WAVE COMMERCIAL RADIO
        COMMUNICATION   L.J.W.van Boetzelaer   (Holland)

On April 23rd,1925, an experiment began in Hilversum which turned out to have far-reaching consquences for the link between The Netherlands and the Dutch East Indies, now Indonesia, causing a huge long-wave transmitter to become obsolete only months after it came into operation.

At the Nederlandsche Seintoestallen Fabriek, now PTI, a young research engineer, L.J.W.van Boetzelaer had just received a new water-cooled 4 kW transmitting triode from the Philips factory at Eindhoven, which had a grid-anode capacity low enough to permit oscillating at 11.5 MHz(!) Working in a humble wooden shed, Mr.van Boetzelaer, after many difficulties, managed to get the primitive transmitter "on the air".

To see how far this transmitter could be heard, it was arranged that the steam ship "Prins der Nederlanden", sailing to the East, would listen in on a daily schedule. With admirable perseverance, van Boetselaer operated the morse key until late at night, hoping to be read. The ship's reactions, coming in slowly via coastal stations, were favourable. Then, in a bold mood, the diligent operator invited Malabar to send a cable if they happened to read the transmissions. The telegram sent back caused great excitement in Hilversum. It flatly stated that reception on 26 metres had been loud and clear since the beginning of the experiment.

SOURCE: Philips Telecommunication Review. Vol.33. No:4 December 1975. page 191.

1925   IONOSPHERE LAYER   E. Appleton   (United Kingdom)

At Cambridge University, British physicist Edward Appleton lays the foundation for the development of radar: he measures the height of the ionosphere and finds that radio waves are reflected from the upper atmosphere to a height of 310 mls (500 km) above ground level.

SOURCE : "The Timetable of Technology", published by Michael Joseph, London, and Marshall Editions, London (November 1982), p.78.

1925   TELEVISION (Mechanical scanning) J.L.Baird (U.K.)

When Baird, in 1923, decided to devote his untried inventive genius to the development of a practical television scheme, the problem seemed to him to be comparatively simple. Two optical exploring devices rotating in syn-chronism, a light-sensitive cell and a controlled varying light source capable of rapid variations in light flux were all that were required, 'and these appeared to be already, to use a Patent-Office term, known to the art'. Baird, however, realised the difficult nature of the problem. "The only ominous cloud on the horizon", he wrote "was that in spite of the apparent simplicity of the task no-one had produced television".

Baird's principal contemporaries in this challenge were C.F.Jenkins of the U.S.A. and D.von Mihaly of Hungary. Other inventors were patenting their ideas on television at this time (1923) but only Jenkins, Mihaly and Baird and a few others were pursuing a practical study of the problem based on the utilization of mechanical scanners.

SOURCE: "The first demonstration of television" by Prof.R.W.Burns. Electronics & Power.(9th October 1975) p. 953.

1925   NOISE (Johnson)   J.B.Johnson (U.K.)

In addition to fluctuation effects produced by vacuum tubes, it was found that random noise signals were generated in metallic resistors made of homogeneous materials. These effects were found to be temperature dependent

and are known as thermal noise. A number of basic contributions to the understanding of thermal noise were made in the 1920's and 1930's among which was
the outstanding paper by Johnson in 1925. The source of thermal conductor
noise was traced to the random excitation of the electron gas in the conductor
in consequence of its existence in an environment of thermally-agitated molecules. The effect is similar to the Brownian movement of particles suspended
in a liquid in which the thermally-agitated molecules of the liquid collide with
the suspended particle and impart to it a certain amount of energy. Since the
particle is cohesive, collision with any one of its molecules sets the entire
particle in motion thereby resulting in random movements observable under a
microscope.

SOURCE: "Noise and Random Processes" by J.R.Ragazzini and S.S.L.Chang
Proc.IRE (May 1962) p.1147/8

SEE ALSO: "Thermal agitation of electricity in conductors" by J.B.Johnson
Phys.Rev. Vol.32. 2nd Series (July 1928) p.97.

1926   SCREENED GRID TUBE   H.J.Round (U.K.)

          By the end of the 1914-1918 war the triode was the only tube in common
use as a detector and amplifier and generator of high-frequency oscillations.
Broadcasting was, in fact, started with the triode, although some of its
limitations were, by that time, well recognised. One major limitation arose
from the inherent electrostatic capacitance between the grid and the anode,
within the valve itself. This gave rise to a coupling between grid and anode
circuits which resulted in uncontrollable, and therefore undesirable, reaction
between the output circuit and the input circuit. The introduction of a screed
grid, between the control grid and the anode, to reduce this inter-electrode
coupling, was first suggested by A.W.Hull. Schottky had earlier suggested a
four-electrode tube, but his suggestion of the introduction of an additional grid
had been to secure an increase of amplification factor. Hull's suggestion, by
contrast, was directed to the reduction of grid-anode capacitance. But it remaine
remained for H.J.Round to bring the screen-grid valve into practical use in
1926.

          Other, and later, versions of the screen-grid valve were developed,
the screen grid being provided with one or more skirts which extended to the
walls of the container bulb. In one form the grid, and in another the anode,
was brought out uniquely at the top of the tube, the other electrodes being
brought out from the base. In this way the undesired capacitance between the
control grid and the anode could be reduced to 0.001 or 0.01$\mu\mu$F.

SOURCE: "Thermionic devices from the development of the triode up to 1939"
by Sir Edward Appleton. IEE Pub. Thermionic Valves 1904-1954. IEE London
(1955) p.22-23.

1926   COPPER OXIDE RECTIFIER   L.O. Grondahl (U.S.A.)
                                          P.H. Geiger

          In the course of an investigation of copper oxide formed on a piece of
copper, during which current was passed through the oxide in a direction at
right angles to the surface of separation, it was observed that the resistance of
the combination was less when the current flowed from the oxide to the copper
than when it flowed in the reverse direction. In the first unit, the ratio of the
resistances in the two directions was about 3 to 1. The phenomenon was so
different in nature from anything that had been observed in other known types of
rectifiers that an intensive study and experimental investigation was undertaken
during which it became more and more evident that the new device has characteristics which make it very probable that it will find general application as a
rectifier.

SOURCE: "A new electronic rectifier" by L.O. Grondahl and P.H. Geiger.
Proc. A.I.E.E. Winter Convention, New York (February 7 - 11, 1927). p.357.

1926   CIRCUITRY (Automatic volume control)   H.A.Wheeler (U.S.A.)

          In 1925 the stage was set for the invention of a practical automatic
volume-control circuit, and on January 2,1926, Wheeler invented his diode AVC
and linear diode-detector circuit. This circuit was first incorporated in the
Philco Model 95 receiver which he designed at the Hazeltine laboratory and

which was announced about September,1929.

Full AVC bias voltage was applied to the first two RF tubes, and, to prevent distortion in the third RF stage, half AVC bias voltage was applied to that stage. The automatic volume-control action was sufficiently gradual to permit accurate tuning by ear, and it was unnecessary to touch the volume control once it was adjusted.

SOURCE: "The development of the art of radio receiving from the early 1920's to the present" by W.O.Swinyard. Proc.IRE (May 1962) p.795.

SEE ALSO: "Automatic volume contrll for radio receiving sets" by H.A. Wheeler. Proc.IRE Vol.16. (Jan.1928) p.30.

1926   FILM SOUND RECORDING
          (Sound-on-disc system)   Warner Bros.   (U.S.A.)

As a result of lagging interest in the motion pictures by the public, Warner Bros. in 1926 decided to test the popularity of sound pictures. To minimize the cost of the venture, this studio arranged to have Western Electric develop the necessary equipment to synchronize disk-recording machines with cameras that were housed in booths to suppress the camera noise. Arrangements were made with the Victor Talking Machine Company to do the recording in their facilities and with their personnel. The Victor Talking Machine Company was a Western Electric licensee, and their studios were equipped with Western Electric recording equipment.

Western Electric developed motor drives for theatre projectors and disc turntables. These were mechanically connected to the same constant speed motor system. Essentially standard public address system amplifiers andloudspeakers were used. The first picture produced was "Don Juan". In October, 1927 "Jazz Singer" followed and was a success.

The public reaction was so enthusiastic that the large theatre chains wanted equipment immediately to play the pictures. Western Electric agreed to lease equipment to them. As a result of the success of the first pictures, Warner Bros. installed disk-recording equipment in their studios. This system was called Vitaphone. It was destined to be supplanted by systems that record-ed the sound as photographic images on the same film the picture was printed on. Having demonstrated the popularity of sound pictures and developed the equipment, the industry proceeded with great speed to convert studios for sound-picture production.

SOURCE: "Film Recording and Reproduction" by M.C.Batsel and G.L. Dimmick. Proc.IRE (May 1962) p.745.

1926   FIXED RESISTOR (Sprayed Metal Film) S.Loewe (Germany)

The basic idea involved is the very old art of decorating chinaware with precious metals. In Germany in 1926, Loewe developed a resistive film by atomizing a liquid solution of platinum resinate by forcing compressed air through it and applying the spray to an insulating base. Heating the film thus formed reduces it to the metal.

SOURCE: "Resistors - a survey of the evolution of the field" J.Marsten. Proc.IRE (May 1962) p.922.

SEE ALSO: S.Loewe. German Patent NO:591,735 (1926) and U.S.Patent 1,717,712 (1926)

1926   TRANSITRON OSCILLATOR   B. van der Pol (Holland)

One of the earliest forms of single-valve trigger circuit was described by van der Pol in 1926. This circuit employs a tetrode in which the screen and grid have a resistance-capacitance coupling, the grid being fed from the high potential source via a high resistance, which forms part of the coupling network.

The term "transitron" has recently come into use to denote any circuit which employs a single pentode valve in such a way that amplification is possible without the phase reversal. The name was originally coined by Brunetti, who defined it as a "retarding-field negative-transconductance device". Generally, these circuits are extremely versatile, and one may produce a sinusoidal, saw-tooth or square wave output from one oscillator by means of very simple switching. It is equally easy to convert the continuous oscillator into a flip-flop by changing the bias potential. When the circuit is arranged as a relaxation oscillator or as a flip-flop it is very valuable as a switching device, and has many applications to time bases and control circuits.

SOURCE: "Time bases" by O.S.Puckle (2nd Ed.) Chapman & Hall London (1951) page 56.& 57.

SEE ALSO: "On relaxation oscillations" by B. van der Pol. Phil.Mag. Vol.2. (1926) p.978.

"The transitron oscillator" by C.Brunetti and E.Weiss. Proc.IRE Vol.27 (1939) p.88.

1926   YAGI AERIAL   H.Yagi (Japan)

Dr.H.Yagi studied under the direction of Prof.Dr.G.H.Barkhausen at the Dresden Technische Hochschule from 1913 to 1914, under the direction of Prof.Dr.J.A.Fleming at University College, London from 1914 to 1915, and under the direction of Prof.Dr.G.W.Pierce at Harvard University, Cambridge, Massachusetts from 1915 to 1916.

In 1926, during his career as a professor at Tohoku University, he invented the VHF directive "Yagi antenna" which was widely put into practical use for domestic television reception. Some years later, as a result of this invention, the Academy of Technical Science in Copenhagen awarded him the Valdemar Poulsen Gold Medal "for his outstanding contributions to radio technique".

SOURCE: "Death of Dr.Hidetsugu Yagi" Telecommunication Journal Vol.43. (V/1976) p.372.

1927   NEGATIVE FEEDBACK AMPLIFIER   H.S.Black (U.S.A.)

In one of the most fundamental discoveries in the history of communications, H.S.Black in 1927 at Bell Laboratories found that by feeding part of an amplifiers output back into its input (negative feedback), it was possible by sacrificing some amplification to achieve stable operation at low distortion.

SOURCE: "Mission Communications - the story of Bell Laboratories" by Prescott C.Mabon. Published by Bell Laboratories Inc., Murray Hill, New Jersey. U.S.A. (1975) p.171-2.

1927   FILM SOUND RECORDING   Fox Movietone News (U.S.A.)
          (Sound-on-film system)

The first commercially successful photographic sound recording system (Fox Movietone News) used a variable intensity method of modulating a beam of light to expose the film negative. The gas-filled lamp known as the Aeo-light had an oxide coated cathode, and its intensity could be modulated over a considerable range by varying the anode voltage, at audio frequencies, between 200 and 400 volts. The Aeo-light was mounted in a tube which entered the camera at the back. Directly against the film was a light restricting slit which passed a beam about a tenth inch long and 0.001 inch high, placed between the picture and the sprocket holes. The Aeo-light could produce a sufficiently high intensity to expose the sensitive negative films used for picture taking. The system worked quite well for news photography, where the sound and pictures were taken simultaneously on the same camera.

SOURCE: "Film Recording and Reproduction" by M.Batsel and G.L. Dimmick. Proc. IRE (May 1962) p.746.

1928   PENTODE TUBE   Tellegen and Holst   (Holland)

The substantial suppression of secondary emission in a tetrode is not an easy matter, particularly where it is desired to operate with high

anode and screen potentials, and so by far the most common method of
suppressing secondary emission is by way of the inclusion of a suppressor
grid, between screen grid and anode, as in the pentode invented by Tellegen
and Holst of the Philips Company, in Holland. The suppressor grid is
maintained at the filament potential. Pentodes, before 1939, had become
extremely popular for both high- and low-frequency amplifications.

SOURCE: "Thermionic devices from the development of the triode up to 1939"
by Sir Edward Appleton. IEE Pub. Thermionic Valves 1904-1954. IEE London
(1955) p. 23-24.

1928   FREQUENCY STANDARDS   J. W. Horton & W. A. Marrison   (U. S. A.)
            (Quartz Clocks)

        The tuning fork was developed to a point at which it gave a stability of
1 part in 10^7 per week and could have been improved still further. By this time,
however, the first quartz clock had been made by Horton and Marrison and it
seemed clear that quartz possessed many advantages. One fundamen tal advantage was the higher frequency of quartz vibrations. Frequencies of many millions of cycles per second were already being used for radio transmissions, and
it was not very convenient to measure them in terms of a standard having such
a low value as 1 kc.

SOURCE: "Frequency and time standards" L. Essen. Proc. IRE (May 1962)
page 1159

SEE ALSO: "Precision determination of frequency" by J. W. Horton and W. A.
Marrison. Proc. IRE Vol. 16. p. 137. (Feb. 1928)

1928   RADIO   (Diversity reception)   H. A. Beverage
                                                   H. O. Peterson (U. S. A.)
                                                   J. B. Moore

        Because of the turbulence and abrupt changes encountered in the HF
medium, special attention had to be given to improved means of reception as it
became apparent that transmitter power increases were not alone sufficient.

        Among the significant techniques developed, one of the most important
is diversity reception, wherein a considerable improvement is obtained due to
the statistical independence in the fading characteristics of two or more paths.
Diversity reception is basic, improving the reception of any type of modulation
at any frequency. H. H. Beverage, H. O. Peterson and J. B. Moore described and
developed a triple-space-diversity system for on-off telegraph reception around
1928. This employed three antennas spaced about 1000 f e et apart. The rectified outputs of three separate receivers were combined across a common load
resistor and the voltage across the resistor keyed a local tone generator. As
long as the voltage was above a certain minimum from any receiver, a properly keyed tone signal was reproduced.

SOURCE: "Radio Receiver -- past and present" by C. Buff. Proc. IRE (May
1962) p. 888.

SEE ALSO: "Diversity receiving system of RCA Communications Inc. for
radiotelegraphy". H. H. Beverage and H. O. Peterson. Proc. IRE. Vol. 19.
pp. 531-561. (April 1931)

1929   COLOUR TV   Bell Laboratories (U. S. A.)

        Colour comes to television. With spectacularly good results, the first
transmission is beamed between Washington and New York. The 50-line system
used by the Bell Telephone Laboratories transmits the three primary colours -
red, blue and green - along three separate channels. Later in the year, the
basis of modern colour TV is laid down when several colour signals are transmitted over a single channel.

SOURCE : "The Timetable of Technology", published by Michael Joseph,
London, and Marshall Editions, London (November 1982), p. 86.

1929   CYCLOTRON   E. O. Laurence   (U. S. A.)

     Laurence used a curved path for the particles, so that the particles could circulate continuously, travelling long distances in a relatively small volume and using the same accelerating system over and over again. An electrically charged particle entering a magnetic field directed at right angles to the motion of the particle, proceeds to move in a circle with constant speed; as the particle speed is increased, the radius of the circle in which the particle moves also increases. Further acceleration occurs at each revolution.

SOURCE: "The Sources of Invention" by J. Jewkes, D. Sawers and R. Stellerman. MacMillan, London(1958.) pp. 290/1

SEE ALSO: "Atomic Slingshot" by Howard Blakeslee. Science Digest, (April, 1949.)

     "Maestro of the Atom" by L. A. Schuler. Scientific American. (August 1940)

1929   MICROWAVE COMMUNICATIONS   A. G. Clavier   (France)

     In 1920 Barkhausen positive-grid oscillator provided a means for the efficient generation of 40-cm waves. This revived the interest in the cent-meter waves. In 1929 Andre G. Clavier, then associated with Laboratoire Central de Telecommunications in Paris, started an experimental project to challenge the then accepted principle that wire or cable circuits should be used in preference to radio whenever physically possible.

     In 1930 a link was started between two terminals in New Jersey using 10-ft. parabolic antennas. Just as testing started, the project was transferred back to France. On March 31st 1931, Clavier and his associates demonstrated that microwave transmission provided a new order of economy, quality, depend-ability and flexibility in communications over a 40-km path between Calais and Dover. The circuit provided both telephone and teleprinter service using 17-6cm waves transmitted in a 4° beam by means of a parabolic reflector 3 meters in diameter with a power output of a fraction of a watt. Andre Clavier went on to establish the first commercial microwave radio link in 1933 from Lympne, England, to St. Inglevert, France.

SOURCE: "Microwave Communications" by J. H. Vogelman. Proc. IRE (May 1962) p. 907.

1930   TRANSISTOR (MOSFET CONCEPT)   J. Lilienfeld (Germany)

     A 1930 patent was issued to Julius Lilienfeld of the University of Leipzig for a device that could be compared to today's MOSFET, or insulated-gate field-effect transistor. The device was reported to provide a means of obtaining amplification in a thin film of copper sulfide. However, a working device was probably never built, since the low mobility of holes in the material and other factors would seem to preclude any amplification.

SOURCE: "Solid State Devices"- "Electronic Design" 24. (Nov. 23, 1972) page 72.

1930   VAN DE GRAAF ACCELERATOR   V. de Graaf   (U. S. A.)

     For nuclear structure research, constant-potential accelerators use the electrostatic belt generator invented by Van de Graaf. About 5.5 million volts were insulated in air between two large generators in 1930 (equipment now in the Boston Museum of Science).

SOURCE: The Encyclopedia of Physics (2nd edition) Editor R. M. Besancon. Van Nostrand Reinhold, (c) Litton Educational Pub. Inc. New York (1974) p. 13.

SEE ALSO: R. J. Van de Graaf, J. G. Trump and W. W. Bruechner "Electrostatic generators for the acceleration of charged particles" Rept. Prog. Phys. 11 p. 1. (1948).

1930   HIGH FIELD SUPERCONDUCTIVITY   W. J. de Haas
                                            J. Voogd
                                                        (Netherlands)

      The story of high field superconductivity began as long ago as 1930, when de Haas and Voogd discovered that resistance was restored in the Pb-Bi eutectic only by magnetic fields as high as 16,000 to 20,000 gauss at 4.2°K. The eutectic alloy had a transition temperature of about 8.8°K. This discovery immediately suggested to its authors the old idea of making a superconducting solenoid with wire of this material, but because of the very low critical current densities that they observed, the idea was soon dropped and in fact lay fallow for over 20 years. Then in 1955 Yntema described a superconducting solenoid wound with niobium wire which produced fields up to 7,000 gauss, but this received little attention. Autler (1960) made a similar coil producing 4,300 gauss, but, the subject didn't really take off until the discovery in 1961 by Kunzler and his co-workers of the remarkable current-carrying properties of the intermetallic compound, Nb_3Sn. This material had been found to have the very high critical temperature of 18.0°K by Matthias et al (1954) and indeed this is still the highest known transition temperature of any material, give or take a few tenths of a degree.

SOURCE: "Materials for Conductive and Resistitive Functions" by G. W. A. Dummer. Hayden Book Co. N. Y., p.141.

SEE ALSO: W. J. de Haas and J. Voogd. Comm. Phys. Lab. University of Leiden. No. 208b. (1930)

      G. B. Yntema "Superconducting Winding for Electromagnetics" Physical Review, 98, 1197. (1955)

      S. H. Autler "Superconducting Electromagnetics" Review of Scientific Instruments, 31, 369 (1960)

      J. E. Kunzler, E. Beuhler F. S. L. Hsu and J. H. Wernick "Superconductivity in Nb_3Sn at High Current Density in a Magnetic Field of 88 kgauss" Physical Review Letters, 6, 89. (1961)

1930's RADIOPHONIC SOUND - MUSIC   P. Grainger (Aus.)

      These techniques first came into real use during the 1950s with the maturation of the magnetic tape recorder although as long ago as the 1930s Percy Grainger, the Australian composer of "Country Gardens" fame, had produced a brief composition based on pure frequencies for the Theremin, an early electronic sound generator. The beginnings were however with 'musique concrete' pioneered in Europe, although as the name suggests the 'music' was made through the manipulation of pre-recorded natural sounds and was in fact orchestrated noise. With further study it became apparent that if more 'musical' sounds were used as the raw material. i.e. sounds with a more ordered harmonic structure, greater malleability was achieved as the timbre changes encountered during pitch changes, due to differing tape speeds on playback, still bore some audible relationship to each other.

SOURCE: "Electronics Engineer's Reference Book" Newnes-Butterworth London (1976) Chap.17. p.17 - 16.

1931   FIXED RESISTOR (Oxide Film)   J. T. Littleton (U.S.A.)

      The seed for this important contribution was provided by Littleton (1931) who developed an iridized, conducting tin-oxide coating for glass insulators. Its resistivity was sufficiently low to equalize potential across the insulator, thereby reducing corona effect, but too high for use in conventional resistors. Mochel modified this film by the addition of antimony oxide which stabilized its electrical properties. By varying the tin-antimony proportions, negative or positive temperature coefficients are obtained.

SOURCE: "Resistors - a survey of the evolution of the field" J. Marston. Proc. IRE (May 1962) p. 922.

SEE ALSO: J. T. Littleton. U.S. Patent No: 2,228,795 (1931)

      J. M. Mochel. U.S. Patent No: 2,564,707 (1947) Reissue 25,556.

1931   <u>STEREOPHONIC SOUND</u>   A.D.Blumlein   (U.K.)
        <u>REPRODUCTION</u>   and Bell Labs.   (U.S.A.)

        Stereophonic reproduction per se was pioneered almost simultaneously
by Blumlein in Great Britain and at the Bell Telephone Laboratories.Blumlein's
contributions are presumed to be described in his patents. He showed a complete system applicable to sound-on-disc motion pictures, including microphone arrays utilizing bidirectional as well as omnidirectional microphones,
transmission circuits, and disc recording systems utilizing simultaneous
lateral and vertical recording. Economic difficulties are believed to have
prevented completion and commercial exploitation of these systems.

        The recognised early systems approach to large audience stereo-
phonic reproduction was a public demonstration of Bell Telephone Laboratories
equipment under the guidance of Dr.Harvey Fletcher on April 27,1933. The
Philadelphia Orchestra was in the Academy of Music in Philadelphia and it was
reproduced in Constitution Hall, Washington, D.C.

SOURCE: "The history of steophonic sound reproduction" by J.K.Hilliard
Proc.IRE (May 1962) p.776.

SEE ALSO: A.D.Blumlein. Brit.Patent No:394,325. Dec.14,1931.also U.S.
Patent No:2,093,540.

        "Perfect transmission and reproduction of symphonic music in auditory
perspective" F.B.Jowett, et al. Bell Telephone Quart. Vol.12. July 1933.
page 150.

1931   <u>C.R.O.CARDIOGRAPH</u>   P.Rijlant (Belgium)

        The first workers to use the cathode-ray tube in bio-electrical work
were Gasser and Erlangers, in their work on the action potentials of nerves.
At the present time, many workers in bio-electrical fields, realizing its
unique properties, are adapting the cathode-ray tube to their own particular
problems; a general account will be found in Holzer's book. Among the first
to adapt it to electrocardiographic work was Rijlant, of Brussels (1931, et seq.)
who has published many papers on the subject. Schmitz, in Germany, and
Matthews (1933) in this country, were also among the first to publish electrocardiograms recorded by the cathode-ray oscillograph (C.R.O.).Matthews
(1934) was able to show that Rijlant's electrocardiograms were inaccurate, and
to point out that the new waves (P_2, T_2, etc.) described by him were really
caused by deficiencies in his amplifer.

SOURCE: "The examination and recording of the human electro-cardiogram by
means of the cathode-ray oscillograph" by D.Robertson. Journal IEE. Vol.81
(1937) p.497.

SEE ALSO: "Some observations on the adaptation of the cathode ray oscillograph
to the recording of Bio-electrical phenomena with special reference to the
electrocardiogram" by D.Robertson. Proc.Royal Soc.of Medicine (Section of
Physical Medicine) Vol.29. (1936) p.593.

        "Cathode ray oscillography in biology and medicine" by W.Holzer.
Published by Maudrich, Vienna (1936)

        "The cathode ray oscillogram of the human heart" by P.Rijlant.
Comptes Rendues des Seances de la Societe de Biologie. Vol.109(1932) p.42.

1931   <u>COMPUTERS (Differential Analyser)</u>   V.Bush (U.S.A.)

Early analogue computer for solving differential equations.

SOURCE: "The differential analyser - a new machine for solving differential
equations" by V.Bush.Journal of the Franklin Institute (Vol.212) 1931. p.477

SEE ALSO: "The computer from Pascal to von Neumann" by H.H.Goldstine.
Princeton University Press (1969) p.88.

1931   RELIABILITY - QUALITY CONTROL CHARTS   W.A.Shewhart (U.S.A.)

          After languishing in libraries for several years, the work of Dodge and
Romig in acceptance sampling and the work of Shewhart on control charts
finally was brought to light during World War II through the nationwide training
programmes sponsored by the Office of Production Research and Development
of the War Production Board.

          Although the underlying concepts were developed by scientific investi-
gators and statisticians in the preceding decades, the genius of Dodge, Romig
and Shewhart lay in their recognition of basic principles as an aid to solving
practical problems, and their ability to recognise and formulate a systematic
approach.

SOURCE: "Treating real data with respect" by J.A.Henry. Quality Progress.
(March 1976) p.18.

SEE ALSO: "Economic Control of Quality of Manufactured Product" by
W.A.Shewhart (1931) D.Van Nostrand Co.Inc. New York.N.Y.

1932   NEUTRON   J. Chadwick (U.K.)

          Some years before, the German physicists Bothe and Becker had dis-
covered some very penetrating radiations obtained by bombarding the light
metal beryllium with particles shot out of polonium. They assumed that the
radiations were of a wave nature. The French physicists Frederic Joliot and
his wife Irene Curie - whose mother Marie Curie had discovered polonium and
named it after her native country - studied these radiations and made a very
striking experiment. They found that if a piece of paraffin wax was placed in
front of them, then the amount of radiation seemed to be increased, and not
decreased by the inter-position of the wax. Chadwick, by further experiment
and interpretation, was able to prove that this paradox could be resolved if the
radiations were not waves, but a new kind of atomic particle without any
electric charge. Thus he discovered the neutron.

SOURCE: "Science at War" by J.G.Crowther and R.Whiddington. H.M.S.O.
London (1947) p.127.

1932   COCKCROFT WALTON ACCELERATOR   J.D. Cockcroft
        (Atom - smasher)   E.D.S. Walton   (U.K.)

          Cockcroft was an electrical engineer from Manchester. He had grad-
uated as an engineer at the Manchester College of Technology and joined the
engineering firm of Metropolitan-Vickers Electrical Company Ltd. After
spending four years in the Army in the war of 1914-1918 he returned to his firm.
He engaged in advanced study with Professor Miles Walker and was presently
awarded a post-graduate scholarship to continue his studies at Cambridge. His
engineering knowledge fitted him to devise powerful electrical apparatus, and
he attacked the problem of devising an electrical machine by which an electrical
field of several hundred thousand volts could be applied to atomic particles, so
that they could be given a very high speed and energy, like those thrown out
naturally by radium. Cockcroft and his colleague Walton succeeded in disin-
tegrating lithium with electrically accelerated protons in 1932, shortly after
Chadwick's discovery of the neutron. This was a great advance, because
electrical machines could be developed, and large streams of atomic projectiles
could be produced at will.

          Cockcroft used protons, the nuclei of hydrogen atoms, in his first
experiments. Each proton released about sixty times as much energy as it
possessed itself. But the number of protons accelerated was relatively
insignificant, and the amount of energy used in producing the accelerating field
was much greater than the total amount released in the atomic disintegrations.
As a machine, Cockcroft's atom-smasher was very inefficient in the engineer's
sense.

SOURCE: "Science at War" by J.G.Crowther and R.Whiddington. H.M.S.O.
London (1947) p.128.

1932   CIRCUITRY (Energy Conserving   A. D. Blumlein   (U.K.)
            Scanning Circuit)

        As with most of the diagrams , Figure 1 (a) is taken from the Patent
Specification, and shows the basic features of the line scan circuit which is now
universal in television receivers, although not brought into a common use until
1946. Figure 1 (b) shows the method of operation, involving three separate
regimes during the cycle.

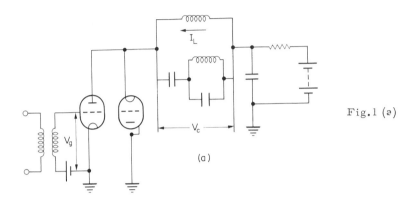

Fig.1 (a)

(a)

        It is now such a well-known circuit that it will not be described
in detail, but it is interesting to compare it with the single LC circuit which
is all that is necessary for a sinusoidal waveform whereas to handle the
saw-tooth waveform it is necessary to provide also the switches in the form
of the valve and diode as shown. The element of symmetry mentioned earlier
can be seen here.

SOURCE: "The World of Alan Blumlein" British Kinematography Sound
and Television. Vol. 50. No. 7. (July 1968) p. 209.

SEE ALSO: British Patent Specification No:400, 976 (1932)

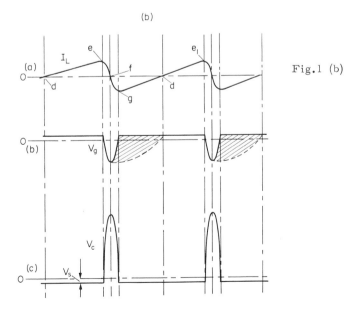

Fig.1 (b)

1932   TRANSMISSION ELECTRON   M.Knoll & E.Ruska (Germany)
              MICROSCOPE

                     The first electron microscope was built at the Technical University
              of Berlin early in 1931. It had two electro-magnetic lenses in series and
              achieved a modest magnification of 17. Improvements were made later. A
              condenser lens was added and an iron shield with a narrow gap built around the
              magnetic lens. Ruska in 1934 was able to demonstrate a magnification of
              12.000.

              SOURCE: "The Encyclopaedia of Physics"(2nd Edition) Editor:R.M.Besancon
              Von Nostrand. Reinhold & Litton Educational Pub.Inc. New York (1974) p.275.

              SEE ALSO: E.Ruska "Uber Fortschritte im Bau und in der Leistung des
              Magnetischen Elektronemikroskops " Z.F.Physik 87. (9 & 10) (1934) p.580.

                     "Origin of the Electron Microscope" by M.M.Freundlich.Science 142.
              (3589) (1963) p.185.

1933   FREQUENCY-MODULATION   E.H.Armstrong   (U.S.A.)

                     The credit for promoting FM as a broadcast service goes to E.H.
              Armstrong. For years he had been seeking a way to reduce static, and
              finally he turned his attention to FM. Toward the end of 1933 he had perfected
              a system of wide-band frequency modulation which seemed to overcome natural
              and many forms of man-made static. In this system the carrier was frequency
              modulated ±75 kc by audio components up to 15 kc.

              SOURCE: "The development of the art of radio receiving from the early 1920's
              to the present" by W.O.Swinyard. Proc.IRE (May 1962) p.797.

              SEE ALSO: "Frequency modulation" by S.W.Seeley. RCA Rev. Vol.5.
              (April 1941) p.468.

                     "Selected papers on frequency modulation"   Edited by J.Klapper.Dover
              Publications Ltd. New York (1970)

              NOTE   The actual invention of frequency modulation goes back to 1902.
              U.S. Patent 785,303. D. Ehret. (Endeavor Review April 1978).

1933   POLYETHYLENE INSULATION   ICI   (U.K.)

                     An outstanding event in the cable world in recent years was the
              discovery of polyethylene in 1933 by Imperial Chemical Industries Ltd. At first
              only minute quantities could be produced, but by 1937 a small amount was made
              available for experimental use. The opportunity was at once taken and, after
              extended research, a mile of submarine cable insulated with Telcothene, a
              synthetic material based on polyethylene, was made by Submarine Cables Ltd,
              in 1939.

              SOURCE: "The story of the submarine cable" Booklet published by
              Submarine Cables Ltd (AEI) London (1960) p.13.

1933   HARD VALVE TIME BASE CIRCUIT   O.S.Puckle (U.K.)

                     In 1933, O.S.Puckle developed a time base which employs a variation
              of the multivibrator as a condenser charging medium. This raised the maxi-
              mum repetition frequency, as compared with that obtainable from a thyratron
              time base, from about 40 Kc/s up to a maximum of about 1 Mc/s.

              SOURCE: "Time Bases" by O.S.Puckle. Chapman & Hall, London (1944) p.30.

              SEE ALSO: "A time base employing hard valves" O.S.Puckle. British Patent
              419198, also Journal of the Television Society Vol.2. (1936) p.147.

1933   RADIO ASTRONOMY   K.G. Jansky   (U.S.A.)

         While looking for the sources of static in overseas radio signals,
K.G. Jansky in 1933 discovered radio energy coming from the stars - thus
launching the science of radio astronomy.

SOURCE: "Mission Communications - the story of Bell Laboratories" by
Prescott C. Mabon. Published by Bell Laboratories Inc., Murray Hill,
New Jersey U.S.A. (1975) p.170.

1933   "IGNITRON" (Mercury-arc Rectifier)   Westinghouse (U.S.A.)

         In 1933 the Westinghouse Company announced its Ignitron. Its potential
value was at once recognised and an active developmental programme soon
commercialised it extensively. Progress in making and applying Ignitrons was
rapid. By the end of 1934 a welding control unit using glass Ignitrons was
installed in a customer's shop.

SOURCE: "Early history of industrial-electronics" W.C. White. Proc. IRE
(May 1962) p.1133.

1934   FREQUENCY STANDARDS (Atomic Clocks)   C.E. Cleeton
                                          and N.A. Williams   (U.S.A.)

         In 1934 Cleeton and Williams at Michigan University excited a spectral
line of ammonia at a frequency of 23,870 Mcs by a source of radio waves generated in the laboratory. The source used by them for exciting the transitions was
a magnetron which generated a fairly wide band of frequencies and the ammonia
was at atmospheric pressure, at which only a very broad resonance effect is
observed. The width was mainly due to the effect of collisions and this can be
reduced by reducing the pressure.

SOURCE: "Frequency and time standards" L. Essen. Proc. IRE (May 1962)
p.1161.

SEE ALSO: "Electromagnetic waves of 1.1 cm wavelength and the absorption
spectrum of ammonia" C.E. Cleeton and N.A. Williams. Phys. Rev. Vol.45.
p.234-237 (February 1934)

1934   TRANS-URANIAN ATOMS   E. Fermi (Italy)

         In 1934, Professor Enrico Fermi in Rome poured out a bewildering
series of discoveries by systematically bombarding atoms of all the elements
with neutrons. He found that several dozen of them could be transmuted by
neutrons, and he obtained particularly interesting results from uranium. This
is the most complicated of the ninety-two different kinds of chemical atoms
found on the earth. These can be placed in an order of complication, depending
on the number of electric charges on the nucleus. The first in the series is
hydrogen, with one positive charge, and therefore known as Atom No.1, and the
last is uranium, with ninety-two charges, and therefore known as Atom No.92.
It is not surprising that Atom No.92 should be naturally radio-active. It might
well be too complicated to be stable. It is in fact an ancestor of radium, whose
atomic number is 88.

        Fermi found that the bombarded uranium produced numerous atoms
with chemical properties quite different from uranium. He concluded that he
had made new atoms, more complicated than uranium atoms, and supposed
that these must be "trans-uranian" atoms, Nos.93, 94 etc. He seemed to have
made a new series of atoms hitherto not found on the earth.

SOURCE: "Science at War" by J.G. Crowther and R. Whiddington. H.M.S.O.
London (1947) p.129.

1934   LIQUID-CRYSTALS   J. Dreyer (U.K.)

              Although the liquid-crystal state was first noted in 1889, it was not
until around 1934 that serious consideration was given to these electro-optical
devices, in the Marconi laboratories, in England. John Dreyer found that their
orderly molecular arrangement could be used to orient dye molecules for
making polarisers - a method still used even though his work was done in the
1940s and patented in 1950. The present explosion in liquid-crystal research
began in the 1960s when the RCA laboratories in the United States began to
investigate them. Its course of development can be charted by counting the US
and British patents that have been granted - one each year in 1936, 1946, 1950,
1951, 1963, 1965 and 1967. Then, suddenly, seven in 1968, 11 in 1969, at least
two in 1970 and more than 11 in 1971. Few companies claim as long-lived an
association with the subject as Marconi and RCA: most have been in the field
for two years or less.

SOURCE: "The fluid state of liquid-crystals" by M. Tobias. New Scientist.
14 December, 1972. p. 651.

1935   SUPERCONDUCTING SWITCH   Casimir-Jonker
                                          and   W. J. de Haas   (Netherlands)

              The idea of using the superconductive transition to switch a small res-
istance into and out of a circuit at will seems to have occurred at about the same
time in the laboratories at Leiden and Toronto in 1935. Casimir-Jonker and
de Haas (1935) developed an apparatus to detect the first trace of resistance in a
superconducting specimen, using a sensitive magnetometer to observe the
change of field external to the cryostat when the current decayed in a superconducting circuit in series with the specimen. A superconducting lead solenoid
around the specimen was used to restore its resistance, and a resistance as
small as 3×10^{-11} ohm produced a decay of current rapid enough to be detected.

              At Toronto, Grayson Smith and Tarr (1935) used a moving-coil magneto-
meter inside the cryostat itself, consisting of fixed lead field coils and a moving
copper coil. A short section of lead wire in series with the field coils acted as
a superconducting switch and could be driven normal by means of a current in a
copper solenoid. The apparatus was used for measuring small persistent
currents, again by observing their decay. Used in this way as a super-conduct-
ing galvanometer, it was capable of detecting currents as small as 10^{-4} amp in
a circuit of self-inductance of 5×10^{-4} henry.

SOURCE: "Materials for Conductive and Resistive Functions" by G. W. A.
Dummer. Hayden Book Co. N. Y., p. 134.

SEE ALSO: 1935 - Casimir-Jonker J. M. and de Haas W. J.   Physica, 2, 935.

              1935 - Smith, H. Grayson and Tarr F. G. A. Transaction of the
Royal Society of Canada, 29, 23.

1935   TRAVELLING WAVE MICROWAVE
              OSCILLATOR (early Magnetron)   A. and O. Heil   (Germany)

              Studies of the classical triode valve in which the anode current is
controlled by the grid had shown that a fundamental difficulty for the highest
frequencies was the excess grid control power needed due to electron inertia.
In 1935 proposals were made by Arsenjewa Heil and O. Heil for avoiding this
limitation and also of avoiding the power dissipation limit of very high frequency
circuits. These proposals were of particular importance since for the first
time, a new mechanism specially suited for the generation of very high frequencies was suggested.

SOURCE: "Microwave valves: A survey of evolution, principles of operation
and basic characteristics" by C. H. Dix and W. E. Willshaw. Journal Brit. IRE
(August 1960) p. 580.

SEE ALSO: "Eine neue methode zur erzeugung kurzer, ungedampfter,
elektromagnetischer Wellen grosser Intensitat" by A. Arsenjewa Heil and
O. Heil. Z. f. Phys. 95. p. 752. 1935.

1935   SCANNING ELECTRON MICROSCOPE   M. Knoll   )
                                               M. von Ardenne)   (Germany)

                                               D. McMullan   )
                                               C. W. Oatley   )   (U. K.)

Postulated by Knoll in 1935 an early form of scanning electron microsc-scope was built by von Ardenne in 1938. However, the intensity of the electron beam at the specimen was very low (about 10^{-13}A) and it was therefore necessary to record the picture over a period of about 20 minutes in order to obtain an image of reasonable density on the photographic film.

Since the image was not visible until the film had been developed, focusing was a difficult proceeding, it being necessary to find the setting by trial and error.

The results with this microscope were inferior to conventional electron microscopes but v. Ardenne pointed out that the scanning microscope should show advantages with thick specimens.

He also proposed that, instead of a photographic recording, the electron beam should be collected by an electrode, amplified and used to modulate a cathode-ray tube. The surfaces of opaque specimens could then be examined in terms of their secondary emitting properties.

A scanning electron microscope designed especially for opaque speciments was made by Zworykin and others in 1942. The specimen was scanned by an electron spot as in v. Ardenne's microscope, the main difference being that electrostatic lenses were used instead of magnetic ones. Some micrographs were published showing a resolution of about 500A but the interpretation of them was inconclusive. It is well known that with primary voltages below a few thousand volts the secondary emission ratio is very dependent on the cleanness of the surface and in a demountable system with oil pumps it is practically impossible to prevent a thin layer of oil forming on the specimen, and this layer plays a significant part in determining the contrasts in the final micrograph. This difficulty has been overcome in the scanning electron microscope at Cambridge and, in addition, a number of other improvements have been incorporated including direct viewing of the picture before recording.

SOURCE: Letter from Dr. D. McMullan dated 16/10/77. Also "The scanning electron microscope and the electron-optical examination of surfaces" by D. McMullan. Electronic Engineering (Feb. 1953) p. 46.

SEE ALSO: "Aufladepotential und Sekundar - emission elektronbestrahlter Oberflachen" by M. Knoll. Z. Tech. Phys. 2. 467 (1935)

"Das Elektron raster Mikroskop" by M. von Ardenne. Z. Tech. Phys. 19. 407-416 (1938)

"An improved scanning electron microscope for opaque specimens" by D. McMullan. Proc. IEE Vol. 100 Part III No: 75 (June 1953) p. 245.

K. C. A. Smith and C. W. Oatley. Brit. J. App. Phys. (1955) Vol. 6. p. 391.

"First international conference on Electron and Ion Beam Science and Technology" Edited by R. Bakish. John Wiley (1965)

NOTE:
In 1957 a team of scientists at Cambridge University made a break-through in electron probe microanalysers which gave Britain a lead in this field that has so far been maintained.

Before the wholly-British development of scanning techniques, specimens had to be moved under static probes and the element distribution plotted laboriously and slowly. Scanning made it possible to display the information on a TV-type viewing system.

The following year, Tube Investments Research Laboratories found the value of scanning X-ray microanalysis, developed by the team at the Cavendish Laboratory was so great, that they built an instrument of their own.

Early in 1959, the Cambridge Instrument Co., entered into an agreement with TI to manufacture such instruments. Production started later that
year and the first Microscan was completed for the UKAEA, Aldermaston, and
shown in the Cambridge Instrument Company's London office at the time of the
Physical Society Exhibtion in January, 1960.

During the same period the company's efforts to improve the resolution
of Microscan led to the merging of their work with that of the University's
Engineering Department where scanning electron microscopes were being
studied. Stereoscan was the result.

This instrument had a field of focus some 300 times greater than any
previous microscope, optical or otherwise, and produced dramatic results of
both rough and delicate surface alike. So revolutionary were the photographs
taken on this instrument, that the company had to arrange a special demonstration before microscopists were convinced that they were true pictures of
the surface.

SOURCE: "From Microscan to Stereoscan....Cambridge keeping Britain in
front" by P. Slater. Electronics Weekly (January 10th, 1968) p.16.

1935   CIRCUITRY (Constant Resistance
        Capacity Stand-off Circuit)   A.D. Blumlein (U.K.)

Figures a(a) and (b) show two versions of the basis of this invention,
namely two-terminal arrangements of an inductor L, a capacitor C, and two
equal resistors R, having the property that the impedance measured between
the two terminals is purely resistive, of value R, at all frequencies, provided
$L/C = R^2$. This property was known, but Blumlein adapted the circuit
(particularly Figure 2(b)) as a means of removing from critical points in a
circuit (e.g. a wide-band amplifier) the stray capacity to earth of, for example,
floating power supplies. The example shown in Figure 3 is the application of
the idea to the filament supply for the cathode follower output valve of the
vision modulator for the original Alexandra Palace transmitter. The 'hardware'
of this is preserved in the Science Museum.

SOURCE: "The Work of Alan Blumlein" British Kinematography Sound and
Television. Vol. 50. No. 7. (July 1968) p. 209.

SEE ALSO: British Patent Specification No: 462, 530 (1935)

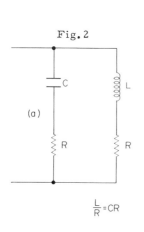

Fig. 2

(a)

$\frac{L}{R} = CR$

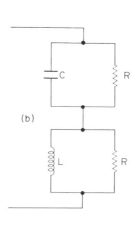

(b)

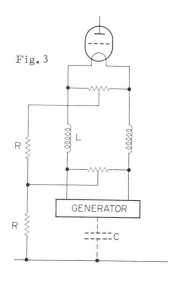

Fig. 3

1935   MULTIPLIER PHOTOTUBES   Zworykin, Morton & Malter (U.S.A.)

One of the most important by-products of television research is the multiplier phototube. When electrons of one or several hundred electron volts energy impinge on a suitably prepared conducting surface, they eject 4 to 10 low-velocity electrons, multiplying the initial current by a corresponding factor of 4 to 10. Repetition of this process leads to current multiplication by an arbitrarily high factor, practically without the addition of amplification noise. If the initial current is derived from a photocathode, the tube output reflects the variation of the light incident on the photocathode with a precision which depends only on the quantum efficiency of the cathode; with proper design, the dispersion in the transit time of the electrons from the cathode to the final collector can be held to quantities of the order of 10^{-10} second.

In the earliest effective multipliers (Zworykin, Morton and Malter 1935) the electrons were guided from dynode to dynode along an approximately cycloidal path by crossed electric and magnetic fields. Purely electrostatic focusing and acceleration systems were developed subsequently by Zworykin and Rajchman and Rajchman and Synder as well as by Larson and Salinger. These may be regarded as the prototypes of present-day multiplier phototubes of RCA and DuMont. The venetian-blind design utilised in the image orthicon is also employed by EMI and RCA for multiplier phototubes. Finally, the early and very simple screen multiplier of Weiss (1936) does without focusing altogether, at the expense of materially lowered multiplication efficiency.

SOURCE: "Beam-deflection and photo devices" by K. Schlesinger and E. C. Ramberg. Proc. IRE (May 1962) p. 1001/2

SEE ALSO: "The secondary emission multiplier - a new electronic device" by V. K. Zworykin, G. A. Morton and L. Malter. Proc. IRE Vol. 24 (March 1936) p. 351.
        "The electrostatic electron multiplier" by - and J. A. Tajchman. Proc. IRE. Vol. 27 (Sept 1939) p. 558.
        "Photocell multiplier tube" by C. C. Larson & H. Salinger Rev. Sci. Instr. Vol. 11 (July 1940) p. 226.

        "On secondary emission multipliers" by G. Weiss. Z. Tech. Physik Vol. 17. (Dec. 1936) p. 623.

1935   TRANSISTOR (Field effect)   O. Heil   (Germany)

In 1935, Oskar Heil of Berlin obtained a British patent on "Improvements in or Relating to Electrical Amplifiers and Other Control Arrangements and Devices". Figure 4 is the inventor's original illustration describing his device. The light area marked 3 is described as a thin layer of a semiconductor such as tellurium, iodine, cuprous oxide, or vanadium pentoxide; 1 and 2 designate ohmic contacts to the semiconductor. A thin metallic layer marked 6 immediately adjacent to but insulated from the semiconductor layer serves as control electrode. Heil describes how a signal on the control electrode modulates the resistance of the semiconductor layer so that an amplified signal may be observed by means of the current meter 5. Using today's experience and language, one might describe this device as a unipolar field-effect transistor with insulated gate.

SOURCE: "The field-effect transistor - an old device with new promise" by J. T. Wallmark. IEEE Spectrum (March 1964) p. 183.

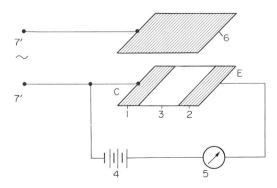

1936   CIRCUITRY (Long-Tailed Pair)   A.D.Blumlein (U.K.)

This now familiar and much used circuit (Figure 5 (a) and (b)) was
first needed in the amplifiers for the original video cable between points in
Central London and Alexandra Palace. The cable was not the now familiar
co-axial type but a shielded pair, and the problem was to obtain the 'push-
pull' signal uncontaminated by 'push-push' interference pick-up. In telephone
practice a transformer serves this purpose, but transformers to handle the
video frequency range were not then available.

The name of the circuit invented to do the job is the name given to
it by Blumlein, and is so descriptive that it has stuck.

Fig. 5 (a)

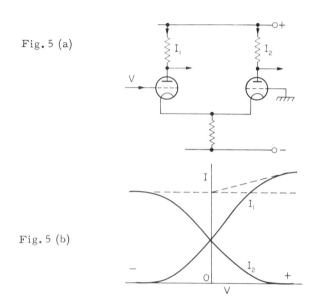

Fig. 5 (b)

SOURCE: "The Work of Alan Blumlein" British Kinematography Sound and
Television. Vol.50. No.7. (July 1968) p.209.

SEE ALSO: British Patent Specification No:482,740 (1936)

1936   COLD CATHODE TRIGGER TUBE   Bell Laboratories (U.S.A.)

Bell Telephone Laboratories announced the first cold cathode trigger
tube in 1936. Using an activated cathode, the 313-A set the pattern for a
number of trigger tubes made in Europe during the next two decades. The
activated cathode led to anode maintaining and critical trigger voltages each
of the order of 70 V.

SOURCE: "A Survey of Cold Cathode Discharge Tubes" by D.M.Neale
The Radio and Electronic Engineer. (February 1964) p.87.

SEE ALSO: "The 313-A vacuum tube" by S.B.Ingram. Bell Lab.Rec.,
p.114-6 (December 1936)

1936   <u>WAVEGUIDES</u>   J.R.Carson, S.P.Meade,
                               S.A.Schelkunoff and   Bell Laboratories (U.S.A.)
                               G.C.Southworth

   In 1936, from the Bell Telephone Laboratories, Carson, Meade and
Schelkunoff published their mathematical theory on "Hyper-Frequency Wave-
Guides" while G.C.Southworth published his experimental results. These
papers provided the basis for the TE01 mode cylindrical waveguide. In that
same year W.L.Barrow of M.I.T. published his work on the "Transmission
of Electromagnetic Waves in Hollow Tubes of Metal". Before 1934 Southworth
had transmitted telegraph and telephone signals at 15.cm wavelengths in a 5-in
diameter hollow metal pipe 875 ft.long with relatively small attentuation.

<u>SOURCE:</u> "Microwave Communications" by J.H.Vogelman. Proc.IRE (May
1962) p.907.

<u>SEE ALSO:</u> (1) "Hyperfrequency waveguides - mathematical theory" by
J.R.Carson, S.P.Meade and S.A.Schelkunoff. Bell Sys.Tech.J.Vol.15
p.310-333(April 1936)

1936   <u>VOCODER</u>   Bell Laboratories (U.S.A.)

   In 1936, Bell Laboratories developed the voice coder, or vocoder, for
analyzing the pitch and energy content of speech waves. With later developments, vocoder output was digitized, encrypted and the digital signal transmitted within a voice channel. The vocoder has been used since World War II
by the U.S.Government for secure communications.

<u>SOURCE:</u> "Mission Communications - the story of Bell Laboratories" by
Prescott C.Mabon. Published by Bell Laboratories Inc., Murray Hill,
New Jersey U.S.A. (1975) p.171.

1937   <u>XEROGRAPHY</u>   Chester Carlson (U.S.A.)

   An individual inventor, Chester Carlson, conceived the idea of
Xerography. This is a new photographic process which in a relatively short
time has found numerous industrial applications. It is completely dry and is
based entirely upon principles of photoconductivity and electrostatics. The
process:

> "employs a plate which consists of a thin photoconductive
> coating on a metallic sheet. This coating can be electrically
> charged in the dark and will hold this charge until exposed to
> light. Thus an electrostatic image can be produced on the
> plate by exposing the plate to an optical image. When the plate
> is dusted with powder particles, the electrostatic image is
> transformed into a powder image which can be transferred to
> paper and fixed by fusing."

   The development of Xerography was turned over to Roland M.
Schaffert, a Battelle research physicist with some previous experience in
printing. For a year he worked alone but after the war Battelle assigned a
few assistants to help him. By the latter part of 1946 two important
developments were completed : a high-vacuum technique for coating plates
with selenium; and a corona discharge wire, both for applying the original
electrostatic charge to the plate and for transferring powder from the plate
to the paper. The most significant contribution was the discovery of a
method to keep the image background from being filled with stray powder.
Thus Battelle improved Xerography to the point where industry became
interested.

<u>SOURCE:</u> "The Sources of Invention" by J.Jewkes, D.Sawers and
R.Stillerman. MacMillan & Co.London (1958) p.405 and 408.

<u>SEE ALSO:</u> "Printing with Powders" Fortune. (June 1949)
    "Xerography - From Fable to Fact" by W.T.Reid.
    "Developments in Xerography" by R.M.Schaffert. The Penrose
     Annual (1954)

1937   <u>POLAR CO-ORDINATE OSCILLOGRAPH</u>   M. von Ardenne   (Germany)
                                                 J. J. Dowling) Eire
                                                 T. G. Bullen)

        A polar co-ordinate oscillograph usually employs a special form of
cathode ray tube or other oscillographic device which has been specially designed for the purpose of depicting oscillograms in polar co-ordinates.

        Von Ardenne and Dowling and Bullen have independently developed
polar co-ordinate cathode ray oscillograph tubes. Von Ardenne's tube and the
circuit employed with it are shown in Fig. 76.

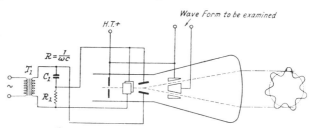

FIG. 76. Von Ardenne's polar co-ordinate oscillograph

        The tube contains two concentric cone-shaped deflectors across which
the potential to be examined is connected. The resultant form of the image is
shown in Fig. 77. This form of cathode ray tube has the advantage that the
final anode potential remains fixed and it is, therefore, possible to obtain larger
deflections without defocusing than is the case when the signal is applied as a
modulation of the final anode potential.

        Von Ardenne has also employed electromagnetic deflection methods
for this purpose.

        The form of the image, as shown in Fig. 77, is not in true polar co-
ordinates in the mathematical sense of the term, but it is difficult to find a
name which truly describes the arrangement.

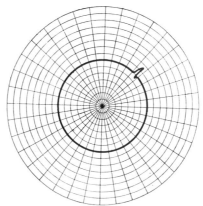

FIG. 77. Polar co-ordinate oscillogram obtained with Von Ardenne's
polar co-ordinate oscillograph

<u>SOURCE:</u>   "Time Bases" by O. S. Puckle. Chapman & Hall (1955) p. 104.

<u>SEE ALSO:</u> "A new polar co-ordinate cathode ray oscillograph with extremely
linear time scale" by M. von Ardenne, Wireless Engineer. Vol. 14(1937) p. 5.

        "Precision measurements with a radial deflection cathode ray
oscillograph" by J. J. Dowling and T. G. Bullen. Proc. Royal Irish Academy.
Section A. Vol. 44 (1937) p. 1.

1937   PULSE CODE MODULATION   A. H. Reeves   (United Kingdom)

       Pulse code modulation, or coded step modulation (which I think would have been an apter name), is a good example of an invention that came too early. I conceived the idea in 1937 while working at the Paris laboratories of International Telephone and Telegraph Corporation.   When PCM was patented in 1938 and in 1942, I knew that no tools then existed that could make it economic for general civilian use.   Only in the last few years, in this semiconductor age, has its commercial value been felt.

       Pulse code modulation was invented mainly for line-of-sight microwave links or link sections, where in 1938 the needed extra bandwidth would have been cheap and easily obtainable, rather than for more limited frequency bands, as in cables, which are now in fact the main fields of application.   It is this change of aim for PCM, for quite sound reasons, that has caused most of the technological difficulties so far in its application.

       It was in the United States during World War II that the next step in PCM's progress was made, by Bell Telephone Laboratories.   In this important stage, a team under Harold S. Black designed a practical PCM system later produced in quantity for the U.S. Army Signal Corps.   Research was also done under Ralph Bown.   It is appropriate that this early Bell work should be stressed, for it was the first time that the principles underlying the new system were translated into hardware.

SOURCE: "The past, present and future of PCM" by A.H. Reeves, IEEE Spectrum, Vol. 3, No. 5, May 1965, p.58.

1937   RADAR AIMING ANTI-AIRCRAFT GUNS   P.E.Pollard   (U.K.)

       The first radar equipment for aiming anti-aircraft guns was devised by Mr.P.E.Pollard in 1937.   It was the basis of the first radar gun-laying equipment, G.L.1, brought into anti-aircraft service in 1939.   It gave range up to 10 miles, with an accuracy of about 25 yards, but no angle of elevation, and was the only equipment of its kind available in the night attacks of 1940-41.

SOURCE: "Science at War" by J.G.Crowther and R. Whiddington. H.M.S.O. London (1947) p.76.

1938   TELEVISION :- SHADOW-MASK TUBE   W.Flechsig (Germany)

       The shadow-mask tube had its genesis in a 1938 conception of a German inventor, Flechsig.   However, at that time, there seemed to be no way to make Flechsig's device in which hundreds of fine wires had to be exact alignment with an equal number of phosphor triads, each with a red-, green-, and blue-emitting phosphor line.   Modifications of the idea were proposed by Goldsmith and by Schroeder, at RCA Laboratories.   Schroeder suggested a hexagonal array of circular holes in a metal mask, together with round phosphor dots and three closely spaced electron beams through a common deflection yoke.   Prior to 1948, some experiments were done on methods of multicolour phosphor deposition, but the basic technology of aligning either holes or wires with phosphor dots or lines appeared well beyond reach.   One of the experimenters on phosphor deposition, and a colleague of Schroeder's, was H.B.Law.   When the RCA crash program to develop a colour tube started in 1949, Law elected to pursue Schroeder's idea.   He then made a key invention, which he called the "lighthouse" this device permitted a photographic process to produce light shadows that were essentially the same as the electron-beam shadows.   Application of any one of several photolithographic techniques then permitted deposition of phosphors in exactly the right place.   Success came rapidly and in a few months Law photographically etched a metal mask with tiny holes, through which the three colour-emitting phosphors could be deposited, shifting the mask slightly for each deposition.   Law's first tube displayed small but remarkably good colour pictures. Application to a larger screen (30-cm diagonal) was made by engineering teams at RCA's Lancaster and Harrison locations, and a single-gun version was developed by R.R.Law (unrelated to H.B.Law but also at RCA's Princeton Laboratories).

SOURCE: "A history of colour television displays" by E.W.Herold. Proc.IRE Vol.64. No.9. (Sept.1976) p.1333.

SEE ALSO: W. Flechsig. German Patent 736575 filed 1938.

"Multi-colour television" A. C. Schroeder, U. S. Patent No:2, 595, 548 filed 1947, issued 1952.

"Picture reproducing apparatus" A. C. Schroeder, U. S. Patent No: 2, 595, 548 filed 1947, issued 1952.

"A three-gun shadow-mask kinescope" by H. B. Law. Proc. IRE Vol. 39 (Oct. 1951) p. 1186.

1938   "GEE" NAVIGATION   R. J. Dippy (U. K.)

Owing to the use of a chart covered with a network or grid of curves, the system was given the code name of Gee. It was invented by Mr. R. J. Dippy and developed by his team at Telecommunications Research Establishment. In practice three stations are actually used. A master station A sends out a series of pulses. If we consider one of these, it will travel past station B, and also to the aircraft P and beyond. It is arranged that when the pulse reaches B, which is called a slave station, a transmitter is then activated which issues a second pulse. A second pulse from A also activates a transmitter in a second slave station at C. Then the cycle repeats.

The aircraft carries one cathode ray tube on which the arrival of the four pulses is recorded. The navigator is provided with a chart covered with two intersecting sets of curves, corresponding to stations A and C, and A and B respectively. This can be super-imposed on a map of the country to be flown over. The target is marked on the map and hence its place in respect to the curves is seen.

SOURCE: "Science at War" by J. G. Crowther and R. Whiddington. HMSO London (1947) p. 54.

SEE ALSO: "Gee - a radio navigational aid" by R. J. Dippy. Proc. IEE. Vol. 93 Pt. IIIA (1946) p. 468.

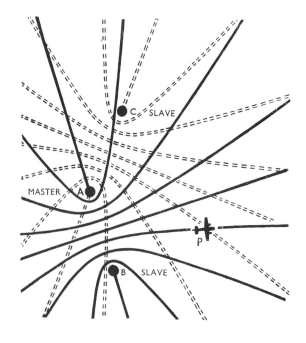

1938   COMPUTERS - (Information Theory)   C. E. Shannon (U. S. A.)

Shows the analysis of complicated circuits for switching could be effected by the use of Boolean algebra.

SOURCE: "A symbolic analysis of relay and switching circuits"
by C.E.Shannon. Trans. A.I.E.E. (1938) Vol.57. p.713.

1938   NUCLEAR FISSION   Fritsch   (Germany)
                                           Meitner

        Late in 1938, Hahn and Dr.F.Strassman, in reviewing the chemical
knowledge of the new substances, recognised that one of them was probably
barium, whose atomic mass and number are only about half of that of uranium.
This meant that they had previously been on a false trail. Frisch and Meitner,
now in Scandinavia, immediately explained the significance of this discovery.
Neutrons did not transmute uranium atoms into new atoms slightly heavier or
more complicated, with a higher atomic number. They split these big uranium
atoms into two roughly equal parts. The splitting could be done in a variety of
ways. Uranium atom No.92 might be split into Barium No.56 + Krypton No.36;
or into Strontium No.38 + Xenon No.54. Dr.J.R.Dunning in America rapidly
repeated the work on fission.

        Here was the explanation of the chemical confusion : a wide variety of
chemically different atoms was being produced by the disintegration. Frisch
and Meitner named this new process of atom-splitting "nuclear fission".
Nothing like this had been seen in heavy atoms before.

SOURCE: "Science at War" by J.G.Crowther and D.Whiddington. H.M.S.O.
London. (1947) p.131

1939   RADIO ALTIMETER   Bell Laboratories   (U.S.A.)

        The radio altimeter, by which a pilot can calculate his height above the
ground, is developed in the USA at the Bell Laboratories. The altimeter
bounces signals off the Earth and measures the time they take to return to the
aircraft. The pilot then uses a calibrated indicator to translate this figure -
the relative altitude - into a figure giving his absolute altitude.

SOURCE : "The Timetable of Technology", published by Michael Joseph,
London, and Marshall Editions, London (November 1982), p.117.

1939   KLYSTRON   W.C.Hahn and Varian Bros. (U.S.A.)

        Perhaps the first great step in understanding the phenomena in micro-
wave tubes came with the invention of the klystron. Bruche and Recknagel
discussed "phase focusing" in 1938 and the work of the Varians, Webster's
theoretical treatment of the klystron and the work of Hahn and Metcalf were
published in 1939.

        With the klystron came a well-thought-out theory of its operation, the
concept of velocity modulation, and a full appreciation of the value of microwave resonators.

SOURCE: "History of the microwave tube art" by J.R.Pierce. Proc.IRE (May
1962) p.979.

SEE ALSO: "High frequency oscillator and amplifier" by R.H. and S.F.Varian
J.App.Phys. Vol.10 (May 1939) p.321.

1939   DOUBLE-BEAM OSCILLOGRAPH   B.C.Fleming-Williams (U.K.)

        A splitter plate is immersed in the beam and divides it into two separ-
ate sections. Two "bucking" wires to which potentials are applied are used in
order to cancel out mutual deflectional interference of the two beams. With this
tube, the two beams are simultaneously deflected in the X axis but they are sep-
ately controlled in the Y direction. Since the Y deflecting potentials are necessarily unbalanced, (because only one plate is available for each beam) it
becomes necessary to employ an anti-trapezium construction for the tube.

        The production of a double beam is much better accomplished by means
of Fleming-Williams' double-beam cathode ray tube than by means of an elec-

tronic switch since, in the former case, the two images appearing are coincident in time. This is not so with the electronic switch and, hence, it is possible for events which are not coincident in time to be assumed to be so.

SOURCE: "Time Bases" by O.S. Puckle (2nd Ed.) Chapman & Hall, London 1951. p. 262.

SEE ALSO: "The double-beam cathode ray oscillograph" by B.C. Fleming-Williams. Electronics & Short Wave World. Vol. 12 (1939) p. 457.

1939   COMPUTERS (Digital)   H.H. Aitken (U.S.A.) and I.B.M.

        Utilizing twentieth-century advances in mechanical and electrical engineering, the Automatic Sequence Controlled Calculator, or Mark I, brought Babbage's ideas into being, giving concrete existence to much more at the same time. The Mark I, an electromechanical calculator 51 feet long and 8 feet high, was built by the International Business Machines Corporation between 1939 and 1944. It could perform any specified sequence of five fundamental operations, addition, subtraction, multiplication, division and reference to tables of previously computed results. The operation of the entire calculator was governed by an automatic sequence mechanism. The machine consisted of 60 registers for constants, 72 adding storage registers, a central multiplying and dividing unit, means of computing the elementary transcendental functions $\log_{10} x$ 10^x and sin x, and three interpolators reading functions coded in perforated paper tapes. The input was in the form of punched cards and switch positions. The output was either punched into cards or printed by electric typewriters.

SOURCE: "The evolution of computing machines and systems" by Serrell, Astrahan, Patterson and Pyrne. Proc. IRE (May 1962) p. 1043.

SEE ALSO: "The computer from Pascal to von Neumann" by H.H. Goldstine. Princeton Univ. Press (1972) p. 118.

        "Proposed automatic calculating machine" by H.H. Aitken. IEEE Spectrum. (August 1964) p. 62.

1939   BELL TELEPHONE LABS "COMPLEX
              COMPUTER"   G. Stibitz et al (U.S.A.)

        It is perhaps a little surprising that it was not until 1937 that Bell Telephone Laboratories investigated the design of calculating devices, although Andrews has stated that from about 1925 the possibility of using relay circuit techniques for such purposes was well accepted there. However, in 1937 George Stibitz started to experiment with relays, and drew up circuit designs for addition, multiplication and division. At first he concentrated on binary arithmetic, together with automatic decimal-binary and binary-decimal conversion, but later turned his attention to a binary-coded decimal number representation. The project became an official one when, prompted by T.C. Fry, Stibitz started to design a calculator capable of multiplying and dividing complex numbers, which was intended to fill a very practical need, namely to facilitate the solution of problems in the design of filter networks, and so started the very important Bell Telephone Laboratories Series of Relay Computers.

        In November, 1938, S.B. Williams took over responsibility for the machine's development and together with Stibitz refined the design of the calculator, whose construction was started in April and completed in October of 1939. The calculator, which became known as the "Complex Number Computer", often shortened to "Complex Computer" and as other calculators were built, the "Model 1" began routine operation in January 1940. Within a short time it was modified so as to provide facilities for the addition and subtraction of complex numbers, and was provided with a second, and then a third teletype control, situated in remote locations. It remained in daily use at Bell Laboratories until 1949.

SOURCE: "The Origins of Digital Computers" Edited by B. Randell. Springer-Verlag, Berlin (1973) p. 238.

SEE ALSO: "Computer" by G.R. Stibitz - "The Origins of Digital Computers" Edited by B. Randell. Springer-Verlag, Berlin (1973) p. 241.

1939   MAGNETRON   J.T.Randall and H.A.H.Boot   (U.K.)

          In the autumn of 1939 the Admiralty asked Professor M.L.Oliphant and
the physics department of the University of Birmingham to develop a high-power
micro-wave transmitter. The majority of the scientists in the laboratory concentrated on the klystron, described by its inventors, R.H. and S.F.Varian of
Stanford University, California, in 1939, which used for the first time closed
resonators, described by W.W.Hansen, also of Stanford, in 1938, for the production of high-frequency power. J.T.Randall and H.A.H.Boot, struck by the
difficulty of getting enough power from the klystron, considered instead applying
the resonator principle to the magnetron, which had been invented by A.W.Hull
of the American General Electric firm in 1921 but which, in its conventional
form, lacked the properties they were seeking. The result was the cavity
magnetron, which proved to be the needed generator, producing high powers on
centimetre wave-lengths.

SOURCE: "The Sources of Invention" by J.Jewkes, D.Sawers and R.Stillerman
MacMillan & Co.London (1958) p.348.

1939   FREQUENCY STANDARDS (Caesium Beam)   I.I.Rabi (U.S.A.)

          The difficulties of bandwidth and low intensity are most easily overcome
by using the atomic beam magnetic resonance method developed at Columbia
University by Rabi and his co-workers. In this method which can be used with
atoms possessing a magnetic dipole moment, a beam of atoms passes to a
detector through a system of magnets and a region of field alternating at the
Bohr frequency. The magnets have a nonuniform field and deflect the atoms in
one direction or the other according to which of the two energy levels they are in.
When the frequency of the RF field is exactly equal to the Bohr frequency and is
of the right amplitude transitions are induced and the deflections in the second
magnet B are the opposite from those in the first magnet, A, and the atoms are
thus focused on the detector.

SOURCE: "Frequency and time standards" by L.Essen.Proc.IRE(May 1962)
p.1162.

1939   LARGE SCREEN TELEVISION PROJECTOR   Fischer (Switzerland)

          The first large-screen television projector was invented by Professor
Fischer at the Swiss Federal Institute of Technology in 1939. At that time,
Prof.Fischer thought that the growth of television would come from the devel-
opment of networks of neighbourhood "television theatres" and he invented the
Eidophor with the capability of projecting TV pictures onto cinema-sized
screens. The earliest Eidophors were cumbersome machines, which could
project only black and white pictures in a darkened or semi-darkened room.
They were not the most reliable of machines and for a number of years the
Eidophor system was little known or used.

          Later the American space programme called for a reliable, high-
performance, large-screen projection system capable of working for long
periods of time, to provide data displays in NASA flight control centres.Gretag
AG, Zurich, a subsidiary of Ciba Geigy and patent holders and manufacturers
of the Eidophor, successfully developed the projector's capability to meet
NASA specifications. The latest Eidophors are able to project full-colour
television pictures onto screens 18m wide.

SOURCE: "Projection television - a review of current practice in large-
screen projectors" by A.Robertson, Wireless World (Sept.1976) p.47.

1940   PLAN POSITION INDICATOR   E.G.Bowen
                                          W.B.Lewis
                                          G.W.A.Dummer   (U.K.)
                                          E.Franklin

          The first radar P.P.I. (Plan Position Indicator) to be used by the
R.A.F. was designed by the authors (G.W.A.Dummer and E.Franklin) in 1940,
and in view of the use since made of this device it may be of interest to recall
early experimental work on radial time bases.

          The possibility of the desirable P.P.I. presentation of radar echoes
had been realised by the pioneers in the early days of radar. It was not, however,

until 1939/40 that the two developments of a radar station using a sufficiently narrow beam and the cathode ray screen with bright and lasting afterglow led to the development of a satisfactory P.P.I.

In 1939/40 work was proceeding on the design of a "radio lighthouse" on a wavelength of 50 cm. It was envisaged that with the narrow beams then obtained on this wavelength it should be possible to rotate a time base in synchronism with the aerial rotation to give a radar "map" of all surrounding aircraft. It was decided that the time base should take one of two forms:-

1.   An inductive or capacitive voltage split of X and Y vectors and recombination on an electrostatically deflected tube.

2.   A mechanically rotated current time base on a magnetically deflected tube.

At that time 12-inch electrostatic tubes were in use on CH and CHL sets and magnetic afterglow tubes were not fully developed; it was therefore decided to adopt the first scheme.

Experimental work was also carried out on "strobing" a portion of the time base, amplifying it and feeding it to another tube. By this means an enlarged P.P.I. was developed. A dim "square" (approximately 1 in. side on a 12 in. tube) was produced on the first tube by partial blackout to identify the area which was being enlarged and this area appeared as full size on the other tube. Owing to the 15° beamwidth of the polar diagram the system was not used, but it is interesting to record that a system of this type was operating in this country in 1941.

The P.P.I. was next adapted for use in the first centimetre AI equipment and afterwards for H₂S and ASV. It has become one of the most widely used presentation systems for radar both in this country and U.S.A. and it seems a far cry from the original 6ft rack to the compact, efficient airborne P.P.I's in use today.

SOURCE: "Radial Time Bases' - How they were Developed for Radar" by G.W.A. Dummer, E. Franklin.   Wireless World (August 1947) p. 287.

SEE ALSO: "Three Steps to Victory" by Sir Robert Watson-Watt.   Odhams Press, London (1957) p. 268.

1940   "OBOE" NAVIGATIONAL SYSTEM   A.H. Reeves (U.K.)

Yet another high-precision method of radar navigation removed the necessity for the pilot to find the target or even know where he was going. All of this could be managed from ground stations without the knowledge of the pilot, who was relieved of much strain and decision, and could devote the whole of his attention to the controls of his aircraft.

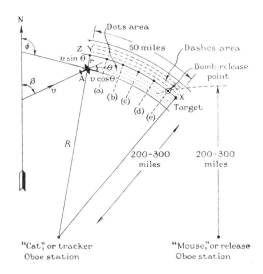

In this system there are two fixed stations, A and B. Station A enables the aircraft P to fly along the circumference of a great circle, whose centre is at A, and whose circumference passes over the target C. This station, which pushes the aircraft this way and that, tracking it, as it were, and making it keep to the circle, is called the Cat station. The stations A and B send pulses which are picked up by the aircraft, magnified and returned. From these responses, the exact distance of P from A is recorded at A, and of P from B is recorded at B. The Cat station emits a signal as a result of its knowledge of the distance PA. If the aircraft has strayed to the right, so that AP is greater than AC, the pilot hears a series of Morse dashes in his earphones. If the aircraft strays to the left, too near to A, he hears a series of dots. But if he keeps exactly on the circle he hears a high-pitched continuous buzz. The Mouse station B watches the aircraft, ready to warn it when it reaches the target and should, as it were, dart down the hole. It gives the pilot a series of warning signals as he comes within range, and then a final signal at the right moment for releasing the bombs. It may even release the bombs without the pilot's intervention at all.

The inventor of this system, called Oboe, was Mr. A. H. Reeves. Like Dippy and Lovell in their developments of Gee and H_2S, he worked in close collaboration with Group Capt. (Now Air Vice-Marshal) D. C. T. Bennett the Pathfinder leader, who tried and adopted it. It was used by Pathfinder aircraft to mark special targets, which could then be attacked by following bombers.

SOURCE: "Science at War" by J. G. Crowther and R. Whiddington. H. M. S. O. London (1947) p. 58.

SEE ALSO: "Oboe - a precision ground controlled blind-bombing system" by F. E. Jones. Proc. IEE Vol. 93 Pt. IIIA (1946) p. 496.

1940   SKIATRON C. R. O.   A. H. Rosenthal (U. S. A.)

The Skiatron, or dark-trace tube, was developed to meet a radar requirement for a large-screen "black on white" picture with a persistance of several seconds.

The Skiatron consists of a magnetically focussed and deflected cathode-ray tube with a screen consisting of a translucent micro-crystalline layer of potassium chloride. This screen is obtained by evaporation of the material in vacuum on to the tube face. An intensity-modulated scanning electron beam produces a picture by causing darkening in the areas bombarded. The picture is episcopially projected using external illumination from mercury-vapour lamps.

SOURCE: "The Skiatron or dark-trace tube" by P. G. R. King and J. F. Gittins. Journal IEE Pt. IIIA Vol. 93 (1946) p. 822.

SEE ALSO: "A system of large-screen television based on certain electron phenonema in crystals" by A. H. Rosenthal, Proc. IRE Vol. 28 (1940) p. 203.

"Photography of cathode ray tube traces" by H. F. Roberts and F. A. Richards. RCA Review Vol. 6 (1941) p. 234.

1941   BETATRON   D. W. Kerst (U. S. A.)

A betatron is an electrical device in which electrons revolve in a vacuum enclosure, in a circular or a spiral orbit normal to a magnetic field, and have their energies continuously increased by the electric force resulting from the variation with time of the magnetic flux enclosed by their orbits. The betatron accelerates electrons to velocities approaching that of light. This magnetic induction accelerator was invented by D. W. Kerst of the University of Illinois in 1941. It is comparable to an ordinary transformer wherein the high voltage winding, or secondary, consists of an evacuated tube in which electrons moving at high velocity form the secondary circuit. The device has been designed to accelerate electrons to 340 million electron-volts.

SOURCE: "Encyclopedic Dictionary of Electronics and Nuclear Engineering" by R. I. Sarbacher. Pitman, London (1959) p. 10.

1941   RADIO "PROXIMITY" FUSE   W.S.Butement (U.K.)

        One of the most brilliant innovations of Army radar was the V-T or
self-acting radio fuse. This was proposed by Butement. It consists of a small
radio transmitter and receiver fitted within a shell. When the shell is fired,
the transmitter emits radio waves. These are reflected from the target. The
time-interval between the emission of the waves and the return of their reflections is a measure of the distance between the shell and its target. By
arranging that the shell explodes when the time-interval falls below a certain
value, the shell is made to explode when it is within a certain distance of the
target, and therefore virtually sure to inflict damage.

        This very ingenious invention was based on an application of the
Doppler principle, and on the use of very rugged radio valves which could be
fired in a shell without being destroyed. Both of these original features were
British inventions, the early successful rugged valves being made by the
research team under Mr.D.I.Lawson of Pye Ltd., while important contributions
were made by the Research Laboratories of the General Electric Company Ltd.
        The later development and manufacture of this fuse were taken over by
American scientists and engineers. Many very difficult problems had to be
solved before it could be produced reliably in quantity. This was done just in
time to meet the flying-bomb menace to London, and at the end of that attack,
nearly 100 per cent of all the flying-bombs approaching London were being shot
down by anti-aircraft guns using V-T fuses invented in England, and developed
and manufactured in the United States.

        Our American allies were able to put 1,500 persons on to the develop-
ment of this fuse : at no time were we able to put more than 50 persons on to
the same task. The Americans performed the prodigy of making 150,000,000
of the special valves for these fuses.

SOURCE: "Science at War" by J.G.Crowther and R.Whiddington. H.M.S.O.
London (1947) p.82.

1941   RADAR (H$_2$S) NAVIGATION SYSTEM   P.I.Dee
                                                     A.C.B.Lovell   (U.K.)
                                                     A.D.Blumlein
                                                         et al

        If you could see air targets with spiral scan A.I. equipments, why not
try to see ground targets with them. Accordingly , in the autumn of 1941 Dee
arranged that this experiment be tried. The aircraft flew from Christchurch
Aerodrome, Hampshire. After four minutes a camp near Stonehenge and the
City of Salisbury were identified on the cathode ray tube.

        Thus it was proved that in the cathode ray tube picture of the general
reflections from the ground of the waves from a 9 centimetre airborne equipment, certain areas of ground could be distinguished from others. The picture
of the mass of echoes from the ground was not just a shimmering confusion, it
had definite features which corresponded to different objects on the ground.

        Specific equipment for scanning the ground was now made and fitted
into a blister or dome on a heavy bomber, in the place of the under-turret.
The blister was made of the synthetic plastic material perspex, which has the
valuable property of being transparent to short radio waves, as well as visible
light. The rotating scanning equipment is thus protected from the rush of the
air. The bomber, Halifax V 9977, flew with the new equipment for the first
time on March 27th,1942, and radar (magic eye) target-finding was born to live
under the name of H$_2$S.

SOURCE: "Science at War" by J.G.Crowther and R.Whiddington. H.M.S.O.
London (1947) p.63.

1941   MICROELECTRONICS   Centralab (U.S.A.)
            (Thick Film Circuits)

        During World War II, the Centralab Div. of Globe-Union Inc.,
developed a ceramic-based circuit for the National Bureau of Standards. This
"printed circuit" used screen-deposited resistor inks and silver paste to

support the miniature circuits in an Army proximity fuse. The PC board that followed stimulated manufacturers to develop components with radial leads and tubular shapes.

SOURCE: "Solid State Devices - Packaging and Materials" by R.L.Goldberg "Electronic Design" 24. (November 23, 1972) p.127.

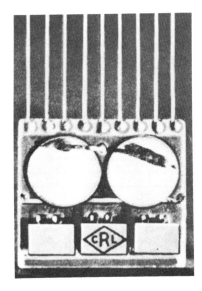 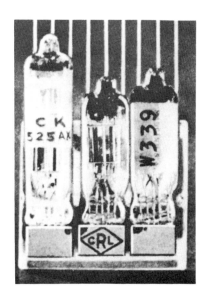

Proximity Fuze Circuits made in 1945.

1942   CIRCUITRY (Miller Integrator)   A.D.Blumlein (U.K.)

          Figure 6 shows this very well-known circuit. The name "Miller" commemorates the man who pointed out (the usually deleterious effect) that gain in a triode valve effectively increases the input capacitance by an amount equal to the product of the gain and the stray grid/anode capacitance. Blumlein, however, first made use of this by deliberately adding an external capacitor to the stray capacitance. Although the date of this patent is 1942, the idea had been used by Blumlein some years earlier, and in fact appears in a somewhat elaborate form in his television frame scan circuit BPS 479, 113 (1936).

SOURCE: "The Work of Alan Blumlein" British Kinematography Sound and Television. Vol.50. No.7. (July 1968) p.211.

SEE ALSO: British Patent Specification No:580, 527 (1942)

NOTE: See "MILLER TIME BASE CIRCUIT" - 1919

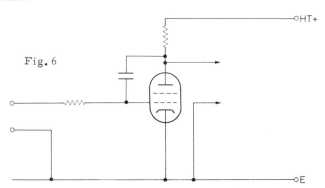

Fig. 6

1942   THE VELODYNE   F. C. Williams
                                      A. M. Uttley   (U.K.)

          The Velodyne is an electro-mechanical system in which a speed of
rotation is held closely proportional to an input voltage by feedback methods.
In such a system the total number of revolutions of the output shaft is a
measure of the time-integral of the input voltage.

          The Velodyne has been applied to the solution of differential equations,
and the T.R.E. differential analyser uses it to solve simultaneous differential
equations of importance in propagation theory; many "simulators" have been
made which obey specific high-order differential equations and have been very
useful, particularly in the design of complicated automatic flight-control
equipment where computation of solutions would have involved prohibitive effort.

SOURCE: "The Velodyne" by F. C. Williams and A. M. Uttley. Proc. IEE
Vol. 93. Pt. IIIA (1946) p. 1256.

1942   SANATRON Linear Time Base Circuit   F.C. Williams
                                                       N.F. Moody   (United
                                                                     Kingdom)

THE SANATRON.   The two valves are arranged in the form of a multivibrator
having one stable and one semi-stable state, the former corresponding with
the quiescent condition of the circuit and the latter having the duration of the
linear discharge of a capacitor C_2.   Operation is initiated by a negative
trigger pulse and the semi-stable state is then maintained until the discharge
of C_2 is completed when the circuit reverts to its quiescent condition.

          As a time base generator the Sanatron leaves little to be desired, a
sweep waveform being available, with amplitudes up to 250 V available from a
300-V supply line and rates from 1 V/sec to 5×10^7 V/sec are readily achieved.
The upper limit of rate may be extended even further by using a modification
of the circuit in which the control of the Miller integrator is applied to the
screen instead of to the suppressor grid.

SOURCE: "Ranging circuits, linear time-base generators and associated
circuits" by F.C. Williams and N.F. Moody.   Abstract of supporting paper
on circuit techniques.   JIEE, p. 320.

SEE ALSO: "Linear time-base generators and associated circuits".
Journal IEE, Vol. 93, Part IIIA, 1946.

1942   PHANTRASTRON Linear Time Base Circuit   F.C. Williams
                                                       N.F. Moody   (United
                                                                     Kingdom)

THE PHANTRASTRON.   The Phantrastron is a circuit which combines the
Miller integrator and trigger properties of the Sanatron in a single valve.
Its main use lies in delayed pulse generation and like the Sanatron, it delivers
a rectangular pulse during the linear sweep.   Due to its simplicity it is widely
used where a somewhat lower order of accuracy and linearity than that provided
by the Sanatron is acceptable.

SOURCE: "Linear time-base generators and associated circuits" by
F.C. Williams and N.F. Moody.   JIEE, Vol. 93, Part IIIA, 1946, p. 1193.

1942   LORAN   MIT   (U.S.A.)

          LORAN (Long Range Air Navigation), which marks the world's air and
sea lanes like streets, goes into operation at four stations between the
Chesapeake Capes and Nova Scotia.   LORAN, a natural successor to the GEE
system developed in Britain and given to the USA in 1940, can at long range,
give an aircraft pilot or ship's captain his position to within a few hundred yards.
The LORAN receiver, developed at MIT, picks up radio signals as pairs of
'pips' on a screen; when the pips intersect, the position is indicated.

SOURCE :   "The Timetable of Technology", published by Michael Joseph,
London, and Marshall Editions, London (November 1982), p. 124.

1943   RELIABILITY (Sequential analysis)   A. Wald (U.S.A.)

A major technical breakthrough occurred in the spring of 1943, when the noted mathematical statistician, Abraham Wald, devised his now celebrated basic theory of sequential analysis for analyzing U.S. war problems. Initial application of the theory in analyzing combat experience demonstrated its value in obtaining reliable conclusions from a minimum of information, swiftly and economically. This seemed like an ideal tool for use in quality control at this crucial period in out history when speed, precision, reliability and economy in production were of the essence. To give the new tool a thorough trial, the military released a multiple sampling plan based on Wald's sequential theory. This plan was first released to a limited number of strategic manufacturing firms for use in acceptance sampling. The early success of the plan and the subsequent widespread demands for its use resulted in removal of its "restricted" classification in 1945.

SOURCE: "The reliability and quality control field from its inception to the present" by C.M. Ryerson. Proc. IRE (May 1962) p. 1326.

1943   TRAVELLING WAVE TUBE   R. Kompfner, A.W. Haeff and
                                J.R. Pierce   (U.S.A.)

It remains, however, for Kompfner to take the decisive step of reasoned approach and effective experiment which gave us the travelling-wave tube and led to a host of related devices.

Kompfner reasoned that an electromagnetic wave on a slow-wave structure, and especially a helix, should interact powerfully with a beam of electrons if the wave velocity and the electron velocity were nearly the same. He built a tube and found a gain of around 10 db for a wave travelling in the same direction as the electrons. A mathematical analysis agreed with this performance.

The travelling-wave tube turned out to be a device which amplifies over unprecedently broad bands, and which can function over an astonishingly wide range of frequencies and powers. A travelling-wave tube with a helix circuit can amplify over a frequency range of more than an octave. More typically, bandwidth is limited to 500-1000 Mc by input and output couplers.

SOURCE: "History of the microwave tube art" by J.R. Pierce. Proc. IRE (May 1962) p. 980.

SEE ALSO: "The travelling wave valve" by R. Kompfner. "Wireless World" Vol. 52. (Nov. 1946) p. 369.

1943   PRINTED WIRING   P. Eisler (U.K.)

This invention relates to the manufacture of electric circuits - such, for instance, as that of a telephone switchboard -- and circuit components - such as inductances, resistances, magnetic cores and their existing windings; and consists in producing them by the methods of the printing art or methods akin to them, methods that is to say, by which the conductor of the circuit is brought into existence in its final form, or a development of that form upon a plane or other surface, instead of being first produced as a linear conductor and afterwards given its three-dimensional form.

A typical instance of the invention comprises the steps of making a drawing of the electric or magnetic circuit, or of a development of it if it is of three dimensions: preparing from that drawing, by any of the well-known methods of the printing art, a printing surface; making an imprint by the aid of the printing surface; and from that imprint producing the conductor.

SOURCE: U.K. Patent No: 639, 178. (February 2nd, 1943)

SEE ALSO: "Technology of Printed Circuits" by P. Eisler. Heywood, London. (1959).

1943   <u>ULTRASONIC RADAR NAVIGATION</u>   G. W. A. Dummer
                        <u>TRAINING DEVICE</u>   A. W. Smart   (U.K.)

        One of the most important of the radar navigation devices designed
during the war was H_2S. This system was used by R.A.F. Bomber Command
for 'blind' navigation over Germany and made possible the accurate and heavy
bombing of Berlin and other targets beyond the effective range of 'Gee" and
'Oboe'. To use this device, training was necessary and navigator/operators
could be trained either by flying training aircraft equipped with H_2S or by
using a synthetic training device on the ground.

        The H_2S training device, originated in February 1943, was the first
to use the new principle of a "miniature" radar system using ultrasonic waves
propagated through water in place of electro-magnetic waves propagated through
air. The long delay time of transmission of ultrasonic waves through a liquid
was used to represent the radar delay times normally encountered in the operation of H_2S. The velocity of ultrasonic waves in water is 1.5×10^5 cm/sec.
and as that of electro-magnetic waves in air is 3×10^{10} cm/sec. the scale on
which the trainer operated was

$$\frac{1.5 \times 10^5}{3 \times 10^{10}} = \frac{1}{200,000}\text{th}$$

of the radar scale. This meant that one "ultrasonic" mile = 0.315in, and a
radar range of fifty miles could be simulated in a physical distance not exceeding
15 inches approximately.

        If an X-cut quartz crystal is pulsed at its resonant frequency under
water and a reflecting object is placed in the path of the transmitted wave, an
echo will be re-radiated from the object in the same way as in radar and picked
up on the crystal.

        The crystal beam was projected on to a glass relief map so that the
projection covered a range of 0.30 miles (0.10 inch approximately), the crystal
being used for both transmission and reception. The crystal was set back 1.25
in. from the axis of rotation to simulate "ground returns" and picture as seen
from an aircraft flying at 20,000ft. On the glass map was reproduced (to a
scale of 1:200,000) a simulation of the area in Germany over which training was
to be effected. The amount of reflection of the radiated energy was proportional
to the 'roughness' of the glass surface, and the map was sand-blasted or etched
to represent land masses. Towns were built up of granules of carborundum
glued to the glass and the sea areas were left as plain glass. The map was
placed at the bottom of a tank of water. The pulsed crystal was then rotated at
60 r.p.m. synchronously with the radial time base, and the signals received
from the map were fed from the crystal to the input stage of the IF amplifier
in the H_2S receiver and thence to the P.P.I.

        The display produced was an excellent simulation of the actual H_2S
picture seen in the air when flying over Germany.

<u>SOURCE</u>: "H_2S trainer - use of ultrasonic reflections from submerged
relief maps" by G. W. A. Dummer. Wireless World (February 1947) p. 65.

<u>SEE ALSO</u>: "Aids to Training - the Design of Radar Synthetic Training Devices
for the R.A.F." by G. W. A. Dummer. Proc. IEE Vol. 96. Pt. III. No. 40
(March 1949) p. 101.

1943   <u>COMPUTERS (ENIAC)</u>   Moore School (U.S.A.)
             <u>(Electronic numerator, integrator</u>
             <u>and computer).</u>

        The ENIAC was developed and built at the Moore School of Electrical
Engineering of the University of Pennsylvania, beginning in 1942 and was completed in 1946. Its principal object was the computation of firing and ballistic
tables for the Aberdeen Proving Ground of the U.S. Army Ordnance Corps.
This computation required the integration of a simple system of ordinary
differential equations involving arbitrary functions.

        This equipment occupied a space 30 x 50 feet and contained 18000
vacuum tubes. The computing elements consisted largely of decade rings,
flip-flops and pentode gates. The input-output system consisted of modified

IBM card readers and punches.

SOURCE: "The evolution of computing machines and systems" by Serrell, Astrahan, Patterson and Pyrne. Proc.IRE (May 1962) p.1044.

SEE ALSO: "The computer from Pascal to von Neumann" by H.H.Goldstine. Princeton Univ.Press (1972) p.117.

1943   MAGNETIC AMPLIFIER   A.S.E.A. (Sweden)
           (TRANSDUCTOR)

        Pioneer work on transductors was carried out by the Swedish firm of A.S.E.A. who devised many techniques involving combinations of saturated reactors and metal rectifiers for supplying controllable d.c. power from single-, three-, and six-phase networks in place of the conventional mercury-arc rectifier systems. Much of this pioneer work is described in the classic work by Uno Lamm.

        Considerable development took place in Germany during the 1939-45 war on transductors as true magnetic amplifiers for various small power applications, mainly in association with servo systems. The V2 rocket incorporated a transductor for controlling the frequency of a 500 c/s, 150 VA alternator.

SOURCE: "The magnetic amplifier" by J.H.Reyner. Rockcliff Pub.Corp. (1950) p.17.

SEE ALSO: "The transductor" by A.U.Lamm. Essalte Aktiebolag, Stockholm (1943)

        "Some fundamentals of a theory of a transductor" by A.U.Lamm AIEE Trans. Vol.66 (1947)

        "Magnetic Amplifier" by A.G.Milnes. Journal IEE. Vol.96. Part I (1949) p.89.

1943   COLOSSUS - CRYPTANALYSIS MACHINE   M. Newman
                                                A. Turing   (United
                                                T.H. Flowers   Kingdom)
                                                A.W.M. Coombs

        As far as building a British stored-program computer, the initial enthusiasm came largely from a group of people who had been involved in code-breaking activity at the government's Bletchley Park establishment. Much of the Bletchley work is still subject to the Official Secrets Act because the methods are (or were) in use by other countries after the war. As far as electronic digital techniques were concerned, the most interesting Bletchley crptanalysis machine was the COLOSSUS, first operational in December 1943 and containing about 1500 thermionic valves. COLOSSUS was not, and had no need to be, a true stored-program computer, but it came quite close to being one. The extraordinary and vital contribution of COLOSSUS to the allied war effort has been told elsewhere.

SOURCE : "The early days of British computers - 1" by S. H. Lavington. Electronics & Power (November/December 1978), p.827.

SEE ALSO : "The Colossus" by B. Randell. University of Newcastle Computer Science Technical Report 90, 1976. Reprinted in condensed form in New Scientist, 1977, 73, p.346-348.

"The Ultra secret" by F.W. Winterbotham. (Wiedenfelt and Nicholson, 1974).

"Colossus : godfather of the computer" by B. Randell. New Scientist. Vol.73 (10 February 1977), p.346.

1944   RELIABILITY - SAMPLING   H.F.Dodge & H.G.Romig (U.S.A.)
           INSPECTION TABLES

        After languishing in libraries for several years, the work of Dodge and Romig in acceptance sampling and the work of Shewhart on control charts finally was brought to light during World War II through the nationwide training programmes sponsored by the Office of Production Research and Development of the War Production Board.

Although the underlying concepts were developed by scientific investigators and statisticians in the preceding decades, the genius of Dodge, Romig and Shewhart lay in their recognition of basic principles as an aid to solving practical problems, and their ability to recognize and formulate a systematic approach.

SOURCE: "Treating real data with respect" by J.A.Henry. Quality Progress. (March 1976) p.18.

SEE ALSO: "Sampling Inspection Tables" by H.F.Dodge and H.G.Romig. 2nd Ed.(1944) John Wiley & Sons, New York, N.Y.

1945   COMPUTERS (Theory)   Von Neumann (U.S.A.)

Basic design of the electronic computer project of the Institute of Advanced Study incorporating ideas underlying essentially all modern machines.

REFERENCE: "Memorandum on the program of the High Speed Computer" by von Neumann. (8th November, 1945)

SEE ALSO: "The computer from Pascal to von Neumann" by H.H.Goldstine. Princeton Univ. Press (1972) p.255.

1945   DECCA Navigation System   W. O'Brien (United Kingdom)
                                          Harvey Schwartz (U.S.A.)

DECCA, internationally regarded as the best navigation system, undergoes crucial tests during the Allies' D-Day landings on the Normandy beaches. Developed by William O'Brien and Harvey Schwartz, in London and Hollywood, DECCA indicates on cockpit dials the position of an aircraft in three dimensions - latitude, longitude and altitude - and is accurate to within a few yards. Waves of radio signals, emitted from two transmitters, collide and are picked up in phase.

SOURCE : "The Timetable of Technology", published by Michael Joseph, London, and Marshall Editions, London (November 1982), p.130.

1945   COMPUTERS (Whirlwind)   Massachusetts Institute of Technology (U.S.A.)

An assignment to build a real-time aircraft simulator was given in 1945 to the Digital Computer Laboratory of the Massachusetts Institute of Technology, at that time a part of the Servomechanisms Laboratory of M.I.T. Beginning in 1947, the major part of the effort was devoted to the design and construction of the electronic digital computer known as 'Whirlwind'. The project was sponsored by the Office of Naval Research and the United States Air Force. The machine was put in operation in March 1951.

'Whirlwind I' was a parallel, synchronous, fixed-point computer utilizing a number length of 15 binary digits plus sign (16 binary digits in all). Physically, it was a large machine containing some 5000 vacuum tubes (mostly single pentodes) and some 11,000 semiconductor diodes. It consisted of an arithmetic "element" including three registers; a control element including central control, storage control, arithmetic control and input-output control; a program counter - a source of synchronizing pulses or master clock supplying 2 megapulses per second to the arithmetic element and 1 megapulse per second to the other circuits; an internal storage element or memory, terminal equipment; and extensive test and marginal checking equipment.

SOURCE: "The evolution of computing machines and systems" Serrell, Astrahan, Patterson and Pyne. Proc.IRE (May 1962) p.1047.

SEE ALSO: "The computer from Pascal to von Neumann" by H.H.Goldstine. Princeton Univ. Press (1972) p.212.

1945   <u>COMMUNICATION (Satellite)</u>   A. C. Clarke (U.K.)

        Early interest in space was concentrated upon the propulsion aspect, and the forthcoming marriage of space and electronics had to await the publication of A. C. Clarke's paper on communications satellites in 1945. The use of a satellite S above the radio horizons of both A and B permitted microwave transmission from A to S, and from S to B, thus bridging the oceans by microwave link. Clarke's paper further pointed out that at an orbital altitude intermediate between a 90-minute SPUTNIK and the 28-day orbit of the moon, there was an orbit taking one day, so that an easterly-launched satellite above the Equator would give the radio engineer an imaginary mast 22 300 miles high on which to place his aerials.

<u>SOURCE</u>: "Electronics in Space" by W. F. Hilton. The Radio and Electronic Engineer. Vol. 45. No: 10 (Oct. 1975) p. 623.

<u>SEE ALSO</u>: "Extra terrestial relays" by A. C. Clarke. Wireless World.Vol. 51. No: 10 (Oct. 1945) p. 305.

1945 -   <u>POTTED CIRCUITS</u>
1950   (U.K.) and (U.S.A.)

        The potting of electrical apparatus in wax or bitumen compounds was carried out for many years, but it is only recently that plastics suitable for this purpose have become available, in the form of cold-polymerizing casting resins. There is no doubt that the small sub-unit is now as essential part of modern electronic equipment, and potting techniques lend themselves to this construction provided that means are available to dissipate the heat developed. The resins are relatively expensive, and mechanically and economically are not attractive for casting exceeding a few inches in major dimensions.

        The casting resins are converted into rigid plastics by the addition of a catalyst and accelerator, without the application of the considerable pressures and temperatures normally associated with the polymerization of thermo-setting resins.

<u>SOURCE</u>: "New constructional techniques" by G. W. A. Dummer and D. L. Johnston. Electronic Engineering(Nov. 1953) p. 456.

<u>SEE ALSO</u>: "How plastics aid miniaturization of electrical assemblies" by R. J. Bibbero and E. B. Chester. Mach. Des. (Oct. 1951) p. 127.

        "Potted Circuits - new development in miniaturization of equipment" Wireless World, 57, 493 (1951)

        "Cast resin embedments of circuit sub- units and components" Elect. Mnfg. 48, 103 (1951).

1946   <u>ACE (AUTOMATIC COMPUTING ENGINE)</u>   A. M. Turing (United Kingdom)

        Turing joined the new mathematics division at NPL, where he immediately set about designing a universal computer with characteristic energy. Turing had written an important theoretical paper on computers in 1936 and, although familiar through personal contact with von Neumann, he had no need or inclination to copy anyone else's design. On the 19th February 1946, he presented to the Executive Committee of NPL probably the first complete design for an electronic stored-program computer, including a cost estimate of £ 11 200. It is likely that Sir Charles Darwin, the NPL Director, thought of Turing's proposal for an Automatic Computing Engine (ACE) in terms of a single national effort that would result in a computer housed at NPL and serving the needs of the whole country.

        The Pilot ACE had a complicated 32 bit instruction format including provision for specifying one of 32 'sources', one of 32 'destinations' and the source of the next instruction. Instructions also specified the duration of a transfer, so that prolonging a transfer over several cycles could give the effect of shifting or multiplying operands by small integers. Many operations in the instruction repertoire were straight-forward transfers, with the remaining instructions providing about a dozen conventional arithmetic or logical functions, including unsigned multiplication. Signed multiplication took just over 2 ms, being performed partly by subroutine. Other orders could be obeyed in as little as 64 μs (1024 μs in the worst case), depending on the position of the next instruction.   Arithmetic was serial, with a 1 μs digit period.   The main store consisted initially of 128 32 bit words in mercury delay lines.   This was extended to 352 words by the end of 1951, and a 4k drum was added in 1954.   Since the NPL already had a large Hollerith punched card calculator, it was sensible to make cards the medium for both input to and output from the Pilot ACE.

The Pilot ACE contained 800 thermionic valves, the processor logic involving type ECC 81 double triodes.   It first ran a program in May 1950.

SOURCE :   "The early days of British computers - 1" by S.H. Lavington. Electronics & Power (November/December 1978), p. 828.

"The early days of British computers - 2" by S. H. Lavington.   Electronics & Power (January 1979), p. 40.

SEE ALSO :   "Proposals for the development in the Mathematics Division of an Automatic Computing Engine (ACE)" by A.M. Turing.   Report E882, Executive Committee, NPL.   Reprinted in April 1972 as NPL Report Com. Sci. 57.

"The other Turing machine" by B.E. Carpenter and R.W. Doran.   Computer J. (1977), 20, p. 269-279.

1946   COMPUTERS (CRT Storage)   F.C. Williams (U.K.)

Storage of pulses on the face of a CRT as a memory device.

REFERENCE: "A Storage System for use with Binary Digital Computers" by F.C. Williams and T. Kilburn. Proc. IEE Vol. 96. Pt. 2. No: 81 (1949) p. 183.

SEE ALSO: "The computer from Pascal to von Neumann" by H.H. Goldstine. Princeton Univ. Press (1972) p. 248.

1947   HIGH QUALITY AMPLIFIER CIRCUIT   D.T.N. Williamson (U.K.)

The requirements of such an amplifier may be listed as :-
(1) Negligible non-linear distortion up to the maximum rated output. (The term "non-linear distortion" includes the production of undesired harmonic frequencies and the intermodulation of component frequencies of the sound wave.) This requires that the dynamic output/input characteristic be linear within close limits up to maximum output at all frequencies within the audible range.

(2) (a) Linear frequency response within the audible frequency spectrum of 10-20,000 c/s.   (b) Constant power handling capacity for neglible non-linear distortion at any frequency within the audible frequency spectrum.

(3) Neglible phase-shift within the audible range.   Although the phase relationship between the component frequencies of a complex steady-state sound does not appear to affect the audible quality of the sound, the same is not true of sounds of a transient nature, the quality of which may be profoundly altered by disturbance of the phase relationship between component frequencies.

(4) Good transient response.   In addition to low phase and frequency distortion, other factors which are essential for the accurate reproduction of   transient wave-forms are the elimination of changes in effective gain due to current and voltage cut-off in any stages, the utmost care in the design of iron-cored components, and the reduction of the number of such components to a minimum.

(5) Low output resistance.   This requirement is concerned with the attainment of good frequency and transient response from the loudspeaker system by ensuring that it has adequate electrical damping.

(6) Adequate power reserve.   The realistic reproduction of orchestral music in an average room requires peak power capabilities of the order of 15-20 watts when the electro-acoustic transducer is a baffle-loaded moving-coil loudspeaker system of normal efficiency.

SOURCE: "Design for a high-quality amplifier" Part I. by D.T.N. Williamson Wireless World (April 1947) p. 118.

"Design for a high-quality amplifier" Part 2. by D.T.N. Williamson
Wireless World (April 1947) p.161.

1947   CHIRP RADAR TECHNIQUES   Bell Laboratories (U.S.A.)

The Chirp or pulse-compression technique for radar originated at
Bell Laboratories in 1947. This technique, in which long, modulated pulses
are transmitted and then compressed upon reception, permitted pulsed radar
systems to have long range and high resolution while avoiding problems associated with generating and transmitting short pulses with high peak powers.

SOURCE: "Mission Communications - the story of Bell Laboratories" by
Prescott C. Mabon. Published by Bell Laboratories Inc., Murray Hill,
New Jersey U.S.A. (1975) p. 179.

1947   ECME (Electronic circuit making equipment)   J.A. Sargrove (U.K.)

The process of John A. Sargrove for the automatic assembly of
electronic apparatus was the first modern approach to automatic operation in
electronic manufacturing. In 1947 he built and operated a machine for the
automatic production of two- and five-tube radio receivers.

The first operation of Sargrove's machine was to prepare the $\frac{1}{4}$-in.
molded-plastic plates by blasting with an abrasive grit to roughen both sides
of the plates simultaneously. The plates were then triple-sprayed with zinc
to form the conducting surface. The spraying machine consisted of eight
nozzles arranged four to a side to allow simultaneous spraying of both sides
of the plate once it was positioned. Materials to form resistance, capacitance
and conductors were sprayed through stencils onto their proper positions on
the plate.

SOURCE: "Electronic equipment design and construction" by G.W.A. Dummer
C. Brunetti and L.K. Lee. McGraw-Hill, New York (1961) p.192-193.

SEE ALSO: "New methods of radio production" by J.A. Sargrove, J. Brit. IRE
Vol. 7(1) (Jan/Feb 1947) p. 2.

"Automatic Receiver Production" Wireless World (April 1947)

Examples of radio sets produced by automatic assembly methods.
(Courtesy - Sargrove Electronics)

1947   COMPUTERS (EDVAC)   University of Pennsylvania (U.S.A.)

The Electronic Discrete Variable Automatic Computer, or EDVAC, was
built at the Moore School (University of Pennsylvania) between 1947 and 1950 for
the Ballistic Research Laboratory at the Aberdeen Proving Ground. It is a

serial, synchronous machine in which all pulses are timed by a master clock operating at 1 mega-pulse per second. It contains some 5900 vacuum tubes, about 1200 semiconductor diodes and utilizes the binary number system with a word length of 44 binary digits.

SOURCE: "The evolution of computing machines and systems" by Serrell, Astrahan, Patterson and Pyne. Proc.IRE (May 1962) p.1046.

SEE ALSO: "The computer from Pascal to von Neumann" by H.H.Goldstine Princeton Univ.Press (1972) p.187.

1947   COMPUTERS (UNIVAC)   P.Eckert   and J.Mauchly (U.S.A.)
          Universal Automatic Computer

The development of the Universal Automatic Computer, or UNIVAC was started about 1947 by Presper Eckert and John Mauchly who founded the Eckert-Mauchly Computer Corporation in December of that year. The first UNIVAC I was built for the U.S.A.Bureau of the Census and was put in operation in the spring of 1951. (The Eckert-Mauchly Corporation later became a subsidiary of Remington Rand, forming the organisation which is now the Remington Rand UNIVAC Division of the Sperry Rand Corporation).

UNIVAC I was a direct descendent of the ENIAC and of the EDVAC in the development of which Eckert and Mauchly had both had an important part at the University of Pennsylvania. It was a serial, synchronous machine operating at a rate of 2.25 megapulses per second. It contained some 5000 tubes and several times as many semiconductor diodes in logic and clamp circuits. One hundred mercury delay lines provided 1000 twelve-decimal-digit words of internal storage. Twelve additional delay lines were used as in ut-output registers. Aside from console switches and an electric typewriter providing small amounts of information, the input-output medium was metal-base magnetic tape. Forty eight UNIVAC I machines were built.

SOURCE: "The evolution of computing machines and systems" Serrell, Astrahan, Patterson and Pyne.   Proc.IRE (May 1962)   p.1048/9.

SEE ALSO: "The computer from Pascal to von Neumann" by H.H.Goldstine. Princeton Univ.Press (1972) p.246.

1947   MOLECULAR BEAM EPITAXY   J.Sosnowski
                                        J.Starkiewicz   (U.K.)
                                        O.Simpson

Molecular beam epitaxy (MBE) is a term used to denote the epitaxial growth of compound semiconductor films by a process involving the reaction of one or more thermal molecular beams with a crystalline surface under high vacuum conditions. MBE is related to vacuum evaporation, but offers much improved control over the incident atomic or molecular fluxes so that sticking coefficient differences may be taken into account, and allows rapid changing of beam speeds. For example, using MBE, it is possible to produce "superlattice" structures consisting of many layers of $G_2 A_s$ and $Al_2 Ga_{1-x} A_s$ with layer thickness as low as 10Å. Since electrically active impurities are added to the growing film with separate beams, the doping profile normal to the surface may be varied and controlled with a special resolution difficult to achieve by more conventional, faster growth techniques.

SOURCE: "Molecular Beam Epitaxy" by A.Y.Cho and J.R.Arthur. J.R.Prog. in Solid State Chemistry. Vol.16, Part 3 (1975) p.157.

SEE ALSO: "Lead sulphide photoconductive cells" by L.Sosnowski, J.Starkiewicz and O.Simpson. Nature. Vol.159 (June 14,1947) p.818.

"The structure and growth of Pbs deposits on rocksalt substrates" by A.J.Elleman and H.Wilman. Proc.Phys.Soc.(London) Vol.61 (1948) p.164.

1948   COMPUTERS (SEAC)   Nat.Bureau of Standards (U.S.A.)

The Standards Electronics Automatic Computer, SEAC, was built by the staff of the Electronic Computer Laboratory of the National Bureau

Standards. The design began in June 1948 and the machine was put in operation in May 1950. It was built under the sponsorship of the Office of the Air Comptroller, Department of the Air Force, principally to carry out mathematical investigations of techniques for solving large logistics programming problems.

SOURCE: "The evolution of computing machines and systems" by Serrell, Astrahan, Patterson and Pyne. Proc. IRE (May 1962) p.1046.

SEE ALSO: "The computer from Pascal to von Neumann" by H.H. Goldstine. Princeton Univ. Press (1972) p. 315.

1948   TRANSISTOR   Bardeen, Brattain and Shockley   (U.S.A.)

        Immediately hostilities ceased in 1945 Shockley organised a group for research on the physics of solids. On testing out experimentally Shockley's ideas, it was discovered that the projected amplifier did not function as Shockley had predicted; something prevented the electric field from penetrating into the interior of the semi-conductor. John Bardeen, a theoretical physicist, formulated a theory concerning the nature of the surface of a semi-conductor which accounted for this lack of penetration of field, and also led to other predictions concerning the electrical properties of semi-conductor surfaces. Experiments were carried out to test the predictions of the theory. In one of these, Walter H. Brattain and R.B. Gibney observed that an electric field would penetrate into the interior if the field was applied through an electrolyte in contact with the surface. Bardeen proposed using an electrolyte in a modified form of Shockley's amplifier in which a suitably prepared small block of silicon was used. He believed that current flowing to a diode contact to the silicon block could be controlled by a voltage applied to an electrolyte surrounding the contact. In the earlier experiments testing Shockley's ideas, thin films with inferior electrical characteristics had been employed. Brattain tried Bardeen's suggested arrangement, and found the amplification as Bardeen had predicted, but the operation was limited to very low frequencies because of the electrolyte. Similar experiments involving germanium were successful, but the sign of the effect was opposite to that predicted. Brattain and Bardeen then conducted experiments in which a rectifying metal contact replaced the electrolyte and discovered that voltage applied to this contact could be used to control, to a small extent, the current flowing to the diode contact. Here again, however, the sign of the effect was opposite to the predicted one. Analysis of these unexpected results by the two scientists led them to the invention of the point-contact transistor, which operates on a completely different principle from the one first proposed. Current flowing to one contact is controlled by current flowing from a second contact, rather than by an externally applied electric field. Brattain and Bardeen used extremely simple equipment, the most expensive piece of apparatus being an oscilloscope.

        The Bell Telephone Laboratories announced the invention in June 1948, and since then, development work has proceeded rapidly. The first point-contact transistor had several limitations: it was noisy, it could not control high amounts of power and it had a limited applicability. Shockley had meanwhile conceived the idea of the junction transistor which was free of many of these defects and most of the transistors now made are of the junction type.

SOURCE: "The Sources of Invention" by J. Jewkes, D. Sawers and R. Stillerman MacMillan & Co. London (1958) p. 400.

SEE ALSO: "The First Five Years of the Transistor" by Mervin Kelly. Bell Telephone Magazine. (Summer 1953).

1948   HOLOGRAPHY   D. Gabor (U.K.)

        With holography, one records not the optically formed image of an object but the object wave itself. This wave is recorded (usually on photographic film) in such a way that a subsequent illumination of this record called a "hologram" reconstructs the original object wave. A visual observation of this reconstructed wavefront then yields a view of the object which is practically indiscernible from the original, including three dimensional parallax effects.

SOURCE: The Encyclopaedia of Physics (2nd Edition) Editor R.M. Besancon. Van Nostrand. Reinhold C. Litton Educational Pub. Inc. New York (1974) p. 426.

SEE ALSO: "Optical Holography" by R.J. Collier, C.B. Burckhardt and L.H. Lin Academic Press. New York (1971)

1948   EDSAC (Electronic Delay Storage
            Automatic Calculator)   M.V. Wilkes   (U.K.)

            The EDSAC (electronic delay storage automatic calculator) is a serial
electronic calculating machine working in the scale of two and using ultrasonic
tanks for storage. The main store consists of 32 tanks, each of which is about
5 ft. long and holds 32 numbers of 17 binary digits, one being a sign digit. This
gives 1024 storage locations in all. It is possible to run two adjacent storage
locations together so as to accommodate a number with 35 binary digits
(including a sign digit); thus at any time the store may contain a mixture of long
and short numbers. Short tanks which can hold one number only are used for
accumulator and multiplier registers in the arithmetical united, and for control
purposes in various parts of the machine.

            A single address code is used in the EDSAC, orders being of the same
length as short numbers.

SOURCE: "The Origins of Digital Computers" Edited by B. Randell. Springer-
Verlag, Berlin (1973) p. 389.

SEE ALSO: "The design of a practical high-speed computing machine" by
M.V. Wilkes, Proc. Roy. Soc. London A195 (1948) p. 274.

1948   TRANSISTORS - Single crystal   G.K. Teal & J.B. Little (U.S.A.)
            fabrication - germanium

            In the latter part of 1948 G.K. Teal and J.B. Little of BTL began experi-
ments to grow germanium single crystals, selecting the pulling technique. They
succeeded in growing large single crystals of germanium of high structural
perfection. They also improved the impurity of the material by repeated recrystallization methods.

            At BTL Teal, working with M. Sparks, devised a unique method for
preparing p-n junctions by modifying his crystal-pulling apparatus to allow
controlled addition of impurities during crystal growth. Using ingots they
prepared single crystals containing p-n junctions and soon afterwards n-p-n
grown-junction transistors which had many of the properties predicted by
Shockley.

SOURCE: "Contributions of materials technology to semiconductor devices"
by R.L. Petritz. Proc. IRE (May 1962) p. 1026.

            "Growth of germanium single crystals" by G.K. Teal and J.B. Little.
Phys. Rev. Vol. 78. p. 647. (June 1950).

            "Growth of silicon single crystals and of single crystal silicon p-n
junctions" Phys. Rev. Vol. 87. p. 190 (July 1952)

1948   COMMUNICATION (Information theory)   C.E. Shannon (U.S.A.)

            The term "Information Theory" is used in the current technical liter-
ature with many different senses. Historically it seems first to have been generally applied to describe the specific mathematical model of communication
systems developed in 1948 by Shannon. In this pioneering paper, Shannon introduced a numerical measure, called by him and others entropy, of the random-
ness or uncertainty associated with a class of messages and showed that this
quantity measures in a real sense the amount of communication facility needed
to transmit with accuracy messages from the given class. He also showed
(quite incidentally to his main argument) that this measure of uncertainty agreed
in certain aspects with the common, vague intuitive notion of the "information
content of a message". He accordingly used the words "information content"
as a synonym for the precisely defined notion of entropy. As a result, his work
and its immediate extensions became known as information theory.

SOURCE: "Information Theory" by B. McMillan and D. Slepian. Proc. IRE
(May 1962) pp. 1151/2.

SEE ALSO: "The mathematical theory of communication" by C.E. Shannon
and W. Weaver. University of Illinois Press, Urbana. (1949).

1948   FILM SOUND RECORDING   RCA and others (U.S.A.)
             (Magnetic film)

             Although magnetic recording was one of the oldest methods known, it
was not until World War II that this form of recording came into its own.
During this period there was developed in Germany a fine grain, low-noise,
magnetic oxide, and a process for uniformly coating it on a thin flexible base
$\frac{1}{4}$ inch in width. Use of this new tape in properly designed recorders and repro-
ducers resulted in sound quality which was higher than had previously been
obtained from either the film or the disk method. Immediately after the war
some of the German recorders were demonstrated in this country, and the
potential impact of magnetic recording on the motion-picture industry was
quickly recognised.

             Early in 1948, oxide coated 35 mm. film became available for use in
motion picture sound recording. It was then possible to convert photographic
sound recorders to combination units capable of recording either magnetic or
photographic sound. Many recorders were converted as quickly as possible and
were tried in motion-picture sound studios for original "takes". The tests of
magnetic recording in the studios were immediately successful, not only
because of its high-quality and large dynamic range, but also because magnetic
film provided a more flexible and a more economical means of recording. Since
re-recording of all original "takes" to a composite photographic negative was
already the accepted practice in the industry, little inconvenience was caused by
the change.

SOURCE: "Film Recording and Reproduction" by M.C. Batsel and G.L. Dimmick
Proc. IRE (May 1962) p. 749.

1949   MICROWIRE   Ulitovsky (U.S.S.R.)

             The microwire process for producing ultrafine wires was invented in
Russia in 1949 by Professor Ulitovsky of the Baykov Institute. It is important
to realise that up to then it was quite difficult to obtain fine insulated wire of
reasonable quality and price.

             Three processes have been developed as advances over the traditional
die drawing techniques:

                    1.   Wollaston process
                    2.   Taylor process
                    3.   Microwire process

             In the Wollaston process wires are compound-drawn; that is, a plati-
num rod encased in a silver tube is drawn through wire dies, being finally subjected to a reduction of between 20 and 40 to one. The silver coated wire so
produced is etched to remove the silver, leaving the fine platinum core behind.
This procedure will give platinum wires down to 0.5μ.

             In the Taylor process a composite rod is made consisting of a glass
tube with a metal core cast into position by being aspirated, pipette-fashion,
from a melting pot. This composite rod is attenuated in a muffle furnace, using
the technique of pulling and stretching as practiced in the glass fibre industry.
The metal usually has a lower melting point than the softening point of the glass,
thus allowing the capillary to be filled with cast metal. Wires of lead, antimony,
bismuth, gold, silver, tin, copper and others were successfully produced down
to 0.25μ.

             In the Microwire process this technique was improved. The basic pro-
cedure consists of melting the core metal by induction in a crucible formed by
the walls of a glass tube extended upwards to a glass feeding device above the
melt. The fact that the process takes place in a vertical line is an improvement
immediately since there is less distortion of the melt and interference with the
capillary by gravity.

SOURCE: "Materials for conductive and resistive functions" by G.W.A.
Dummer. Hayden Book Co. N.Y. p. 62.

SEE ALSO: "Glass-coated microwire" by Herbert Wagner. Wire and Wire
Productions (June 1964)

"Microwire. A new engineering material" by R. G. S. Clarke
Electronic Components (Sept. 1963)

"The structure of copper microwire" by E. M. Nadgorny and
B. I. Smirnov. Fizika Tverdogo Tele. Vol. 2 (12) (1960) pp. 3048-3049.

1949   DIP-SOLDERING OF PRINTED CIRCUITS   S. F. Danko & Abramson
                                                                    (U. S. A.)

        When Danko and Abramson of the Army Signal Corps invented dip
soldering in 1949, a new era of automation came into being.

SOURCE: "Packaging and Materials" by R. L. Goldberg. Electronic Design 24.
(Nov. 23, 1972). p. 126.

SEE ALSO: "Autosembly of miniature military equipment" by S. F. Danko and
S. J. Lanzalotti. Electronics 24 (7) (July 1951) p. 94.

        "Printed circuits and microelectronics" by S. F. Danko. Proc. IRE
(May 1962) p. 937.

        Autosembly. U. S. Patent No:2756485 assigned to U. S. Army. July 31,
1956.

1949   COLD CATHODE STEPPING TUBE   Remington Rand (U. S. A.)

        The first published account of a multi-cathode stepping tube descibed
a tube developed in America by Remington Rand. The first tubes to be widely
used, however, were made in England by S. T. C. and by Ericsson Telephones.
For some ten years following its introduction in 1949, the Ericsson 'Dekatron'
became practically synonymous with the cold cathode stepping tube. The
original double-pulse 'Dekatron' was followed in 1952 by the single-pulse tube
operating up to 2o kc/s. Routing guides were added in 1955 to simplify the
construction of reversible scalers. In 1962 auxiliary-anode tubes were introduced from which numerical indicators could be driven directly.

SOURCE: "A Survey of Cold Cathode Discharge Tubes" by D. M. Neale
The Radio and Electronic Engineer. (February 1964) p. 87.

SEE ALSO: "Poly-cathode glow tube for counters and calculators" by J. J. Lamb
and J. A. Brustman. Electronics 22. No. 11 p. 92-6. (November 1949)

1949/50   ION IMPLANTATION IN SEMICONDUCTORS   R. S. Ohl,   W. Shockley (U. S. A.)

        Ion implantation is a technique for modifying the properties of solids
by injecting (implanting) charged atoms (ions) into them. The ions alter the
electrical, optical, chemical, magnetic and mechanical properties of a solid by
the interactions they have with the solid both as they slow down and by their
presence after they have come to rest.

        The implantation idea isn't a new one - work at Bell Laboratories
during the late 1940s and early 1950s by Russel S. Ohl and William Shockley
pioneered the application of ion implantation to semiconductor device fabrication.

SOURCE: "Ion implantation" by W. C. Brown and A. U. MacRae, Bell Laboratories Research (Nov. 1975) p. 389.

SEE ALSO: "Forming semiconductive devices by ionic bombardment"
W. Shockley. U. S. Patent No:2787564 (28 Oct. 1954)

        "Ion implantation in semiconductor device technology" by J. Stephen.
The Radio Electronic Engineer. Vol. 42. No. 6. (June 1972) p. 265.

1950   VIDICON - TV Camera Tube   RCA (U.S.A.)

        The Vidicon, the first TV camera tube to use the principle of photo-
conductivity by which light of different intensities produces a change in electrical
activity, is devised in the USA by RCA. After further developments, the
Vidicon proves significantly more adaptable, more sensitive - and cheaper -
than previous TV cameras.

        SOURCE :   "The Timetable of Technology", published by Michael Joseph,
London, and Marshall Editions, London (November 1982), p.138.

1950   HAMMING CODE   R. W. Hamming (U.S.A.)

        It was simple human frustration that led Dr. Richard W. Hamming, then
a research mathematician at Bell Laboratories, to devise the first method for
correcting machine-caused errors in digital computers.

        Hamming's technique - which was a result of his research in pure
mathematics - enabled computers to spot electrical errors in the data and
instructions and tell where they occurred. And, for the first time, it enabled
computers to correct those errors and go right on solving problems without
interruption.

        April 20, 1980, marks the 30th anniversary of Hamming's pioneering
work, which has evolved into a new field of research called error-correcting
codes.

        SOURCE: "Bell Labs marks 30th anniversary of computer error-correcting
codes". Bell Laboratories Record (May 1980), p.152.

        SEE ALSO :   "Applying the Hamming code to microprocessor-based systems"
by Ernst L. Wall.   Electronics (November 22, 1979), p.103.

1950   COMPUTERS (IBM 650)   I.B.M. (U.S.A.)

        The IBM 650, an intermediate-size, vacuum-tube computer was con-
sidered a workhorse of the industry during the late 1950s. Development began
in 1949 and the first installation was made late in 1954. Over a thousand 650's
have been in service since then.

        The 650 operates serially by character on words of 10 decimal digits
plus sign. A 2-out-of-5 decimal representation in storage is translated into a
biquinary code in the operating registers, allowing a fixed-count check to
detect the presence of more or less than 2 bits per character. The main store
of the 650 is a 12,500 rpm 2000-word magnetic drum. A two-address instruction format accommodates, as part of each instruction, the location of the next
programme step. This format allows the programmer to place instructions
anywhere on the program drum, and makes it possible for him to minimise
access times to successive instructions.

        SOURCE: " The evolution of computing machines and systems" by Serrell,
Astrahan, Patterson and Pyne. Proc. IRE (May 1962) p.1050.

        SEE ALSO: "The computer from Pascal to von Neumann" by H.H.Goldstine.
Princeton Univ. Press(1972) p.330.

1950   PIN DIODE   Jun-ichi NISHIZAWA (Japan)

        The p-i-n sometimes called as the "pin" diode in which an intrinsic or
a high resistivity layer is sandwiched between a p layer and an n layer was
invented by J. Nishizawa in 1950. In the same year several manufacturing
methods of the p-i-n diode, including the thermal diffusion, the chemical treatment, the anodic oxidation, the ion implantation and the nuclear transmutation
by the bombardment of high energy perticle were also invented by J. Nishizawa.
In 1958, his research group successfully realized the Si p-i-n diode, in which
the characteristics were 2300V reverse breakdown voltage and 1.5V forward
voltage drop at 100A, by using the alloying method and the elaborate simple
surface passivation technology prior to the Westinghouse Company. Also the
4000V p-i-n diode was realized by his research group with the development of
the original high purity epitaxial growth technology using $SiCl_4$ and the hydrogen
system.

The p-i-n diode features a high reverse breakdown voltage up to ten thousands volts, low forward voltage drop and small junction capacitance. It is used all over the world for various application fields such as the industrial high power systems, which includes the famous Japanese bullet train, owing to its efficient high power handling capability and also in consumer electronics products such as TV.

The p-i-n diode is also used as a low loss switching device, a phase shifter and an attenuator in microwave circuits in radar systems and communication equipments.

The p-i-n photo diode also invented by J. Nishizawa is an excellent optical detector and is widely used for the recent optical communication system. The fundamental patent also included the application of the high resistivity layer into a transistor as pnip or npin type.

SOURCE : "Semiconductor device having the high resistivity region" by J. Nishizawa and Y. Watanabe. Japanese Patent No. 205068 (Application date : December 20, 1950).

SEE ALSO : "Chemical surface treatment method of semiconductor device" by J. Nishizawa and Y. Watanabe. Japanese Patent No. 221722 (Application date : September 11, 1950).

ALSO Patent Nos. 221695, 223246, 226589, 229685, 235980, 236731 and 226859 (1950).

1950   COMPUTERS (IBM 701)   I.B.M. (U.S.A.)

The development of the IBM 701 Data Processing System began at the end of 1950. A model was operating late in 1951 and the first production machine was delivered at the end of 1952. The heart of the IBM 701 system was a 36-bit single, address, binary, parallel, synchronous processor employing vacuum-tube flip-flops and diode logic at a rate of one megapulse per second. Multiple pluggable circuit packages were used. The arithmetic registers employed a recirculating-pulse bit-storage circuit, developed for the NORC in which a combination of diode gating and pulse delay made it possible to store, shift right, or shift left with one triode per bit. Computation was governed by a single-address stored program of two 18-bit instructions per 36-bit word.

SOURCE: "The evolution of computing machines and systems" by Serrell, Aastrahan, Patterson and Pyne. Proc. IRE (May 1962) p. 1050.

SEE ALSO: "The computer from Pascal to von Neumann" by H.H.Goldstine.

1950's   THERMO-COMPRESSION BONDING   O.L.Anderson
                                          H.Christensen   (U.S.A.)
                                          and P.Andreatch

In the 1950's, O.L.Anderson, Howard Christensen and Peter Andreatch of Bell Laboratories, discovered a new bonding technique particularly useful for connecting transistors to other elements in electronic circuits. The technique, pressing the connecting wire to the transistor mounting at low heat levels, provides a firm bond without introducing undesired electrical properties and has been widely used throughout the electronics industry. It is particularly advantageous in avoiding contamination, thus achieving long life and reliability.

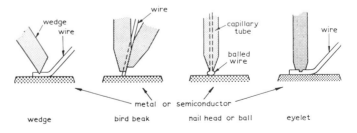

SOURCE: "Mission Communications - the story of Bell Laboratories" by Prescott C.Mabon. Published by Bell Laboratories Inc., Murray Hill, New Jersey U.S.A. (1975) p.173.

1950's   MODEM (MODulation & DEModulation)   MIT
                                                  Bell Labs. (U.S.A.)

         The modem as we know it today is a product of research done for the
defence department by MIT Lincoln Laboratory and Bell Telephone Laboratories
in the 1950's. Encouraged by advances in data processing technology, scientists
at these research centres concentrated on finding ways to improve the speed and
the accuracy of data transmission using existing facilities of the nation's vast
telephone network. A great deal of study went into understanding telephone line
parameters and deriving new modulation techniques that could work efficiently
with these line parameters.

         This early work paid off handsomely. In less than 20 years, reliable
transmission speeds jumped 700 percent, from 1200 baud in the late 1950's to
9600 bps in the early 70's.

SOURCE: "Where are MODEM's going" by J.L.Holsinger. Telecommunications
(June 1977) p. 12.

1950's   FERREED SWITCH   Bell Laboratories (U.S.A.)

         The ferreed switch, invented at Bell Laboratories in the late 1950's
comprises two or four sealed-in-glass contacts and is controlled by magnetized
wire coils. Ferreed switches, used to switch phone calls in most electronic
switching systems, are smaller, faster operating and require less power than
older switching devices.

SOURCE: "Mission Communications - the story of Bell Laboratories" by
Prescott C.Mabon. Published by Bell Laboratories Inc., Murray Hill, New
Jersey U.S.A. (1975) p. 177.

1950's   APL (A Programming Language)   K. Iverson (U.S.A.)

         APL, which stands for A Programming Language, was developed in the
late 1950's by Kenneth Iverson, a professor at Harvard University. He
invented this simple, elegant notational system to fill a need for a pithy way to
represent mathematical expressions, describe and analyze various topics in
data processing, and teach his classes.

         In 1960, Iverson joined IBM Corp. There, with the help of Adin
Falkoff and other interested researchers, an interpretive version of the
language was adapted for the System/360. In 1973, IBM released APLSV.
The appended SV stands for Shared Variables - a means whereby a number of
users may communicate information. More recently, the language has surfaced
on less expensive machines: IBM's 5100 series of business computers, Digital
Equipment Corp.'s DEC system 2020, Hewlett-Packard Co.'s 3000, and the
newly introduced Interactive Computer Systems Inc.'s System 900, to name a
few of the computers in question.

         APL's primitive functions, of which there are about 60, fall into two
categories, scalar and mixed. Scalar functions can be used with scalar
arguments and arrays on an item-by-item basis. Mixed functions apply to
arrays with various ranks and may produce results that vary from the original
arguments in rank and shape. The scalar functions can be subclassified as
monadic and dyadic, which are defined for one and two arguments, respectively.
The primitive operators, which currently number five, modify the action of
scalar dyadic functions and some mixed functions, resulting in a great number
of new functions.

         APL uses alphanumerics, Greek letters, and some uncommon mathe-
matical symbols to represent the functions and operators. These make APL
programs appear cryptic to the beginner; in fact, the language is easy to learn.
With a little practice, powerful routines can be generated with a few simple key
strokes.

1951   QUALITY CONTROL   J.M.Juran (U.S.A.)

         First edition of "Quality Control Handbook" written by J.M.Juran,
published by McGraw-Hill, New York.

         An authorative treatise on all aspects of quality control.

1951   AUTOMATIC CIRCUIT ASSEMBLY   Nat.Bureau of Standards (U.S.A.)
          "TINKERTOY" system

          In 1950 the Navy Bureau of Aeronautics asked the National Bureau
of Standards to study further automation of circuit assembly. The process
that followed in 1951 - developed by Robert Henry of the Bureau of Standards -
was dubbed Project Tinkertoy. It provided for the automatic assembly and
inspection of circuit components, and it led to the first modular package.

          The system started with individual components mounted on steatite
ceramtic wafers 7/8-inch square by 1/16-inch thick. The components were
machine-printed or mounted over printed wiring. Four to six wafers were
then automatically selected, stacked and mechanically and electrically joined
by machine-soldered riser wires, which were attached at notches along the
sides of each wafer. The resulting module generally had a tube socket on the
top wafer.

          Though this modular approach to packaging was used for production
items, it faded in the late 50s as the transistor began to replace the vacuum
tube.

Pie section modular build-up using Tinkertoy modules.
                                (Courtesy - Sanders Associates, Inc).

SOURCE: "Solid State Devices - Packaging and Materials" by R.L.Goldberg.
"Electronic Design" 24. (November 23,1972) p.126.

1951   MICROPROGRAMMING COMPUTERS   M.V.Wilkes (United
                                                            Kingdom)

          Credit for the original microprogramming concept is generally given to
Britisher M. V. Wilkes of Cambridge University's Mathematical Laboratory.
In a paper he presented at Manchester University's Computer Conference in
July 1951, Wilkes discussed "The Best Way to Design an Automatic Calculating

Machine". Wilkes' intention, ironically enough, was to simplify the design of a hardwires machine. Today microprogramming is used to replace hardwired logic altogether.

As Samir Husson notes in his book "Microprogramming: Principles and Practices", the technique attracted little attention before the 1960's because it was too expensive to implement. The first commercial microprogrammed computer was IBM's 7950, introduced in 1961. Other early appearances of microprogramming were in such machines as the IBM System/360, the RCA Spectra/70, and the Honeywell H4200.

A variety of memory technologies were used for the read-only memories needed for the microinstruction store. Among these were traditional ferrite cores, cores cut into an E-shape to create a transformer memory, and arrays of diodes or capacitors. A novel approach was taken by IBM, which devised a card capacitor ROM for several of its System/360 models.

Microprogramming came to minicomputers early in the 1970's in units such as Microdata's Micro 800, which used diode arrays as ROMs, and the later 3200, Hewlett-Packard's 2100 family, and Digital Equipment's PDP-11 line. The increasing availability of low-cost and fast semiconductor ROMs, however, has recently caused the technique to boom in popularity.

Virtually all mainframes and minicomputers today are microprogrammed. Many of the machines now use random-access memory to store the microinstructions, making it easier for the user or the manufacturer to change the machine. Also, the ease with which microprogramming allows one machine to emulate another has resulted in its widespread use in the IBM-compatible computer area.

SOURCE: "How Microprogramming started" by A. Durniak. Electronics, (November 9, 1978), p.126.

1952   ALLOYED TRANSISTOR   R.C.A.   (U.S.A.)

In 1952 an announcement was made by the Radio Corporation of America that they had successfully made transistors by heating small quantities of impurities, on the surface of germanium chips, so as to make alloy regions with the germanium. The resulting penetration of the chips was made from both sides.

SOURCE: "Semiconductors for Engineers" by D.F. Dunster. Business Books Limited, London, p.20.

1952   MICROELECTRONICS   G.W.A.Dummer   (U.K.)
        (Integrated Circuit Concept)

In a paper read at the IRE Symposium in Washington D.C. on May 5th 1952, entitled: "Electronic components in Great Britain" G.W.A.Dummer stated:-

"At this stage, I would like to take a peep into the future. With the advent of the transistor and the work in semiconductors generally, it seems now possible to envisage electronic equipment in a solid block with no connecting wires. The block may consist of layers of insulating, conducting, rectifying and amplifying materials, the electrical functions being connected directly by cutting out areas of the various layers".

SOURCE: Proc.IRE Symposium on "Progress in Quality Electronic Components" Washington D.C. (May 1952) p.19.

SEE ALSO: "A History of Microelectronics Development at the Royal Radar Establishment" by G.W.A.Dummer. "Microelectronics and Reliability" Vol.4. (1965) p.193.

"Solid Circuits" - Wireless World (November, 1957)

"The Semiconductor Story - 3, Solid Circuits - a New Concept" by K.J.Dean and S.White. Wireless World (March 1973) p.137.

"The genesis of the integrated circuit" by M.F.Wolff. IEEE Spectrum (Aug. 1976) p.45.

1952   ZONE MELTING of GERMANIUM
            AND SILICON   W. G. Pfann (U.S.A.)

                    W. G. Pfann discovered a simple method for repeating the action of
            normal melting and freezing, which avoided handling the material between each
            operation. This resulted in material of extremely high purity which was then
            grown into single crystals by the pulling technique. Pfann also developed the
            zone levelling technique, which distributes impurities uniformly through a rod.
            He grew single crystals in his zone levelling apparatus using seeding techniques.
            The combination of zone levelling and horizontal growth of single crystals has
            become the standard technique used in today's transistor manufacturing operat-
            ions.

            SOURCE: "Contributions of materials technology to semiconductor devices"
            by R. L. Petriz. Proc. IRE (May 1962) p. 1027.

            SEE ALSO: "Segregation of two solutes, with particuar reference to semi-
            conductors" by W. G. Pfann. J. Metals Vol. 4. (August 1952) p. 861.

                    "Techniques of zone melting and crystal growing" Solid State Physics
            Academic Press, New York. Vol. 4. (1957) p. 423.

1952   DARLINGTON PAIRS, DIRECT-CONNECTED
            TRANSISTOR CIRCUIT
                                                           S. Darlington (U.S.A.)

                    In one illustrative embodiment of this invention, a translating device
            comprises a pair of similar junction transistors , the collector zones of which
            are electrically integral and the base zone of one of which is tied directly to
            the emitter zone of the other. Individual connections are provided to the other
            emitter and base zones. The device constitutes an equivalent single transistor
            having emitter and collector resistances substantially equal to those of one of
            the component transistors, but having a current multiplication factor substant-
            ially greater than that of either of the components.

            SOURCE: "Semiconductor Signal Translating Device" U. S. Patent No:2, 663, 806
            Bell Telephone Laboratories(dated May 9th, 1952).

1952   DIGITAL VOLTMETER   A. Kay (U.S.A.)

                    The digital revolution started in 1952, when Andy Kay unveiled the
            first digital voltmeter. The Model 419 was crude, compared with today's DVMs.
            But both the idea and Non Linear Systems - the company formed around the
            idea - took off like a rocket. Today, almost every instrument from signal
            generators to multimeters to scopes is digitized, thanks to Andy Kay and to
            the commercial digital readout tube, introduced by Burroughs (then Haydu)
            just one year before. (Burroughs' familiar Nixie tube actually had a rival in
            its early days - the Inditron, which was developed by National Union Radio
            Corpn. and which did not survive).

            SOURCE: "Solid State Devices - Instruments" by S. Runyon. "Electronic
            Design" 24.(November 23, 1972) p. 102.

1952   NEGATIVE - FEEDBACK TONE   P. J. Baxandall (U.K.)
            CONTROL CIRCUIT

                    The circuit to be described is the outcome of a prolonged investigation
            of tone-control circuits of the continuously-adjustable type, and provides in-
            dependent control of bass and treble response by means of two potentiometers,
            without the need for switches to change over from "lift" to "cut". Unusual
            features are the wide range of control available, and the fact that a level res-
            ponse is obtained with both potentiometers at mid-setting. The treble-response
            curves are of almost constant shape, being shifted along the frequency axis
            when the control is operated, and there is practically no tendency for the curves
            to "flatten off" towards the upper limit of the audio range. The shape of the
            bass-response curves, though not constant, varies less than with most contin-
            uously-adjustable circuits.

            SOURCE: "Negative-feedback tone control" by P. J. Baxandall, Wireless World
            (October 1952) p. 402.

1952   <u>COMPUTERS (SAGE)</u>   I.B.M.   M.I.T.(Lincoln Labs)
                                                      (U.S.A.)

          Descendents of M.I.T's "Whirlwind I" and of the IBM 701, the AN/FSQ
-7 air defence computers for the SAGE system began, in 1952, as a co-operative
IBM-M.I.T. Lincoln Laboratories effort based upon previous studies and specifications by Lincoln Laboratories. SAGE, a real-time communication-based
digital computer control system, accepts radar data over phone lines, processes,
displays information for operator decisions and guides interception weapons.
The first engineering model of the computer was delivered by IBM in 1955, and
production deliveries began in June 1956.

<u>SOURCE</u>: "The evolution of computing machines and systems" by Serrell,
Astrahan, Patterson and Pyne. Proc.IRE (May 1962) p.1051.

1952   <u>TRANSISTORS (Single crystal</u>   G.K.Teal and E.Buehler (U.S.A.)
              <u>fabrication) - Silicon</u>

          Large single crystals of silicon and silicon p-n junctions were pre-
pared by Teal and Buehler by an extension of the pulling technique developed
for germanium. These crystals were used by G.L.Pearson to prepare p-n
junction diodes by the alloy method.

<u>SOURCE</u>: "Contributions of materials technology to semiconductor devices"
by R.L.Petritz. Proc.IRE (May 1962) p.1028.

<u>SEE ALSO</u>: "Growth of silicon single crystals and of single crystal p-n
junctions" by G.K.Teal and E.Buehler. Phys.Rev. Vol.87 (July 1952) p.190.

1953   <u>TRANSISTOR (Unijunction)</u> UJT   G.E.C. (U.S.A.)

          Engineers at General Electric's electronics advanced semiconductor
laboratory in Syracuse, N.Y., experimented in the early 1950s with germanium
alloy tetrode devices in search of a semiconductor that could be used at
frequencies higher than the 5-megahertz operating level of existing conventional
bipolar transitors. The tetrode structures, it was found, produced a transverse
electric field that boosted the device's cutoff frequency and lessened the semiconductor's input impedance - the limiting factor regarding frequency.

          But in 1953, while examining the waveforms on the structure's
terminals I.A.Lesk of the laboratory noticed that an oscillatory signal was
present on the tetrode's emitter. And when the collector supply was removed,
the oscillations persisted for a while. The researchers realised that they had
stumbled on a new switching-type device.

          The GE engineers added to the development tetrode models, a
double-based diode structure that was being studied because of a negative-
resistance property.

<u>SOURCE</u>: "Solid State - a switch in time" by W.R.Spofford Jr., and
R.A. Stasior. Electronics. (February 19th,1968) page 118.

1953   <u>TRANSISTOR (Surface barrier)</u>   Philco (U.S.A.)

          Advancing the early trend toward higher frequencies, Philco
developed the jet-etching technique in 1953. Here electrochemical machining
was used to fabricate the necessary thin base layers. A major product of
this process was the surface-barrier transistor, which boosted the upper
frequency limit of transistors into the megahertz region.

<u>SOURCE</u>: "Solid State Devices - The Processes" E.A.Torrero.
"Electronic Design" 24. (November 23,1972.) p.73.

1953   COMPUTERS IBM 704, 709 and 7090   I.B.M. (U.S.A.)

      The development of the IBM 704, descended from the 701, began in November 1953 and the first system was delivered in January 1956. The 704 featured higher speed, magnetic-core memory, floating point and indexing. It was followed in 1958 by the 709, featuring simultaneous read, write and compute by means of Data Synchronizer units that allowed the input-output channels to operate independently, as well as several special operations, including a table look--up instruction and indirect addressing. A 32,768-word core memory was installed on a 704 in April, 1957.

      The IBM 7090, the first units of which were delivered in 1959, is a transistorised system compatible with the 709. About 5-times faster than the 709 on typical problems, the 7090 incorporates a 2.18 -μsec core memory and improved magnetic-tape units.

SOURCE: "The evolution of computing machines and systems" by Serrell, Astrahan, Patterson and Pyne. Proc. IRE (May 1962) p.1052.

SEE ALSO: "The computer from Pascal to von Neumann" by H.H. Goldstine. Princeton Univ. Press (1972).

1953   AUTOMATIC ASSEMBLY SYSTEMS   Autofab (General Mills)   (U.S.A.)
                                                 IBM Machine (IBM)   (U.S.A.)
                                                 United Shoe Machinery Co.   (U.S.A.)
                                                 GE/Signal Corps.   (U.S.A.)
                                                 Mini-Mech (Melpar)   (U.S.A.)

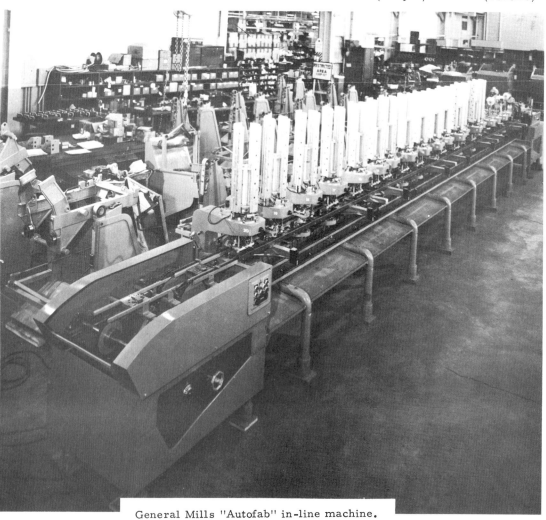

General Mills "Autofab" in-line machine.

In attempting to develop standards which would allow unrestricted electronics design and production processes sufficiently flexible to permit rapid transition from design to manufacturing, the following conclusions are reached:-

1. The use of a printed-circuit board is basic to practically all automatic approaches.

2. A generic grid pattern to govern circuit layout must be used in order to prevent obsolescence of tooling (0.1 and 0.025 in. are commonly accepted).

3. In conjunction with the standard grid pattern, standard mounting dimensions for all components and parts must be used.

The use of single-head component-insertion machines will satisfy these fundamental requirements. The flexibility of such a machine permits the insertion of a variety of components by recycling the printed-circuit modules through the unit for each component. When higher volumes are obtained, several machines can be utilized. By this evolutionary method one is able to create any degree of automation which the manufacturer wishes to attain. It is possible to stop with the production of the printed circuit and to use one or several component-attaching or inserting machines. Dip-soldering can be mechanised or performed manually. It would be impractical to consider taking the giant step into completely automatic production at the beginning of the factory's use of automation principles. Even for mass producers, it is necessary to develop specific procedures for particular requirements documented by many years of experience in the step-by-step evolution toward a mechanised operation.

SOURCE: "Electronic equipment design and construction" by G.W.A. Dummer, C. Brunetti and L.K. Lee. McGraw-Hill, New York (1961) p.185-186.

SEE ALSO: Proceedings of Symposium on Automatic Production of Electronic Equipment, sponsored by Stanford Research Institute and the U.S. Air Force (April, 1954.)

| 1953 | MASER (Microwave amplification by stimulated emission of radiation) | C.H. Townes and J. Weber | (U.S.A.) |
| | | N.G. Basov and A.M. Prokhorov | (U.S.S.R.) |

The first clear recognition of the possibility of amplification of electromagnetic radiation by stimulated emission seems to have been by a Russian, Fabrikant, who filed a patent in 1951 (although it was not published until 1959) and who had discussed various aspects of his thesis of 1940. However, his attempts to produce optical amplification in caesium were unsuccessful.

The first statement in the open literature about amplification was by Weber in 1953, followed by the detailed proposals of Basov and Prokhorov for a beam-type maser in 1954. However, the real excitement was caused by the short article of Gordon, Zeiger and Townes, in the same year, announcing the operation of the first maser using ammonia. Townes had conceived the required experimental arrangement three years earlier, based on his experience in microwave spectroscopy. In the years immediately following many other techniques were studied, but the only one to give any degree of practical success was the three-level maser of Bloembergen which resulted in the ruby maser amplifier.

SOURCE: "Lasers and optical electronics" by W.A. Gambling. The Radio & Electronic Engineer. Vol. 45 No:10 (October 1975) p.538.

SEE ALSO: "Mekhanizm izlucheniya gazovogo razryada" by V.A. Fabrikant in Elektronnye i ionnye pribory (Electron and ion devices); Trudy Vsesoyuznogo Elektrotekhnicheskogo Instituta (Proceedings of the All-Union Electrotechnical Institute), Vol. 41, (1940); pp.236.

"Evolution of masers and lasers" by B.A. Lengyel. V.A. Fabrikant. Am. J. Phys. 34. p.903 (1966)

"Amplification of microwave radiation by substances not in thermal equilibrium"by J. Weber. IRE Trans. on Electron Devices. No:PGED-3, p. 1-4 (June 1953)

"Application of molecular beams to radio spectroscopic studies of rotation spectra of molecules" by N. G. Basov and A. M. Prokhorov. J. Exp. Theor. Phys. (U.S.S.R.) 27, pp. 431-8 (1954)

"Molecular microwave oscillator and new hyperfine structure in the microwave spectrum of NH_3" by J. P. Gordon, H. J. Zeiger and C. H. Townes. Phys. Rev. 95. pp. 282-4 (1st July) 1954).

"Proposals for a new type solid-state maser" by N. Bloembergen Phys. Rev. 104. pp. 324-7, (15th October, 1956)

"Forgotten inventor emerges from epic patent battle with claim to laser" Science Vol. 198 (28th October, 1977) p. 379.

1953   CONNECTION TECHNIQUES   R. F. Mallina et al (U.S.A.)
           (Wire wrapped joints)

The mechanical basis of the wire wrapped joint was investigated very extensively by workers at Bell Telephone Laboratories 20 years ago. Their very full analysis of the joining system is still considered to be essentially correct. The work includes photoelastic observations on a wrapped joint model to investigate strain patterns produced by wrapping, and the study of stress relaxation as a function of time and temperature.

The wire wrapped joint consists of a wire which is tightly wrapped around a sharp cornered terminal. Sufficient deformation is engendered in the many notches created by the terminal in the wrapping wire to create metal to metal interfaces with a high level of integrity. The wrapping wire, which is bent several times during wrapping before final positioning in the wrap, is under a high level of tensile stress during wrapping. The tensile strain which is caused remains in the wire after wrapping because the stretched wire is locked by the notches formed in it.

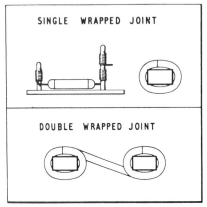
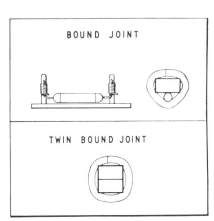

TYPICAL   MATERIALS

WRAPPING WIRE   POST
14-32 S.W.G. SINGLE CORE TINNED COPPER,   BRASS (NICKEL FLASHED OR ELECTRO TINNED)
COPPER/NICKEL IRON (COPPER CORED)   NICKEL SILVER — MONEL METAL — PHOSPHOR BRONZE

SOURCE: "Wire wrapped joints - a review" by P. M. A. Sollars. Electro-component Science and Technology. (1974) Vol. 1. p. 17.

SEE ALSO: "Solderless wrapped connections" The Bell System Technical Journal (May 1953)   Introduction: J. W. McRae

                    Part 1   Structure & Tools - R. F. Mallina

                    Part 2   Necessary conditions for obtaining a
                                 permanent connection - W. P. Mason and
                                 T. F. Osmer.

                    Part 3   Evaluation & Performance tests -
                                 R. H. Van Horn.

1953   TRANSI<u>STORS (Floating zone refining</u>   P.H.Keck, R.Emeis and   (U.S.A.)
        <u>of silicon)</u>   H.C.Theurer

            An improved silicon purification technique was developed which pro-
    duced material of sufficient quality that alloy silicon transistors could be fab-
    ricated with good yields. This was a novel variation of zone refining called
    "floating zone refining" developed by P.H.Keck and independently by R.Emeis
    and H.C.Theurer. This operation employs a vertical system and uses surface
    tension to support a stable liquid zone formed by induction heating. Hence, the
    crucible is completely eliminated. Silicon with thousands of ohm-cm resistivity
    and minority carrier lifetimes of greater than 100 μsec can be produced by this
    method.

    SOURCE: "Contributions of materials technology to semiconductor devices"
    by R.L.Petritz. Proc.IRE (May 1962) p.1028.

    SEE ALSO: "Crystallisation of silicon from a floating liquid zone" by P.H.Keck
    and M.J.E.Golay. Phys.Rev. Vol.89 (March 1953) p.1297.

            "Growing single crystals without a crucible" Z.Naturforsch. Vol.9A
    (January 1954) p.67.

            "Removal of boron from silicon by hydrogen water vapour treatment"
    J.Metals Vol.8. (October 1956) p.1316.

1954   TRAN<u>SISTOR (Interdigitated)</u> N.H.Fletcher (U.S.A.)

            The father of the interdigitated transistor is N.H.Fletcher, an
    engineer with Transistor Products Inc. When, in 1954, he hit upon the idea of
    elongated emitter areas, Fletcher was seeking a means to increase the power
    handling capability of devices, not a way to boost their cut-off frequency levels.

            His discoveries were applied by other firms to most transistor types
    over the next decade, but his own company realised few benefits from his work.

    SOURCE: "Solid state - fingers in the die" by J.E.Tatum. Electronics.
    (February 19,1968) p.94.

    SEE ALSO: "Some aspects of the design of power transistors" by N.H.
    Fletcher. Proc.IRE (May 1955) p.551.

1954   TRANSISTOR RADIO SET   Regency (U.S.A.)

            In 1954 the first transistor radio, the Regency, appeared on the
    market. Although not a commercial success, it introduced the transistor
    into the consumer market and gave transistor makers the impetus they
    needed to develop mass production techniques. That, coupled with an
    awakening of interest by the military, increased transistor sales meteor-
    ically in the mid-fifties.

    SOURCE: "Silicon, Germanium & Silver - the transistor's 25th anniversay"
    by C.P.Kocher. The Electronic Engineer,(November 1972) p.30.

1954   SOLAR BATTERY   D.M.Chapin, C.S.Fuller
                               and G.L.Pearson   (U.S.A.)

            As an outgrowth of work on transistors, Bell Laboratories scientists
    D.M.Chapin, C.S.Fuller and G.L.Pearson in 1954 invented the silicon solar
    battery - an efficient device for converting sunlight directly into electricity.
    Arrays of these devices are used to power satellites and as energy sources
    for other uses.

    SOURCE: "Mission Communications - the story of Bell Laboratories" by
    Prescott C.Mabon. Published by Bell Laboratories Inc., Murray Hill,
    New Jersey, U.S.A. (1975) p.172.

1955   CRYOTRON   D. Buck (U.S.A.)

            A super-conducting switching element was first examined by de Haas
and Casimir-Jonker in 1935; however, it was not until 1955 that Buck demonstrated a practical device which he called the cryotron. The basic principle of
the cryotron depends on the existence of a critical magnetic field above which
the superconducting metal becomes a normal conductor. The original cryotron
utilized a small tantalum rod wound with a niobium wire. At liquid helium
temperatures the tantalum wire has a critical field of the order of several
hundred gauss, whereas that of niobium is of the order of 2000 gauss. Con-
squently, when the niobium wire is pulsed with a suitable current, the magnetic field that it creates is sufficient to destroy the superconductivity in the
tantalum but not in itself. The current in the niobium wire can be smaller than
that in the tantalum wire so that a small current can control a larger one, thus
producing a current gain in the device.

SOURCE: "Solid State Devices other than Semiconductors" by B. Lax and
J.G. Mavroides. Proc IRE (May 1962) p. 1016.

SEE ALSO: "The cryotron - a superconductive computer component" by D. Buck
Proc. IRE Vol. 44 (April 1956) p. 482.

1955   INFRA-RED EMISSION FROM GALLIUM
            ARSENIDE SEMICONDUCTORS   R. Braunstein (U.S.A.)

            Radiation produced by carrier injection has been observed from GaSb,
GaAs, InP and the Ge;Si alloys at room temperature and 77°K. The spectral
distributions of the radiation are maximum at energies close to the best estim-
ates of the band gaps of these materials; consequently, the evidence is that the
radiation is due to the direct recombination of electron-hole pairs.

SOURCE: "Radiative transitions in semiconductors" by R. Braunstein. Phys.
Rev. Vol. 99. (1955) p. 1892.

1956   DIFFUSION PROCESS   C.S. Fuller and H. Reis (U.S.A.)

            The next major advance in device technology was the diffusion process.
Research on the diffusion of III-V impurities into germanium and silicon by
Fuller at the BTL, and by Dunlap at GE laid the foundation for transistor fabrication using diffusion as a key process step. The BTL was the first to fully
integrate these results into germanium and silicon transistors.

            Diffusion techniques have proved to be one of the best controlled meth-
ods for preparing p-n junctions. Because the common doping impurities diffuse
very slowly in semiconductors at rates which can be varied by adjusting temperatures, close control and reproducibility of the impurity distributions can
be achieved. Hence, control over the electrical parameters of the resulting
devices may be maintained. The ability to form base regions only a fraction of
a micron thick allows very high-frequency transistors to be fabricated.

SOURCE: "Contributions of materials technology to semiconductor devices"
by R.L. Petritz. Proc. IRE (May 1962) p. 1029.

SEE ALSO: "Diffusion processes in germanium and silicon" by H. Reis and
C.S. Fuller. Chaper 6 of "Control of composition in semiconductors by freezing
methods" Ed. N.B. Hannay. Reinhold Pub. Corp. New York (1959).

1956   SOLID ELECTROLYTE CAPACITOR   D.A. McLean
                                               F.S. Power   (U.S.A.)

            From the time transistors began to be produced commercially, the need
for a solid electrolytic capacitor as a coupling capacitor or a bypass capacitor
for electronic equipment has increased. Since McLean and Power announced the
development of the tantalum solid electrolytic capacitor in 1956, studies of this
capacitor began to appear with great frequency in literature through the world
and applications of the solid capacitor gradually began to spread.

SOURCE: "Miniaturised aluminium solid electrolyte capacitors using a highly effective enlargement technique" by K. Hirata and T. Yamasaki. IEEE Trans. on Parts, Hybrids & Packaging. Vol. PHP-12 No. 3 (Sept. 1976) p. 217.

SEE ALSO: "Tantalum solid electrolyte capacitor" by D. A. McLean and F. S. Power. Proc. IRE Vol. 44 (July 1956) p. 872.

1956   VALVES - VAPOUR COOLING   C. Beutheret (France)

The first vapour-cooled tubes were made by Beutheret who used an anode with teeth approximately 10 mm square tapering to 5 mm square over 20 mm protruding from the surface. The object of the teeth was to stabilise the anode temperature and prevent a sudden catastrophic increase in anode temperature known as calefraction.

SOURCE: "Electronic Engineer's Reference Book" Newnes-Butterworth London (1976) Chapter 7. p. 7-47.

SEE ALSO: "The Vaportron Technique" by C. Beutheret. Rev. Tech. Thomson-CSF Vol. 24. (1956)

1956   "FLOWSOLDERING" OF PRINTED
        CIRCUITS   Fry's Metal Foundries Ltd. (U.K.)

In the Flowsolder dipping unit, developed by Fry's Metal Foundries Ltd, which avoids some of the difficulties inherent in the conventional flat dip-soldering of printed circuits, a stationary wave of molten solder is created by pumping the metal upwards through a rectangular nozzle and the pre-fluxed circuit panels are passed through the crest of the wave. It is claimed that this unit, which is used with specially developed fluxes, facilitates the soldering of printed circuits, free from faulty joints or bridging.

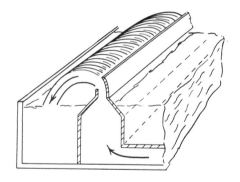 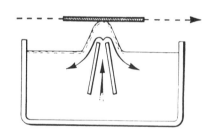

TWO SYSTEMS OF WAVE SOLDERING

SOURCE: "Flowsolder Method of Soldering Printed Circuits" R. Strauss & A. F. C. Barnes. Electronic Engineering. Vol. 28. No. 345 (Nov. 1956) pp. 494-496.

1956   SEMICONDUCTOR DIODE
        JUNCTION CAPACITOR   L. J. Giacoletto and J. O'Connell (U.S.A.)

A semiconductor junction when biased in the reverse (non-conducting) direction is a capacitance which can be varied by the bias voltage. When biased in this direction the mobile charge carriers are moved away from the junction, leaving uncompensated fixed charges in a region near the junction. The width, and hence the electrical charge of this space-charge layer, depends on the applied voltage, thus giving rise to a junction transition capacitance.

SOURCE: "History, Present Status and Future Developments of Electronic Components" by P. S. Darnell, IRE Transactions on Component Parts. (September 1958)   p. 128.

SEE ALSO: "A variable capacitor germanium junction diode for U.H.F." by L. J. Giacoletto and J. O'Connell. R. C. A. Review. Vol. 17. (March 1956) p. 68.

1956   TRANSATLANTIC TELEPHONE CABLE   (TAT-1)   U.K./U.S.A.

                    The 25th September, 1956 was an auspicious date for international tele-
              communications being the day that the first transatlantic telephone cable (TAT-1)
              entered service.

                    The history of the Atlantic cables started in 1858 with the laying of the
              first telegraph cable which didn't have a very long life but did prove the
              feasibility of the operation. By 1956 there were 28 transatlantic telegraph
              cables.

                    As far as a telephone service was concerned the main problem lay
              with the fact that repeaters were necessary to amplify the already weak signals
              on their long journey. By 1920 repeatered telephone links were in the course
              of preliminary experiment. In 1946 a repeatered link was laid between the
              United Kingdom and mainland Europe. With this impetus development work was
              brought to the point where it was felt that the repeaters would exhibit the 20-
              year life required of them for deep water service.

                    Consequently, TAT-1 runs from Oban in Scotland under the Atlantic to
              Clarenville in Newfoundland, using 51 American made repeaters. Then the
              line goes overland to Terenceville where again it takes to the water for its
              journey to Sydney Mines in Nova Scotia. In the latter run the cable uses 16
              British made repeaters, the first two of which are buried in Newfoundland.

                    For the cable's 36 telephone channels two cables are used, one for
              each direction, in the transatlantic section while from Clarenville the cable
              becomes a 60-circuit single cable system, connecting Newfoundland to Canada.

                    In the main section one telephone channel had been reserved as a
              United Kingdom - Canada telegraph link. From Sydney Mines, the circuit takes
              a short radio hop to Spruce Lake where it is divided to allow circuits to be
              directed to Montreal and New York.

                    Although in 1956 the 36-channel cable was hailed as an historical
              achievement only 20 years later the latest in the family of TAT cables, TAT-6,
              sports over 100 times the capacity while TAT-7 and TAT-8 are already being
              considered.

              SOURCE: "TAT-1" 20 years old. ITU Telecommunication Journal. Vol.43 -
              XII/1976 p.734.

1956   MAGNETIC MATERIAL -   F. Bertaut and
              YTTRIUM IRON GARNET   F. Forrat   (France)

                    The crystallographic structure of yttrium iron garnet (YIG) was
              discovered by Bertaut and Forrat in 1956 and very soon afterwards large bulk
              crystals (several centimetres in one dimension) began to be grown by the
              molten flux technique. Yttrium iron garnet was found to be cubic, to be
              essentially an insulator, to have a saturation magnetization of 1750 gauss, and,
              unlike the previously available ferrite materials, to have a ferrimagnetic
              resonance linewidth of the order of 1 oersted at 10 GHz.

              SOURCE: "Epitaxial magnetic garnets" by J.H. Collins and A.B. Smith.
              The Radio and Electronic Engineer. Vol.45. No:12 (Dec.1975) p.707.

              SEE ALSO: "Structure of ferrimagnetic ferrites of rare earths" by F. Bertaut
              and F. Forrat. C.R. Acad. Sci. Paris. 242, p.382 (1956)

1956   TRANSISTORIZED COMPUTER   Bell Laboratories (U.S.A.)

                    Bell Telephone Laboratories in New York, the place where the transistor
              was invented in 1947, builds the Leprachaun, the first experimental transis-
              torized computer. The on-off switching transistor fathers a new breed of more
              reliable, more economical machines. IBM, Philco and General Electric
              quickly follow suit with 'second generation' computers.

              SOURCE :   "The Timetable of Technology", published by Michael Joseph,
              London, and Marshall Editions, London (November 1982), p.153.

1956   RADIO PAGING   Multitone (U.K.)

        Concurrently with the development of two-way radio communication, there has been a remarkable development of radio-paging equipment and services. The first paging systems were established in the mid-fifties using a magnetic loop around the building to be served and operated on very low frequencies around 70 kHz. One of the first of these systems was installed by the Multitone Company at St. Thomas's Hospital, London in 1956.

        Later the technique changed to v.h.f. radiating system using frequencies in the 27,150 and 450 MHz bands. Development is now very widespread with over 2000 systems and 100,000 paging receivers in use in Britain alone. The recievers involved are very small, weighing only a few ounces.

SOURCE: "Fifty years of mobile radio" by J.R. Brinkley. The Radio and Electronic Engineer. Vol. 45. No:10 (Oct. 1975) p. 556.

1957   FULL FREQUENCY-RANGE ELECTROSTATIC LOUDSPEAKER

                                            P. J. Walker   (United Kingdom)

        Much work was done on electrostatic loudspeakers in the 1920's and 30's, but all early workers seem to have made the mistake of assuming the basic principle to be necessarily that the force between two capacitor plates is proportional to the square of the voltage between them. Inventors sought to overcome the non-linearity distortion inherent in this square-law relationship by employing push-pull arrangements, in which the diaphragm, at a high and constant DC voltage, was placed between a pair of fixed perforated electrodes across which the audio programme voltage was applied. But linearity was only achieved if the vibration amplitude was kept small compared with the plate spacing, and this in practice confined all these early electrostatic loudspeakers to high audio frequencies only.

        Professor F.V. Hunt of Harvard University seems to have been the first to appreciate that the above non-linearity could theoretically be totally removed, even for very large diaphragm amplitudes, if the electrically-conductive diaphragm had a constant electric charge on it instead of being held at a constant voltage. The principle is then simply that the force on a constant charge is proportional to the product of the charge and the strength of the electric field in which it is situated. There is a tacit assumption, however, that all parts of the diaphragm move equally. This does not happen in practice, causing charge to move about on the diaphragm surface, thus preventing each part of the diaphragm from operating under the desired constant-charge condition that gives low distortion.

        P.J. Walker and D.T.N. Williamson discovered that this latter major cause of distortion could be removed by making the diaphragm of very thin plastic film, treated to be sufficiently conductive to allow it to be charged up in a reasonable time, but not sufficiently to allow any significant moving about of charge during a low-frequency audio cycle. It then became possible to achieve large acoustic outputs at bass frequencies with very low distortion. The plastic diaphragm also greatly reduced the danger of spark damage occurring at high signal levels, and led ultimately, after many other engineering problems had been solved, to the marketing of the first successful full-frequency-range electrostatic loudspeaker in 1957.

        Recent improvements by P.J. Walker have been largely concerned with making the polar radiation characteristic of the loudspeaker vary less with frequency than in the earlier designs, and in a smooth and controlled manner. The electrode area is divided into a number of annular sections, the signals being fed to the outer sections via a delay line. The wavefront radiated is then as if from a point source located behind the loudspeaker, but since the actual radiating area is quite large, high volume levels can be produced.

SOURCE: Communication from P.J. Baxandall, Malvern (22nd July 1982).

SEE ALSO: Walker, P.J. "Wide Range Electrostatic Loudspeakers", Wireless World, May, June, August, 1955.

Walker, P.J. "New Developments in Electrostatic Loudspeakers",
J. Audio Eng. Soc., November 1980.

Walker, P.J. and Williamson, D.T.N., "Improvements Relating to Electrostatic Loudspeakers", British Patent No. 815,978. (Application dated
20th July 1954. Complete filed 19th October 1955. Complete published
8th July 1959).

Hunt, F.V., "Electroacoustics", Chapter 6. Harvard University Press/
John Wiley (1954).

1957   PLUMBICON TV CAMERA TUBE   Philips   (Holland)

Philips introduce the Plumbicon TV camera tube, a greatly improved
version of its predecessor, the Vidicon. In Britain the BBC bases all its plans
for the development of colour TV on the new tube, which soon becomes universal
in colour TV design.

SOURCE : "The Timetable of Technology", published by Michael Joseph,
London, and Marshall Editions, London (November 1982), p. 157.

1957   PLATED-WIRE MEMORIES   U.F.Gianole (U.S.A.)

The plated-wire memory uses the principle of the direction of magnetization in a material to store digital information. The original concept of
the wire memory was invented in 1957 by U.F.Gianole of Bell Laboratories.
Plated-wire memories require no standby power, are non-volatile, inexpensive
to manufacture and will work in a high electrical noise environment.

SOURCE: "Mission Communications - the story of Bell Laboratories" by
Prescott C.Mabon. Published by Bell Laboratories Inc., Murray Hill,
New Jersey U.S.A (1975) p.179.

1957   RESISTORS - (Nickel-chromium   R.H.Alderton
           thin film)   and F.Ashworth   (U.K.)

It is well known that nichrome is the material most used today in thin-
film resistors. Alderton and Ashworth stressed the importance of the following
parameters: the source temperature, the degree of vacuum maintaining in the
system, and the temperature of the receiving surface during deposition from
the vapour phase. They state that stable films can only be made if the substrate
temperature is greater than $350^{\circ}C$ and the vacuum in the system better than
10^{-4} torr. They also stated that the maximum surface resistivity that produced
stable films was $300\Omega/\square$. These results are still used today as a guide in the
production of nichrome films. Alderton and Ashworth measured the reistivity
as a function of thickness and obtained a temperature coefficient of resistance
from 100-200ppm/$^{\circ}C$.

SOURCE: "Resistive thin films and thin film resistors - History, Science and
Technology" by J.A.Bennett. Electronic Components (Sept.1964) p.748.

SEE ALSO: "Vacuum deposited films of a nickel-chromium alloy" by R.H.
Alderton and F.Ashworth. Brit. Journ. Applied Physics. Vol.8 (1957) p.205.

1957   SPUTNIK 1 Satellite   (U.S.S.R)

Launched 4th October,1957. First artificial satellite. Study of ionosphere,
radio wave propagation. Batteries. Transmitted for 21 days. Decayed on
4th January,1958.

SOURCE: Table of Artificial Satellites Launched Between 1957 and 1976.
International Telecommunication Union. Geneva (1977)

1957   SPUTNIK 2 Satellite   (U.S.S.R)

Launched 3rd November,1957. Study of ultraviolet rays and Xrays from the
sun. Study of cosmic rays. Medico-biological study of the dog Laika. Trans-

mitted for seven days. Decayed on 14th April, 1958.

SOURCE: Table of Artificial Satellites Launched Between 1957 and 1976 International Telecommunication Union. Geneva (1977)

1957   TRANSISTORS (Oxide masking process)   C.J.Frosch (U.S.A.)

Another important technological advance in this period was the development of oxide masking for silicon by C.J.Frosch of BTL. He observed that a thermally grown oxide on silicon impeded the diffusion of certain impurities, including boron and phosphorus. This technique, coupled with photographic masking against etching, provides a powerful tool for silicon processing.

SOURCE: Contributions of materials technology to semiconductor devices" by R.L.Petritz. Proc.IRE (May 1962) p.1030.

SEE ALSO: "Surface protection and selective masking during diffusion in silicon" by C.J.Frosch and L.Derick. J.Electrochem.Soc.Vol. 104 (Sept.1957) p.547.

1958   TECHNETRONField Effect (FET)   Stanislas Teszner (France)

The first commercial FET was produced in France in 1958, by Stanislas Teszner, a Polish scientist employed by CFTH, a General Electric Company affiliate. Called the Technetron, Teszner's device was a germanium alloy semiconductor. It had a transonductance of 80 micromhos, a pinchoff of 35 volts, a gate leakage current of 4 microamps, and a low gate capacitance of 0.9 picofarads. The low trans-conductance and high leakage severely limited its applications. But the high pinchoff voltage was closer to the operating levels of some tubes, and its gate capacitance permitted it to be operated at a few megahertz.

SOURCE: "Solid state - an old-timer comes of age" by J.M.Cohen. Electronics. (February 19th, 1968. p.123.

1958   PEDESTAL PULLING OF SILICON   W.C.Dash (U.S.A.)

The "pedestal" method was devised to avoid oxygen contamination and at the same time achieve the high perfection attainable with the Czochralski technique. In this method the melt is an inductively heated mound held on top of a solid silicon support by surface tension and electromagnetic levitation. The support is sectored to inhibit electromagnetic coupling to the pedestal. A seed is inserted and the growing crystal is withdrawn at a rate which may vary from 3 cm per minute at the start to 3 or 4 mm per minute during the major part of its growth.

SOURCE: "Growth of silicon crystals free from dislocations" by W.C.Dash. J.App.Physics. Vol.30. No:4 (April 1959) p.459.

SEE ALSO: W.C.Dash J.App.Phys. Vol.29 (1958) p.736.

1958   TUNNEL DIODE   L.Esaki (Japan) now (U.S.A.)

First in chronological sequence came the tunnel diode, first described by Esaki in 1958. Again with hindsight, what a beautifully simple idea - to form a p-n junction between two such highly-doped regions that, in equilibrium, the continuity of the Fermi level across the junction would result in an energy barrier to the flow of carriers in the "forward" direction. The device thus presents a high impedance at low forward bias, progresses through a region of negative impedance and then into a fairly normal "forward" region of positive resistance.

SOURCE: "Semiconductor Devices - portrait of a technological explosion" by I.M.Mackintosh. The Radio and Electronic Engineer. Vol.45. No:10 (Oct.1975) p.517.

SEE ALSO: "New phenomenon in narrow germanium p-n junctions"
by L. Esaki. Phys. Rev. Vol. 109 (1958) p. 603.

1958   FIELD-EFFECT VARISTOR   Bell Laboratories (U.S.A.)

        This device, closely related in principle to the field-effect
transistor, has a constant-current feature which makes it ideally suited for a
current regulator in circuits where either the load or supply voltage varies
over wide limits. It can also be used as a current limiter or pulse shaper.
Its ac impedance is very high, making it useful as a coupling choke or as an
ac switch. This device is based on the field effect principles developed by
Shockley, Dacey and Ross.

SOURCE: "History, Present Status and Future Development of Electronic
Components" by P.S. Darnell. IRE Transactions on Components Parts.
(September, 1958) p. 128.

SEE ALSO: "A Field Effect Varistor" Bell Labs. Record. Vol. 36.
(April, 1958.) p. 150.

1958   VIDEO TAPE RECORDER   Ampex   (U.S.A.)

        The first battery of 'video'tape recorders, a system called Ampex,
was installed in the largest American television studios early in 1958. This
system used tape moving at a speed of 200 inches per second but only half an
inch wide; the recording being done on three tracks, two for storing the video
signals and one for sound. A special machine for the cutting and editing of the
tape had to be devised as it could be edited visually like cine film. Today, the
majority of television programmes are recorded on videotape before transmission, in black-and-white as well as in colour; cine-film material shot for
television can also easily be transferred on to videotape. Whether live, film
or video, the viewer cannot detect the difference; the quality is equally high.

        An important development of video recording is the 'canned' television
programme for homes and schools, There are three rival systems. Two work
with 'cassettes', which are inserted into a special replay unit plugged into the
television set; one system uses Ampex tape, the other 8 mm. film with two
parallel tracks for sound and vision, electronically recorded. The third system
a British-German venture, works with a fast-rotating (1, 500 r. p. m.) disc and
a pickup, providing a monochrome or colour programme of up to 12 minutes.

SOURCE: "A History of Invention" by E. Larsen.   J. M. Dent & Sons, London.
and Roy Publishers, New York (1971) p. 330.

1958   EXPLORER 1 Satellite   (U.S.A.)

Launched 1st February 1958. Measurement of cosmic radiation and micro-
meteorites. Discovery of the Van Allen radiation belt. Batteries. Transmitted
up to 23rd May 1958. Decayed on 31st March, 1970.

SOURCE: Table of Artificial Satellites Launched Between 1957 and 1976.
International Telecommunication Union. Geneva (1977)

1958   VANGUARD-1 Satellite   (U.S.A.)

Launched 7th March, 1958. Part of the International Geophysical Year Programme. Permitted the discovery of the "pear-shaped" earth. Studied the earth
and measured the "far out" density of the atmosphere. Batteries and solar
cells. Transmitted until 12th February, 1965.

SOURCE: Table of Artificial Satellites Launched Between 1957 and 1976.
International Telecommunication Union. Geneva (1977)

1958   AUTOMATIC CIRCUIT ASSEMBLY   U.S. Army Signal Corps. (U.S.A.)
         "Micro-module" system.

    By 1957 the goal for packaging had shifted from automation to
miniaturization. Working with the Army Signal Corps., RCA suggested an
approach that was similar to Tinkertoy's but with smaller wafers. Using
wafers 310 mils square, spaced 10 mils apart, RCA encapsulated the
assembled module with an epoxy resin to increase mechanical strength and
provide environmental protection.

    With RCA as the prime contractor for an $18-million contract,
the Signal Corps promoted micromodule as a standard package. A Signal
Corps team headed by Daniel Elders, Stan Danko and Weldon Lane, established
a continuing development programme for micromodule.

    The micromodule approach combined high packaging density,
machine assembly and modular design. It was the first attempt at functional
modular replacement, where the entire module was treated as a single component. The programme established a compact universal packaging system
using standard-shaped parts. But just as micromodule was gaining popularity
in the early 60s, the IC deflated its chance of achieving sufficient volume for
a competitive price.

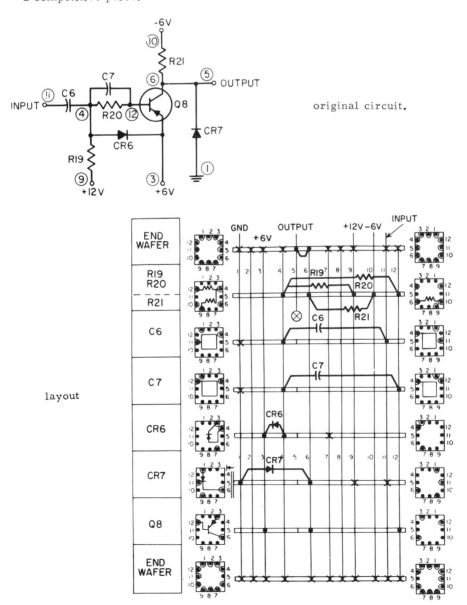

original circuit.

layout

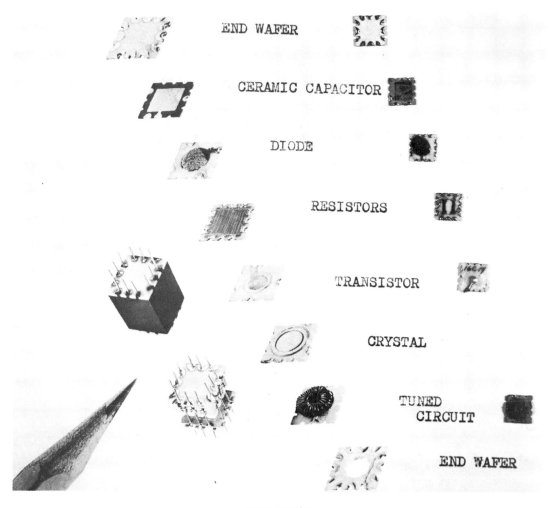

END WAFER

CERAMIC CAPACITOR

DIODE

RESISTORS

TRANSISTOR

CRYSTAL

TUNED
CIRCUIT

END WAFER

components

SOURCE: "Solid State Devices - Packaging and Materials" by R. L. Goldberg.
"Electronic Design" 24 .(November 23, 1972)   p. 126/7

SEE ALSO: Signal Corps.   Contract DA-36-039-SC-76968 RCA Camden. N. J.

1958   PIONEER-1 Satellite   (U. S. A.)

Launched 11th October, 1958.   Moon probe.   Failed to reach the moon. Sent data
for 43 hours. Batteries.

Decayed on 12th October, 1958 coming back to earth and burning in the atmosphere.

SOURCE: Table of Artificial Satellites Launched Between 1957 and 1976.
International Telecommunication Union.   Geneva (1977)

1958   SCORE Satellite   (U. S. A.)

Launched 18th December, 1958. Signal Communication by Orbiting Relay Equipment. First communication satellite. Transmitted taped messages for 13 days.

Decayed on 21st January, 1959.

SOURCE: Table of Artificial Satellites Launched Between 1957 and 1976.
International Telecommunication Union.   Geneva (1977)

1958   MOSSBAUER EFFECT   R.L.Mossbauer   (Germany)

        The Mossbauer effect is the phenomenon of recoiless resonance
fluorescence of gamma rays from nuclei bound in solids. The extreme sharp-
ness of the recoiless gamma transitions and the relative ease and accuracy in
observing small energy differences made the Mossbauer effect an important
tool in nuclear physics, solid state physics etc.

SOURCE: The Encyclopaedia of Physics. (2nd Edition) Editor:R.M.Besancon.
Von Nostrand. Reinhold Ⓒ Litton Educational Pub.Inc.

1958   LASER (Light amplification by   A.L.Schalow and C.H.Townes (U.S.A.)
              stimulated emission of radiation).

        Because of the great interest aroused by masers it was not until
1957 that further serious attention was given to the idea of producing an optical
version of the maser.

        In their classic article of 1958 on the principles of laser action
Schalow and Townes suggested potassium vapour as a possible medium and
much effort was devoted to it but with no success. The reasons for this
failure were rather puzzling, especially as other works later found caesium
vapour to behave as predicted. Another medium under consideration was ruby
(Cr^{3+} in Al_2O_3) although an internal report at Bell Telephone Laboratories
concluded that the existing material was much too poor to give any hope of
success and the experts of the time expected that the first laser would be
based on a gas. Great was the surprise and general jubilation, therefore, when
Maiman, who had persevered with ruby, achieved laser action in 1960. Maiman's
own jubilation was short-lived as the manuscript which he prepared announcing
his remarkable result was rejected by Physical Review Letters and an historic
scoop of scientific journalism was achieved by the journals "Nature" and
"British Communications and Electronics" which carried the first announcement
in the established scientific literature.

        Some months later the helium/neon laser was successfully operated
and there followed over the next few years a tremendous explosion of publications on laser transitions in hundreds of different materials and on the
properties of laser devices.

SOURCE: "Lasers and optical electronics" by W.A.Gambling. The Radio and
Electronic Engineer. Vol.45. No:10 (Oct.1975) p.539.

SEE ALSO: "Infrared and optical masers" by A.L.Schalow and C.H.Townes
Phys.Rev. Vol.112 (Dec.15,1958) p.1940.

        "Infrared and optical masers" in Quantum Electronics. C.H.Townes Ed.
Columbia Univ.Press. New York 1960.

        "Stimulated optical production in ruby" by T.H.Maiman. Nature.
Vol.187 (6th August,1960) p.493.

1959   PLANAR PROCESS   J.A.Hoerni (U.S.A.)

        At Fairchild, Dr.Jean A.Hoerni, a physicist, was trying to develop
a family of double-diffused silicon mesa transistors. But instead of mounting
the base layer on top of the collector, the traditional mesa approach, Hoerni
diffused it down into the collector and protected the base-collector junction on
the top surface with a layer of boron-and-phosphorous diffused silicon oxide.
This first planar transistor was less brittle than the mesa and far more
reliable - dust or other foreign matter could not contaminate the protected
p-n junction. In 1959 Fairchild started marketing planar transistors and
shortly thereafter applied the planar technique to the new integrated circuits.

SOURCE: "Silicon, Germanium & Silver - the transistor's 25th anniversary"
by C.P.Kocher. The Electrical Engineer.(November 1972) p.30.

1959   INTRINSIC 10μ PHOTOCONDUCTORS   W. D. Lawson
        Cadmium telluride and Mercury Telluride.   S. Nielsen   (U. K.)
                                                   E. H. Putley
                                                   A. S. Young

                The elements mercury, cadmium and tellurium have been purified and
        crystals of the compounds CdTe and HgTe, and of the mixed compounds CdTe-
        HgTe have been prepared. X-ray and cooling-curve data have established that
        coefficient and conductivity measurements show that HgTe is a semiconductor
        with a very low activation energy (0.01 eV) and a high mobility ratio (∿ 100).
        HgTe is opaque to infrared radiation out to a wavelength of 38μ. but the mixed
        crystals show absorption edges which vary in position with composition from
        0.8μ in pure CdTe to 13μ in crystals containing 90 per cent HgTe. Photo-
        conductivity has been observed in filamentary detectors made from the mixed
        crystals.

        SOURCE: "Preparation and properties of HgTe and mixed crystals of
        HgTe-CdTe" by W. D. Lawson, S. Nielsen, E. H. Putley and A. S. Young.
        J. Phys. Chem. Solids. Vol. 9 (1959) p. 325.

1959   THIN FILM CIRCUITS - TANTALUM   Bell Laboratories (U. S. A.)

                Tantalum film circuitry is a single material technology in that
        capacitors, resistors and elementary interconnections are all derived
        from tantalum. Use of tantalum for this purpose is based on its chemical
        and structural stability, and on its capability of being anodized to form
        dielectrics for capacitors, and to protect and adjust resistors.

                In addition, to the general value of tantalum film circuitry, tantalum
        film resistors and resistance networks, especially when sputtered in nitrogen,
        have independent interest as exceptional circuit elements.

        SOURCE: "Developments in tantalum nitride resistors" by D. A. McLean
        N. Schwartz and E. D. Tidd. IEEE International Convention (March 26, 1964).

        SEE ALSO: "Microcircuity with Refractory Metals" by D. A. McLean
        IRE Wescon Convention Record. Vol. 3. Part 6, pp. 87-91 (1959)

1959   MICROELECTRONICS   J. S. Kilby (U. S. A.)
        (INTEGRATED CIRCUIT-PATENT)

                "It is, therefore, a principal object of this invention to provide a novel
        miniaturised electronic circuit fabricated from a body of semiconductor
        material containing a diffused p-n junction wherein all components of the
        electronic circuit are completely integrated into the body of semiconductor
        material."

        SOURCE: U. S. Patent No: 3, 138, 743 filed February 6th, 1959 (J. S. Kilby)

        SEE ALSO: U. S. Patent No: 3, 261, 081 patented July 19th, 1966. (J. S. Kilby and
        Texas Instruments).

1959   LUNIK-1 (Mechta) Satellite   (U. S. S. R)

        Launched 2nd January, 1959. In solar orbit. Moon probe passed within 600 km
        of moon. Equipment for studying circumterrestrial and circumlunar space. No
        magnetic field detected when passing close to the moon. Emission of a sodium
        vapour cloud. Batteries.

        SOURCE: Table of Artificial Satellites Launched Between 1957 and 1976.
        International Telecommunication Union. Geneva (1977)

1959   DISCOVERER-1 Satellite   (U. S. A.)

        Launched 28th February, 1959. Stabilization defective: mission not fully
        accomplished. Decayed early March 1959.

        SOURCE: Table of Artificial Satellites Launched Between 1957 and 1976.
        International Telecommunication Union. Geneva (1977)

1960   <u>NEURISTOR</u>   H. D. Crane (U. S. A.)

          A neuristor may be visualized as a channel having energy available
to it everywhere along its length. The line or channel includes certain trigger-
able (active) processes arranged so that when any section of line is triggered,
the locally available energy is itself converted into "trigger form". Thus,
successive neighbouring sections of line are activated. A signal propagating in
this manner is generally referred to as a discharge, since it is continually
regenerated by discharging available energy into the line as it propagates.

          <u>SOURCE</u>: "Neuristor - a novel design and system concept" by H. D. Crane.
<u>Proc. IRE</u> Vol. 50 (Oct. 1962) p. 2048.

          <u>SEE ALSO</u>: "Neuristor studies" by H. D. Crane. Stanford Electronics Lab.
<u>Tech. Rept.</u> No: 1056-2 AD240306 (July 11, 1960)

               "An integrable MOS Neuristor line" by C. Kulkarni-Kohli and R. W.
Newcombe. Proc. IEEE (Nov. 1976) p. 1630.

1960   <u>FEMITRON - Field Emission</u>   W. P. Dyke (U. S. A.)
              <u>Microwave Amplifier)</u>

          Microwave devices incorporating field-emission cathodes have been
described by Charbonnier et al., and Dyke. The term "femitron" was, in fact,
first used by Dyke to describe a microwave amplifier resembling a 2-cavity
klystron but incorporating a field emission cathode in the input-cavity gap. The
femitron and derivatives of it, in particular frequency-multiplying devices, have
also been investigated by Fontana and Shaw.

          <u>SOURCE</u>: "Field emission microwave amplifier : a reappraisal" by A. J.
<u>Sangster</u>. IEE Solid-State & Electron Devices. Vol. 1. No: 5 (Sept. 1977) p. 151.

          <u>SEE ALSO</u>: "Basic and applied studies of field emission as microwave
<u>frequencies</u>" by F. M. Charbonnier, J. P. Barbour, L. E. Garrett and W. P. Dyke.
Proc. IEEE (1963) 51. pp. 991-1004.

               "Field emission, a newly practical electron source" by W. P. Dyke.
IRE Trans. (1960) 4. pp. 38. 45.

               "Microwave devices with field emission cathodes" by J. R. Fontana and
H. J. Shaw. Trans. Amer. Inst. Engrs. (1962) 81. pp. 43-48.

               "The carbon fibre field emitter" by F. S. Baker et al. J. Appl. Phys.
(1974) 7. pp. 2105-2115.

1960   <u>TELEPHONE ELECTRONIC SWITCHING SYSTEM</u>   Bell Labs (U. S. A.)

          Historically, electronic switching systems for telephone communicat-
ions began in the U. S. and saw their pioneering development undertaken there.
The world's first electronic switching field experiment took place in Morris, Ill.
(1960), and the first production system was placed in service in the United States
(No: 101 ESS in 1963).

          While the development of electronic switching is proceeding rapidly in
many countries, the bulk of installed electronic equipment is now found in the
U. S. More than 82 percent of the world's telephone lines that are switched
electronically, as well as more than 48 percent of the world's electronic central
office systems, are located in the United States. One of the most significant
developments made possible by electronic switching has been stored program
control flexibility, and here the U. S. accounts for over 80 percent of the world's
switching entities.

          <u>SOURCE</u>:   "ESS: Minimonster" by A. E. Joel Jnr. IEEE Spectrum (Aug. 1976)
p. 33.

          <u>SEE ALSO</u>: "Morris electronic telephone exchange" by W. Keister, R. W.
<u>Ketchledge</u> and C. A. Lovell. Proc. IEEE 107 Suppl. No: 20 (Nov. 1960) p. 257.

               "Electronic PBX telephone switching systems (ESS101)" by W. A. Depp
and M. A. Townsend. IEEE Trans. Communications. Vol. 83 (July 1964) p. 329.

1960   SUB-MILLIMETER PHOTO-   E.H.Putley   (U.K.)
            CONDUCTIVE DETECTOR (n-type InSb)

            Photoconductivity has been observed using a cryostat fitted with a light
pipe so that a specimen* could be illuminated with radiation of wavelength between
0.1 and 4.0 mm.   The sources of radiation were a mercury lamp and grating
spectrometer covering the range 0.1 to 1.4 mm. and a Philips DX151 klystron
and harmonic generator operating at 2mm and 4mm.   The light-pipe was fitted
with a black paper filter at room temperature and a black polythene filter in the
helium to remove short-wave radiation.

            The sample dimensions were 0.5 cm x 0.5 cm x 1.0 cm and indium
electrodes were applied to opposite 0.5 cm x 1.0 cm faces.   The magnetic field
was applied at right angles to the direction of current flow and of the incident
radiation.   The radiation was directed along the long direction of the sample.
For the majority of these experiments the radiation was modulated at 800 c/s
and detected using a tuned amplifier and phase sensitive detector.

            When the temperature was reduced to below 1.5^{O}K and a magnetic in-
duction of 6-8000 gauss applied, the sample resistance was about 10-30Ω.   The
sample was able to detect the applied radiation, the minimum detectable energy
per unit bandwidth being approximately 5×10^{-10}W at 0.5 mm, 5×10^{-11}W at
2mm and 10^{-10}W at 4 mm.

            SOURCE: "Impurity Photoconductivity in n-type InSb" Proc.Phys.Soc.Vol.76
            Pt.5 No:491 (1 Nov.1960) p.802.

            * n-type indium antimonide.

1960   COMPUTERS (CD 1604)   Control-Data Corporation (U.S.A.)

            The CDC 1604 is a general-purpose data-processing system manu-
factured by the Control Data Corporation.   The first installation was made in
January 1960.

            The entire system includes some 100,000 diodes and 25,000 transistors.
The internal number system is the binary, with a word length of 48 bits.   There
are 62 24-bit one-address instructions (2 per word) each including a six-bit
operation code, a three-bit index and 15 bits for the address.   Indirect addressing is built-in and six index registers are provided. Arithmetic is performed
with fixed or floating point in the parallel synchronous mode concurrently with
other operations.   Addition requires 4.8 to 9.6 μsec, multiplication 25.2 to
63.6 μsec. including storage access.   32,768 words of magnetic-core storage
are provided.   Input-output equipment includes paper-tape typewriter, punched
cards, magnetic tape (up to 24 units) and a 667-1000 lines-per-minute printer.

            SOURCE: Serrell, Astrahan, Patterson and Pyne "The evolution of computing
            machines and systems" Proc.IRE (May 1962) p.1054.

1960   PRINTED WIRING MULTILAYER BOARDS   Photocircuits Corp.(U.S.A.)

            Miniaturized replacement for back-panel wiring in computers may be
accomplished by printed circuit sandwiches produced in many layers and lami-
nated together under heat and pressure.   The components are being made by
Photocircuits Corporation, 31 Sea Cliff Ave., Glen Cove, New York.

            The manufacturer believes the new development will have uses in cir-
cuits having multiple crossovers and complicated interconnections among
closely spaced component leads.

            Connections between different levels in a multilayered circuit are made
through use of Tuf-Plate plated-through-holes. A typical six-layered printed-
circuit sandwich measures only 0.026 in. in thickness compared with a thickness of 0.062 in. for a conventional single circuit board.

            SOURCE: "On the market - P.C.sandwich six-layered unit" Electronics.
            (April 8,1960) p.90.

1960   COMPUTERS (HONEYWELL 800)   Honeywell (U.S.A.)

The Honeywell 800 is a general purpose data -processing system capable of running as many as 8 distinct programmes simultaneously without special instructions. The first installation was made in December 1960.

The system includes 30,000 diodes and 6000 transistors, excluding peripheral equipment. The internal number structure is binary and binary-coded decimal with a word length of 48 bits, or 12 decimal digits. These 48 bits are assignable to numerical, alphanumerical or pure binary information. There are 59 basic instructions, each consisting of a twelve -bit operation core and 3 twelve -bit addresses. Eight index registers are available for each of the 8 programmes which can be run concurrently. Other special -purpose registers are available. Arithmetic is performed in a synchronous parallel-serial-parallel mode, concurrently with other operations. Addition requires 24 μsec, multiplication 162 μsec including storage access. Up to 32,000 words of magnetic -core storage can be used. Input-output equipment includes punched cards, paper tape and a 900 lines -per -minute printer. Up to 64 magnetic tape units can be connected to the system.

SOURCE: Serrell, Astrahan, Patterson and Pyne "The evolution of computing machines and systems" Proc. IRE (May 1962) p.1054.

1960   CIRCUITRY (Linear Integrated Circuits)   Various (U.S.A.)

Linear ICs came into their own in the 1960s. Starting with op amps, linear monolithics grew steadily in complexity and functions.

Monolithic op amps were first introduced in the early 1960s. At least two manufacturers - Texas Instruments and Westinghouse - were selling models. Then Fairchild, in 1964, came out with the 702, the result of the first collaboration between Bob Widlar and Dave Talbert. The 702 found limited acceptance - more significantly, its development led to the 709, one of the biggest success stories in an industry accustomed to them.

The 709 was a revolution of sorts. Rather than translate a discrete design into a monolithic form, the standard approach, Widlar played the linear microcircuit game by a different set of rules: Use transistors and diodes - even matched transistors and diodes - with impunity, but use resistors and capacitors - particularly those of large value - only where necessary. Even where use of a big resistor seemed inevitable, Widlar put a dc-biased transistor in its place. He exploited the monolith's natural ability to produce matched resistors and only assumed loose absolute values.

SOURCE: "Integrated Circuits" by E.A. Torrero - "Electronic Design" 24. (November 23, 1972) p. 77.

1960   LIGHT EMITTING DIODE (LED)   J.W. Allen and
                                             P.E. Gibbons   (U.K.)

It has been known for some time that rectifying contacts to GaP emit light when current is passed through them (Wolff et al, 1955). Experiments at this laboratory and elsewhere suggest that the electronic transitions involved in this electroluminescence are different for the two directions of current flow. We consider alloyed or point-contact junctions on n-type GaP. Then the light emitted with forward bias has a spectrum which is a comparatively narrow band, the position of the band depending on the impurities present in the GaP. If the junction is biased in the reverse direction the current flowing is small until a certain voltage is reached. Beyond this voltage the current increases rapidly and orange light is emitted which has a very broad spectrum extending from the infra-red down to the absorption edge or beyond (Loebner and Poor, 1959). It would seem that electroluminescence in the forward direction is due to radiative recombination of injected carriers via impurity levels, while that in the reverse direction is due to emission by 'hot' carriers produced by an avalanche break-down (Chynoweth and McKay, 1956).

SOURCE: "Breakdown and light emission in gallium phosphide diodes" by J.W. Allen and P.E. Gibbons. Journal of Electronics Vol. VII No. 6. (Dec. 1959) p. 518.

1960   COMPUTERS - UNIVAC Solid State 80/90   Sperry Corporation   (U.S.A.)

            The UNIVAC Solid State 80/90 was designed as a medium-sized data-
processing system. The term "solid-state" refers to the use of "Ferractor"
magnetic amplifiers and transistors. The system consists of a central
processor, a read-punch unit, a 450 card-per-minute card reader and a 600-
lines-per-minute printer. The card equipment can be obtained for either the
80-column or the 90-column punched-card system. The first installation was
made in January 1960.

            SOURCE: Serrell, Astrahan, Patterson & Pyne "The evolution of computing
machines and systems" Proc. IRE (May 1962) p. 1053.

1960   TIROS-1 Satellite   (U.S.A.)

Launched 1st April, 1960. First meteorological satellite. Sent 22952 photos up
to 17th June 1960. 9000 solar cells. Batteries.

SOURCE: Table of Artificial Satellites Launched Between 1957 and 1976.
International Telecommunication Union. Geneva (1977)

1960   ECHO-1 Satellite   (U.S.A.)

Launched 12th August, 1960. Passive telecommunication satellite. Relayed
voice and television signals. 70 solar cells and batteries.

SOURCE: Table of Artificial Satellites Launched Between 1957 and 1976.
International Telecommunication Union. Geneva (1977)

1960   COURIER-1B Satellite   (U.S.A.)

Launched 4th October, 1960. First active repeater communication satellite.
Operated for 17 days. 19 152 solar cells. Batteries.

SOURCE: Table of Artificial Satellites Launched Between 1957 and 1976.
International Telecommunication Union. Geneva (1977)

1960   TRANSIT -1B Satellite   (U.S.A.)

Launched 13th April, 1960. First navigation satellite. Transmitted until 12th
July 1960. Solar cells. Batteries. Decayed on 5th October, 1967.

SOURCE: Table of Artificial Satellites Launched Between 1957 and 1976.
International Telecommunication Union. Geneva (1977)

1960   EPITAXIAL CRYSTAL GROWTH   H.H. Loor, H. Christensen   (U.S.A.)
                                           J.J. Kleimock & H.C. Theurer
            Until 1960 the semiconductor industry followed a pattern of starting
with a crystal as pure as needed in the initial stage, and each step added impurities in a controlled manner. In June 1960, the Bell Telephone Laboratories
announced a new method of fabricating transistors using epitaxial single
crystals grown from the gas phase with controlled impurity levels. This
technique had been studied at a number of laboratories, but it was not until the
1960 announcement that the potential was fully grasped by the semiconductor
industry. Its unique advantage is the ability to grow very thin regions of
controlled purity.

            SOURCE: "Contributions of materials technology to semiconductor devices"
by R.L. Petritx. Proc IRE (May 1962) p. 1030.

            SEE ALSO: "New advances in diffused devices" by H.H. Loor, H. Christensen,
J.J. Kleimock and H.C. Theurer. Presented at the IRE/AIEE Solid State Device
Research Corp. Pittsburgh Pa. (June 1960)

            "Epitaxy - a fresh approach to semiconductor circuit design" by
Materials Dept. Motorola S/C Products Division. International Electronics
(March 1964) p. 24.

1960-
1964

CIRCUITRY (Logic Circuits)   Various (U.S.A.)

        Much of the early activity was involved with digital logic families.
Almost from the beginning, a host of semiconductor manufacturers were
attempting to establish the dominance of one logic family over the other -
or were second-sourcing the strong suit of a competitor.

        At the start resistor-transistor logic (RTL) seemed to be the way
to go: Fairchild and Texas Instruments were strongly promoting it. Then
diode-transistor logic (DTL) appeared in 1962 from the recently formed
Signetics, and it took off. Transistor-transistor logic (TTL) emerged in
Sylvania's Universal High Level Logic (SUHL) in 1963 and again, more
permanently, in Texas Instruments' 5400 series in 1964.

SOURCE: "Integrated Circuits" by E.A.Torrero - "Electronic Design" 24.
(November 23, 1972)   p. 76.

1961   TRANSFERRED ELECTRON EFFECT   B.K.Ridley
                                               T.B.Watkins (U.K.)

        The possibility of obtaining negative resistance effects in a new way
in semiconductors is discussed. The principle of the method is to heat carriers in a high mobility sub-band with an electric field so that they transfer
when they have a high enough "temperature" to a higher energy low mobility
sub-band.

SOURCE: "The possibility of negative resistance effects in semiconductors"
by B.K.Ridley and T.B.Watkins. Proc.Physical Soc. Vol. LXXVIII (1961) p. 293.

SEE ALSO: H.Kromer. Phys.Rev. Vol. 109 (1958) p.1856.

1961   TRANSFERRED ELECTRONIC DEVICE   C.Hilsum (U.K.)

        In some semiconductors the conduction band system has two minima
separated by only a small energy, and the lower minimum has associated with
it a smaller electron effective mass than the upper minimum. At high electric
fields it should be possible to transfer electrons to the upper minimum where
they will have a power mobility. The conductivity of a homogeneous crystal bar
can therefore decrease as the field is increased and it is conceivable that a
differential negative resistance could be obtained. The conditions needed for
obtaining negative resistance are examined, and calculations made for GaSb
and semi-insulating GaAs. It appears that negative resistances should be observable in both these materials.

SOURCE: "Transferred electron amplifiers and oscillators" by C.Hilsum.
Proc. of the IRE Vol. 50. No:2 (February 1962) p. 185.

SEE ALSO: "Proposed negative mass microwave amplifier" by H.Kromer.
Phys.Rev. Vol.109 (March 1958) p.1856.

        "Indium Phosphide: a semiconductor for microwave devices" by
H.D.Rees and K.W.Gray. IEE Solid State & Electronic Devices, Vol.1
No: 1 (September 1976) p.1.

        "Three-level oscillator: a new form of transferred-electron device"
by C.Hilsum and H.D.Rees. Electron.Lett. Vol. 6 (1960) p. 277.

1961   LIQUID PHASE EPITAXY   H.Nelson (U.S.A.)

        An apparatus and procedures have been developed for the epitaxial
growth of GaAs and Ge from the liquid state. The resulting technology has been
found to posess advantages over vapour-phase epitaxy in some applications
demanding highly doped epitaxial films and high-quality p-n junctions at the sub-
strate-film interface. In this connection, it is an important feature of the liquid
phase process that chemical impurities and mechanical damage of the substrate
are removed when material is initially dissolved from the substrate surface
prior to epitaxial growth. A clean interface p-n junction is thus obtained. Since
liquid-phase epitaxy also favours the achievement of high doping and a steep con-
centration gradient at the p-n junctions, the process has proved itself eminently
suitable for application in the manufacture of Ge tunnel diodes. In its applica-

tion to the fabrication of the GaAs laser diode, it is an additional advantage of the liquid-phase process that the interface p-n junction is formed on a (100) crystal plane. As a consequence, this p-n junction is perfectly planar and also perpendicular to the (110) cleavage planes of the wafer. An optimum geometry (plane-parallel ends perpemdicular to a perfectly flat p-n junction) is thus insured for diodes cleaved from (100) oriented GaAs wafers whose p-n junction has been formed by liquid-phase epitaxy.

SOURCE: "Epitaxial growth from the liquid state and its application to the fabrication of tunnel and laser diodes" by H. Nelson. RCA Review (December 1963) p. 603.

SEE ALSO: "Epitaxial growth from the liquid phase" by H. Nelson. Solid State Device Conference, Stanford University (June 26, 1961)

     "Properties and applications III-V compound films deposited by liquid phase epitaxy" by H. Kressel and H. Nelson. Physics of Thin Films. Vol. 7 Academic Press. N.Y. (1973).

1961   VENUS-1 Satellite   (U.S.S.R)

Launched 12th February 1961. Automatic interplanetary station. Reached Venus in the second half of May 1961. Minimum distance to Venus 100 000 km. Investigation of interplanetary ionized gas and of solar corpuscular radiation. Investigation of the radiation belts and of space radiation. Magnetic measurements. Investigation of meteoristic dust. Solar cells, batteries.

SOURCE: Table of Artificial Satellites Launched Between 1957 and 1976. International Telecommunication Union. Geneva (1977)

1961   VOSTOK-1 Satellite   (U.S.S.R)

Launched 12th April, 1961. First manned satellite. Pilot Yuri Gagarin. Returned to earth in the USSR after one orbit and 1.8 hours in space on 12th April, 1961 near Smelovka, 800 km south-east of Moscow.

SOURCE: Table of Artificial Satellites Launched Between 1957 and 1976. International Telecommunication Union. Geneva (1977)

1961   ELECTRONIC CLOCK   P. Vogel & Cie   (Switzerland)

     According to the present invention there is provided an electronic clock comprising no macroscopic moving parts which comprises an oscillator for delivering electrical pulses at a given frequency, distributing means arranged to be controlled by said oscillator for delivering at outputs thereof the pulses delivered by the oscillator, a counting device arranged to be controlled by the oscillator for delivering signals of a frequency of n cycles per hours, where n is an integral factor of 60, and 1 cycle per minute, an electronic switch arranged to be controlled by the distributing means for delivering signals corresponding to the state of the counting device said signals being associated with hours and minutes successively, to a distribution matrix for controlling a display device.

SOURCE: British Patent Specification No :995, 546 "Improvements in or relating to electronic clocks" Application made in the U.S.A. (No:94832) on March 10th, 1961.

1961   MERCURY-ATLAS-4 Satellite   (U.S.A.)

Launched 13th September, 1961. Test of a cabin with a dummy on board. Checking of ground equipment performance (tracking stations). Cabin recovered in the Atlantic 260 km east of the Bermudas after 1st orbit on 13th September, 1961. Batteries.

SOURCE: Table of Artificial Satellites Launched Between 1957 and 1976. International Telecommunication Union. Geneva (1977).

1961   OSCAR-1 Satellite   (U.S.A.)

Launched 12th December, 1961. Orbital Satellite Carrying Amateur Radio. Transmitted for 18 days.

Decayed on 31st January 1962.

SOURCE: Table of Artificial Satellites Launched Between 1957 and 1976. International Telecommunication Union. Geneva (1977)

1961   MINICOMPUTER   Digital Equipment Inc. (U.S.A.)

It is generally accepted that the first minicomputer was designed by Digital Equipment in 1961 - a 12-bit 4K word memory machine selling for approximately £15,000. Many applications were found for a machine of this type, and the market blossomed rapidly with a number of manufacturers designing specialised machines with 8, 12 or 16 bits and varying memory sizes. The direct descendants of such machines, with increased power are still available at prices around £1,500, i.e. a reduction of 10:1 over ten years. This price reduction, and/or increase in performance, was made possible by the introduction of integrated circuits, MSI and SI logic, which allowed computers to become physically smaller while at the same time increasing performance. Since those early days when machines had very little software and peripheral support great strides have been made with the addition of extras, such as disks and magnetic tapes which allow the provision of operating systems running high level languages.

SOURCE: "How minicomputers can produce an integrated solution to the running of a business" by I. Evans. Paper read at Seminex, London (March 25th, 1976).

1962   SATELLITE (Telstar I)   Various (U.S.A.)

The first earth satellite was launched by the USSR on the 4th October, 1957. Telstar I, the first communication satellite, successfully transmitted high-definition television pictures across the Atlantic on the 10th July 1962, and its successors promise a new form of global communication by sound and vision. Telstar I, now silent, has orbited the Earth about 17000 times and is expected to remain in orbit for some 200 years; the 170lb satellite was powered by nickel-cadmium batteries, recharged by 3600 solar cells, and contained 1064 transistors and a single electron tube (a travelling-wave tube for amplifying signals). By the 23rd July, 1962, 16 European countries were exchanging live television with the United States and Telstar 2, launched on the 7th May 1963, paved the way for the world's first commercial communication satellite, Early Bird.

SOURCE: "The Scope of Modern Electronics" by F.A.Benson. Electronics & Power (January 1969) p. 13.

1962   MERCURY-ATLAS-6   (U.S.A.)
           "FRIENDSHIP 7" Satellite

Launched 20th February 1962. Investigation of man's capability in space. First United States manned spacecraft: astronaut John H. Glenn Jr. Capsule recovered on 20th February 1962 after three orbits and 4.6-hour lifetime. Batteries.

SOURCE: Table of Artificial Satellites Launched Between 1957 and 1976. International Telecommunication Union. Geneva (1977)

1962   OSO-1 Satellite   (U.S.A.)

Launched 7th March, 1962. Orbiting Solar Observatory. Measurement of solar electromagnetic radiation in the ultraviolet, X-ray and gamma-ray regions. Investigation of dust particles in space. Transmitted data on 75 solar flares until 6th August, 1963. 1860 solar cells.

SOURCE: Table of Artificial Satellites Launched Between 1957 and 1976.
International Telecommunication Union. Geneva (1977)

1962   RELAY-1___Satellite   (U.S.A.)

Launched 13th December, 1962. Active telecommunication satellite to test
microwave communications. Measurement of energy levels of space radiation.
Study of radiation effect on solar cells and electronic components. Transmission
of one television broadcast. 12 simultaneous 2-way phone calls or 144 tele-
printer circuits. 8215 solar cells, batteries. Experiments conducted until
February 1965.

SOURCE: Table of Artificial Satellites Launched Between 1957 and 1976.
International Telecommunication Union. Geneva (1977)

1962   MOS (Metal-oxide-semiconductor)   S.R.Hofstein
            INTEGRATED CIRCUIT   F.P.Heiman (U.S.A.)

        Who made the first MOS integrated circuit. Undoubtedly, Drs. Steven
R. Hofstein and Frederick P. Heiman, who working under the direction of Thomas
O. Stanley, head of the Integrated Electronics Group at the RCA Electronic
Research Laboratory in Princeton, N.J., were the first to succeed in late 1962
in integrating a multipurpose logic block of 16 MOS transistors into a silicon
chip, 50 x 50 mils. They reported their success at the 1962 Electron Devices
Meeting in October, 1962.

SOURCE: "The first MOS" by A. Socolovsky. The Electronic Engineer.
(February 1970) p. 56.

1962   LED (LIGHT EMITTING DIODE)   N. Holonyak (U.S.A.)

        Nick Holonyak, Jr (F) a member of the faculty at the University of
Illinois at Urbana-Champaign, Ill., received the 1975 John Scott Award for his
inventions leading to the first practical light-emitting diode (LED).

        Dr. Holonyak received the bachelor's degree in 1950, the master's
degree in 1951 and the doctorate in 1954, all from the University of Illinois.
In 1962, while at the General Electric Advanced Semiconductor Laboratory,
Syracuse, N.Y., he invented the first gallium arsenide phosphide LED and
the first semiconductor laser to operate in the visible spectrum. Since
joining the University of Illinois faculty in 1963 he has continued his research
with semiconductors, diodes and other solid-state devices.

SOURCE: IEEE Spectrum. (October 1975) p. 106.

1962   "DUANE" RELIABILITY GROWTH THEORY   J.T. Duane (U.S.A.)

        Historically, the subject of reliability growth theory of electronic
systems has received an abundance of attention and concern.   Beginning with
J.T. Duane in 1962, the literature on this subject has proliferated.

        Duane and other investigators developed reliability growth analysis
techniques based on actual data and used these data to test their models.
However, these efforts had no statistically developed theory of inherent or
analytical design reliability to fix their initial or starting points for their
growth curves.   However, in the structural reliability analysis, data relative
to inherent or analytical reliability has been available and evolving since 1955.
This data became available with the publication of Jablecki of data obtained
from first time static structural tests of major aircraft structural subsystems
under relatively controlled conditions at Wright-Patterson Air Force Base in
the years 1940 to 1949.

SOURCE: "A Theory of Reliability Growth in Structural Systems" by
Halsey B. Chenoweth. 1980 Proceedings Annual Reliability and Maintainability

Symposium, p. 106.

SEE ALSO: Duane, J.T. "Learning Curve Approach to Reliability
Monitoring", IEEE Transactions on Aerospace, Vol. 2, 1964, p. 563 - 566.

Jablecki, L.S., "Analysis of Premature Structural Failures in Static Tested
Aircraft", Dissertation, 1955, Die Eidgenossichen Technesche Hochschule,
Zurich, Switzerland.

1962   JOSEPHSON EFFECT   B.D.Josephson (U.K.)

          In spite of the fact that the history of the Josephson effect is quite long,
it is attributed to Josephson (1962) for its theoretical prediction. Prior experimental results published by Hahn and Meissner (1932) by Dietrich (1952) and
by Giaver and Megerle (1961) have been given without definite conclusions or
in doubt about effects, so that they could not be decisive for discovery. The
experimental confirmation of the Josephson absorption effect (known as the
a.c. Josephson effect) is attributed to Shapiro (1962, 1963) for first published
results. The current discontinuities in the current-voltage characteristic of
the Josephson junction, introduced by the macroscopic quantum absorption
effect, thus became the generally accepted fact in physics. In addition Janson
et al. (1965) described the Josephson emission effect (frequently termed as the
d.c.Josephson effect). Further, first-order discoveries based on the Josephson
effect have been : the Mercerau effect or macroscopis quantum interference,
frequency multiplication (Mercerau et al.1964), frequency mixing (Grimes and
Shapiro 1966), e/h ratio measurements (Langenberg et al.1966), followed by a
series of other applications. The crucial paper in Josephson voltage intro -
duction is published by Finnegan et al.(1971). This paper presents clear and
firm experimental evidence that the Josephson voltage stability exceeds the
best Weston-cell batteries used as national primary voltage standards.

SOURCE: An analysis of the inflexion point structure of Josephson absorption
effect current steps" by Ranko Mutabzija. Int. J.Electronics. (1977) Vol:42
No:3 page 241

SEE ALSO: "New superconducting devices" by B.D.Josephson. Wireless World
(October 1966) p.484.

          I.Dietrich (1952) Z.Phys. 133, 499.
          T.F.Finnegan, A.Denenstein & D.N.Langenberg (1971) Phys.Rev.
                                              B.4. 1487.
          I.Giaver and K.Megerle (1961) Phys.Rev. 122, 1101.
          C.C.Grimes and S.Shapiro (1966) Phys.Rev. 169,186.
          R.Hahn and W.Meissner (1932) Z.Phys. 74, 715.
          B.D.Josephson (1962) Physics Lett. 1, 251.
          D.N.Langenberg, W.H.Parker & B.N.Taylor (1966) Phys.Rev.150,186.
          J.E.Mercerau, R.C.Jaklevic, J.J.Lambe & A.H.Silver (1964)
                    Phys.Rev.Lett, 12, 274.
          S.Shapiro (1963) Phys.Rev.Lett., 11, 80.(1967) Phys.Rev., 169, 186.

1962   ELECTRONIC WATCH   P.Vogel & Cie   (Switzerland)

          An electronic timepiece, comprising an oscillator unit; a frequency
divider unit and a time display unit, in which each unit comprises or consists
of a layer of semi-conductive material, the layers being sandwiched together
and having their interfaces insulated from each other except at selected points
at which the units are electrically connected together.

SOURCE: British Patent Specification No:1,057,453 "Electronic Timepieces"
Application made in Switzerland (No:13423) on November 16th,1962.

1962   MICROELECTRONICS   Y.Tao (U.S.A.)
          (Flat-pack)

          With the emergence of the IC as the modern circuit element of the
early '60s, transistor packages were found to lack sufficient heat sinking and
adequate interconnections. To dissipate heat and provide a standard package
size, Yung Tao created the flatpack in 1962 while at Texas Instruments. It was
1/4 by 1/8 inch and originally had 10 leads.

SOURCE: "Solid State Devices - Packaging and Materials" R.L.Goldberg
"Electronic Design" 24, (November 23rd,1972) p.127.

1962   SEMICONDUCTOR LASER   R.N.Hall   J.D.Kingsley (U.S.A)
                                                     G.E.Fenner   T.J.Soltys
                                                          and R.O.Carlson

              Coherent infrared radiation has been observed from forward biased
GaAs p-n junctions. Evidence for this behaviour is based upon the sharply
beamed radiation pattern of the emitted light, upon the observation of a threshold current beyond which the intensity of the beam increases abruptly, and upon
the pronounced narrowing of the spectral distribution of this beam beyond threshold. The stimulated emission is believed to occur as the result of transitions
between states of equal wave number in the condition and valence bands.

SOURCE: "Coherent light emission from GaAs junctions" by R.N.Hall,
G.E.Fenner, J.D.Kingsley, T.J.Soltys and R.O.Carlson. Phys.Rev.Letters.
Vol.9.No:9 (November 1st, 1962) p.366.

1962   SEMICONDUCTOR LASER   M.I.Nathan
                                                     G.Lasher   (U.S.A.)

              A characteristic effect of stimulated emission of radiation in a
fluorescing material is the narrowing of the emission line as the excitation is
increased. We have observed such narrowing of an emission line from a for-
ward-biased GaAs p-n junction. As the injection current is increased, the
emission line at 77°K narrows by a factor of more than 20 to a width of less than
kT/5. We believe that this narrowing is direct evidence for the occurrence of
stimulated emission.

SOURCE: "Stimulated emission of radiation from GaAs p-n junctions" by
M.I.Nathan, W.P.Dumke, G.Burns, F.H.Dill Jr., and G.Lasher. App.Phys.
Letters. Vol.1. No:1 (1st November 1962) p.62.

1962   MARINER-2 Satellite   (U.S.A.)

Launched 27th August 1962. Data on interplanetary space during trip to Venus,
survey of the planet; magnetic fields, charged particle distribution and intensity
flux and momentum of cosmic dust, flow and density of solar plasma and energy
of its particles. Flew by the planet and scanned it on 14th December 1962. Contact lost on 3rd January 1963 at 87 390 000 km from earth. 9800 solar cells
(222 W) batteries.

SOURCE: Table of Artificial Satellites Launched Between 1957 and 1976.
International Telecommunication Union. Geneva (1977)

1962   ALOUETTE-1 Satellite   (Canada)

Launched 29th September,1962. Ionospheric top-side sounder launched by an
American vehicle. Measurement of radio noise originating in outer space and
within ionosphere, determination of electron density at satellite altitude,
detection of audio frequency signals in VLF radio spectrum, observation of
cosmic rays and measurement of primary cosmic ray particles. 6500 solar
cells, batteries.

SOURCE: Table of Artificial Satellites Launched Between 1957 and 1976.
International Telecommunication Union. Geneva (1977)

1962   MARS-1 Satellite   (U.S.S.R.)

Launched 1st November,1962. Long-term space exploration. Establishment of
interplanetary space radiocommunications. Lost earth lock at 106 Mkm. Passed
within 193 000 km of planet. Solar cells.

SOURCE: Table of Artificial Satellites Launched Between 1957 and 1976.
International Telecommunication Union. Geneva (1977)

1962   ARIEL-1 Satellite   (U.K.)

Launched 26th April,1962. Ionspheric satellite launched by United States rocket. Transmitted ionospheric, X-ray and cosmic data until November 1964. Solar cells.

Decayed on 24th May 1976.

SOURCE: Table of Artificial Satellites Launched Between 1957 and 1976. International Telecommunication Union. Geneva (1977)

1963   INK JET PRINTING PROCESS   R.G. Sweet (U.S.A.)

        In the early 1960's, Sweet developed a method of forming, charging and electrostatically deflecting a high-speed stream of small ink drops to produce high frequency oscillograph traces in a direct-writing, signal-recording system. Each drop is given an electrostatic charge that is a function of the instantaneous value of the electrical input signal to be recorded. The drop is then deflected from its normal path by an amount that depends on the magnitude of its charge and in a direction that is a function of the polarity of the charge. As deflected drops are deposited on a strip of moving chart paper, a trace is formed that is representative of the input signal.

        Lewis and Brown extended Sweet's technique to permit the printing of characters. Character images are stored in binary form in a character generator. An encoded signal addresses the character generator to select a desired character. The binary image of that character is then used to generate the drop charging signals necessary to deflect drops to the appropriate character matrix positions.

SOURCE: "Application of ink jet technology to a word processing output printer" by W.L. Buehler, J.D. Hill, T.H. Williams and J.W. Woods. IBM Journal of Research & Development (January 1977) p. 2.

SEE ALSO: "High frequency recording with electrostatically deflected ink jets" by R.G. Sweet. Stanford Electronics Laboratory Technical Report No: 1772-1. Stanford University, CA (1964)

"High frequency recording with electrostatically deflected ink jets" by R.G. Sweet. Rev. Sci. Inst. 36, 131 (1965).

    "Fluid Droplet Recorder" by R.G. Sweet. U.S. Patent 3,576,275 (1971)

    "Electrically operated character printer" by A.M. Lewis and A.D. Brown. U.S. Patent 3,298,030 (1967)

1963   GUNN DIODE OSCILLATOR   J.B. Gunn (U.S.A.)

        The observation is described of a new phenomenon in the electrical conductivity of certain III-V semiconductors. When the applied electric field exceeds a critical value, oscillations of extremely high frequency appear in the specimen current.

SOURCE: "Microwave oscillations of current in III-V-semiconductors" by J.B. Gunn. Solid State Communications. Vol.1 (1963) p. 88.

1963   VELA-1 Satellite   (U.S.A.)

Launched 17th October, 1963. Experimental nuclear test detection satellite. 1300 solar cells (70-90 W) batteries.

SOURCE: Table of Artificial Satellites Launched Between 1957 and 1976. International Telecommunication Union. Geneva (1977)

1963   SYNCOM-1 Satellite   (U.S.A.)

Launched 14th February 1963. Active telecommunication satellite. Successfully injected into a near synchronous orbit but then lost contact with ground station. 3840 solar cells (135 W) battery.

SOURCE: Table of Artificial Satellites Launched Between 1957 and 1976. International Telecommunication Union. Geneva (1977)

1963   ELECTRONIC CALCULATOR   Bell Punch Co. (U.K.)

The first electronic calculators, containing discrete semiconductor
components wired to printed circuit boards, were produced in 1963 by a
British firm, the Bell Punch Company. The machine was produced under
licence in America and in Japan, where the advantage of cheaper Japanese
labour for the hundreds of connections required led to a Japanese domination
in the manufacture of calculators throughout the sixties. Integrated circuitry
was, of course, the perfect technology for the calculator, and MOS - slower
but more compact and cheaper than bipolar integration - the most appropriate
of the integrated circuit technologies. By the second half of the sixties,
calculators using MOS integrated circuits were available.

The first American company to make calculators was a firm called
Universal Data Machines, operating from a warehouse in Chicago. The
company bought chips from Texas Instruments and, using cheap immigrant
labour from Vietnam and South America, assembled five or six thousand
calculators a week for sale through a local department store. Probably the
second company to enter what was to become a particularly vicious race was
the Canadian firm, Commodore, newly moved from Toronto to Silicon Valley.
Commodore also used a Texas Instruments MOS chip, but adopted a technology
developed by a component supplier, Bowmar, for making a particularly compact calculator. Bowmar had chosen not to make calculators itself and had
found no interest in its technology among the established manufacturers of
electro-mechanical calculators. Although these first mass-produced calculators dropped rapidly in price from about $100 in 1971 to $40 or $50 the
following year, the profits of these small entrepreneurial companies remained
high.

The situation had to change as it became staringly obvious where the
profits in the exploding new market lay. By 1972 Bowmar was struggling to
get back into the business it had earlier farmed out and was joined by other
semiconductor manufacturers, including Texas Instruments. The calculator
provides perhaps the best example of rapid vertical integration in the semiconductor industry, but if small firms had not demonstrated the viability of the
new product, it is doubtful whether such integration would ever have taken place
or, indeed, whether the calculator would ever have gained the acceptance it has.

SOURCE: "Revolution in miniature - The history & impact of semiconductor
electronics" by E. Braun and S. Macdonald. To be published by Cambridge
University Press.

SEE ALSO: "Coming of age in the Calculator business" by N. Valery, New
Scientist (Calculator supplement) 68, 975,(1975) pp. ii-iv.

ALSO: "Electronic Calculator" "Wireless World" 78, 1442 (1972) p. 357.

1963   ION PLATING OF PLASTIC & METALS   .D.H. Mattox (U.S.A.)

World-wide interest in ion plating stems from the new characteristics
or colours the ion plated coats have when compared with films produced by other
coating techniques. Among these properties are:-

(1)   Excellent adhesion of incompatible substrates -
           such as metals on plastics.

(2)   Irregular surfaces of many types such as screw threads,
           bearings and tubes can be coated in one operation because
           of the good "throwing power" obtained in the process.

(3)   Compact pin-hole-free structures are formed with out-
           standing friction and wear characteristics and soldering
           to the metal coat is no problem.

(4)   Oxides and ceramic coating is possible with reactive ion
           plating.

(5)   High corrosion resistant coating can be produced.

(6)   High rate production of coatings is possible because the
soft vacuum required means that long pump down times are
not necessary. However, when good adhesion to metals is
required, the metal substrate must be ion etched for periods up to 30 minutes to remove the oxide before deposition
commences.

Although the technique was invented in 1963 in the United States by
D. M. Mattox of Sandia Laboratories, New Mexico, the true potential of ion
plating had not been appreciated until quite recently. In the beginning, the
process is just like a conventional evaporation one, in that the bell jar is evacuated first with a rotary pump and then a diffusion pump. After the pressure
reaches about 10^{-5} torr, argon is admitted through a needle valve until the
pressure rises to about 2×10^{-2} torr. This soft vacuum is kept constant by
controlling the needle valve and partially opening the baffle valve to the diffusion pump.

In the case of a metal, ion etching of the substrate is carried out first
by striking a discharge between the substrate and top and base plate. A negative voltage of from 1kV to 6kV is applied to the substrate, and the argon ion
discharge remains so long as the cathode voltage or the argon pressure is not
allowed to fall too low. The bombardment of the substrate with neutral and
ionised argon atoms sputters off metal oxide and metal and etches the surface.
When cleaning and etching is complete, the metal source filament is switched
on and the metal is evaporated into the argon discharge.

SOURCE: "Ion plating - coat of many colours" New Scientist. (9th June, 1977)
page 588.

1963   SURFACE ACOUSTIC WAVE DEVICES   J. H. Rowen   (U.S.A.)
                                          and E. K. Sittig

"I filed a patent application describing a number of Surface Acoustic
Wave devices in December 1963 and described these structures in a post-
deadline paper presented at the 1964 Symposium on Sonics and Ultrasonics in
Santa Monica, California - October 14th-16th 1964. Dr. Ehrhardt Sittig, who
worked in my department at that time, constructed working models of these
devices and subsequently filed an application describing an interdigital electrode
structure for balanced (vs. unbalanced) excitation of Rayleigh surface waves on
such devices. I believe these efforts, which predate Prof. R. M. White's work
by at least three years, constitute the original invention of surface acoustic
wave devices".

SOURCE: Letter from Bell Laboratories, Murray Hill, N. J. dated 20th July, 1977.

SEE ALSO: "Tapped ultrasonic delay line and uses therefor" - J. H. Rowen
U. S. A. Patent No: 3, 289, 114 dated November 29, 1966.
"Elastic wave delay device" - E. K. Sittig.
U. S. A. Patent No: 3, 360, 749 dated December 26, 1967.

NOTE: Lord Rayleigh first described the equations governing the propagation
of surface elastic plane waves along the stress-free boundary of a semi-
infinite, isotropic and perfectly elastic solid. Prof. White's paper, in 1967, is
summarised as follows:

Surface elastic-wave propagation, transduction and amplification
(in a piezoelectric semiconductor) are discussed with emphasis on
characteristics useful in electronic devices. Computed curves show
the dependence on distance from the surface of the elastic and the
electric fields associated with surface elastic-wave propagation in
cadmium sulfide. The interaction impedance, relating the external
electric field to power flow, is computed for propagation on the basal
plane of CdS and found to be low in comparison with values characteristic of electromagnetic slow-wave circuits. Amplification with a continuous drift field in cadmium sulfide is reported and differences
between surface and bulk-wave amplifiers are discussed. Some operating characteristics and fabrication techniques for making electrode
transducers on piezoelectric crystals are given, together with experimental results on several passive surface-wave devices.

SOURCE "Surface elastic wave propagation and amplification" by R.M. White
IEEE Trans. on Electron Devices. ED-14 No.4 (April, 1967) p.181.

SEE ALSO:

"On waves propagated along the plane surface of an elastic solid"
by Lord Rayleigh. Proc. London Math. Soc. Vol:17 (November 1885) p.4.

"Surface waves in anisotropic media" by J.L.Synge. Proc.
Royal Irish Acad. (Dublin) Volume A58. (November 1956) p.13.

"Surface waves in anisotropic elastic media" by V.T. Buchwold.
Nature. Volume 191 (August 26th,1961) p.899.

"Design of surface wave delay lines with interdigital transducers"
by W.R.Smith, H.M.Gerard, J.H.Collins, T.M.Reader and H.J.Shaw. IEEE
Trans. on Microwave Theory & Techniques. Vol.MTT17 No:11 (Nov.1969)
page 865.

"Passive Interdigital Devices using surface acoustic waves" IEE
Reprint Series 2. Editor D.P.Morgan. Peter Peregrinus (May 1976)

1963   SILICON ON SAPPHIRE TECHNOLOGY   Various   (U.S.A.)

The technology of silicon-on-insulating substrates, specifically
silicon on sapphire, dates back to the beginning of practical MOS technology
in 1963. The technology is known by different abbreviations, such as: SIS,
ESFI (epitaxial silicon films on insulators), SOS, SOSL (silicon on spinel), etc.

The principal advantage of SOS circuitry is the inherent dielectric
isolation, both dc and ac. The absence of silicon, except in the active device
areas, significantly reduces parasitic capacitance between lines and essentially eliminates the parasitic capacitance to the substrate. Diffusion of device electrodes through the silicon film to the sapphire reduces electrode capacitance by several orders of magnitude because of the reduction in junction
area. This significant reduction in electrode and interelectrode capacitance
enables many devices to achieve their maximum band-width and frequency response; it allows for very high speeds, minimum speed-power products (below
0.5 pJ) on SOS CMOS and for very high frequency linear elements ($f_r > 2.10^9$ Hz)
such as dual-gate MOSFET's (tetrodes).

SOURCE: "Recent SOS technology advances and applications" by R.S.Ronen
and F.B.Micheletti. Solid State Technology (August 1975) p.39.

SEE ALSO: Early publications on SOS Technology e.g. Material and Devices,
include: H.M.Manasevit and W.I.Simpson, "Single crystal silicon-on-Sapphire
substrate" J.Appl.Phys. 35, 1349 (1964) : C.W.Mueller and P.H.Robinson,
"Grown-film silicon transistors on sapphire" Proc. IEEE 52, 1487 (1964).

1964   NIMBUS-1 Satellite   (U.S.A.)

Launched 28th August 1964. Meteorological satellite to achieve a precise continuously earth-pointing orientation to evaluate the advanced vidicon camera
system (AVCS) to provide improved pictures of local clouds by means of automatic picture transmission (APT), and to evaluate the high resolution infrared
radiometer (HRIR) system for mapping global night-time cloud cover. 10 500
solar cells (450 W) batteries, 27 000 cloud cover photos returned until 23rd
September 1964.

SOURCE: Table of Artificial Satellites Launched Between 1957 and 1976.
International Telecommunication Union. Geneva (1977)

1964   VOSKHOD-1 Satellite   (U.S.S.R)

Launched 12th October,1964. Manned spacecraft. First three-man crew:
V.Komarov, K.Feokistov, B.Yegorov. Landed after 16 orbits (24.3 hours)
305 km northwest of Kustanaya, Kazakhstan.

SOURCE: Table of Artificial Satellites Launched Between 1957 and 1976.
International Telecommunication Union. Geneva (1977)

1964   <u>PACKET-SWITCHING-COMMUNICATIONS</u>   P. Baran (U.S.A.)

  Packet-switching demands a different kind of communications network from the normal telecommunications pattern. The channels, whether wire or radio, may be the same; but the switching points and exchanges have "intelligence" - some form of computing device - which can accept a packet, look at it, and send it on its way according to the address and instructions it carries.

  There is, however, also another major difference. The traditional communications network is essentially serial. To make a connection between two subscribers to such a system requires that the connection be established for the duration of the call. This means that the right switches have to be opened/closed and kept so for that duration.

  Such a requirement is not necessary with a packet-switched system or network. In this latter case, you transmit your data to the network, which then takes over, either sending it on or holding it until the addressee's receiver facilities are free and able to take it. The speed of transmission thus becomes a function of the weight of loading within the network. At the conceptual level, this is quite a radical approach to telecommunications.

  To have a fail-safe network, in the terms that Baran proposed, there should be "over connection". In other words, there should be not just one path in or out for a packet, but several. What would then determine the routing is the availability of a channel at a particular time. With this sort of network, the reliability can be far lower than would be necessary for a "normal" linear communications system.

  People have been trying to build such networks now for around 10 years. Today, though no one can be certain how many are being planned or built, most of the communications networks of which we have high expectations are packet-switched - among them the experimental Arpanet in the United States, the commercial Telenet network (also American), Europe's interbank, Swift network, Euronet, and the European Informatics Network.

SOURCE: "Packet-switching's unsung hero" by R. Malik. New Scientist (8th September, 1977) p. 606.

1964   <u>GEMINI-1 Satellite</u>   (U.S.A.)

Launched 8th April, 1964. Testing of the GEMINI launch vehicle compatibility and the structural integrity of the GEMINI spacecraft. The satellite re-entered the atmosphere and disintegrated on 12th April 1964. Batteries.

SOURCE: Table of Artificial Satellites Launched Between 1957 and 1976. International Telecommunication Union. Geneva (1977)

1964   <u>"IMPATT" DIODE</u>   R. L. Johnston and B. C. deLoach (U.S.A.)

  In 1964, Bell Laboratories scientists R. L. Johnston and B. C. DeLoach discovered the IMPATT (IMPact Avalanche Transit Time) diode, subsequently shown to operate by an effect proposed earlier by W. T. Read Jr., also of the Laboratories. IMPATT diodes - semiconductor devices that generate microwaves directly when a DC voltage is applied to them - are becoming increasingly important in the design of microwave systems because of their high reliability and low cost.

SOURCE: "Mission Communications - the story of Bell Laboratories" by Prescott C. Mabon. Published by Bell Laboratories Inc., Murray Hill, New Jersey, U.S.A. (1975) p. 173.

SEE ALSO: "A proposed high frequency, negative resistance diode" by W. T. Read. Bell Syst. Tech. J. Vol. 37 (1958) p. 401.

1964   TRANSISTOR MODELLING   H.K.Gummel (U.S.A.)

    Since the original paper by Gummel in 1964, a great deal of literature
has appeared on the subject of fundamental, or exact, transistor modelling.
Gummel was the first to solve the semiconductor partial difference equations
with no basic simplifications in their one-dimensional steady-state form. His
integral formulation appeared in an improved form in the work of De Mari, who
then went on to tackle the time-dependent 1-dimensional system. This required
the use of a finite-difference formulation from which he could obtain a current-
driven transient solution for a diode. A simple spatial discretization was used
and solutions were obtained for two implicit time integration methods, a Crank-
Nicolson scheme and a pure implicit first-order scheme. This was followed by
the analysis of a Read diode by Gummel and Scharfetter, also using a 1-dimen-
sional implicit scheme, but introducing a new and important spatial finite-
difference formulation.

SOURCE: "Fundamental one-dimensional analysis of transistors" by A.M.
Stark. Philips Research Reports Supplements (1976) No:4. p.1.

SEE ALSO: "A self-consistent iterative scheme for one-dimensional steady
state transistor calaculations" by H.K.Gummel. IEEE Trans. ED-11 p.455-465
(1964)

    "An accurate numerical steady state one-dimensional solution of the
p-n junction" by A.de Mari. Solid State Electronics 11. p.33-58 (1968)

    "An accurate numerical one-dimensional solution of the p-n junction under
arbitrary transient conditions" by A.de Mari. Solid-State Electronics 11.
pp.1021-1053. (1968)

    "Finite difference methods for partial difference equations" by
G.E.Forsythe and W.R.Wasow. J.Wiley & Sons Inc., (1970) pp.101 et seq.

    "Large-signal analysis of a silicon Read diode oscillator" by D.L.
Scharfetter and H.K.Gummel. IEEE Trans. ED-16. 64-77. (1969).

1964   TRANSISTOR (Overlay)   R.C.A. (U.S.A.)

    The overlay transistors, first introduced in 1964, was developed at
RCA under a contract from the Army Electronics Command, Ft.Monmouth,
New Jersey, as a direct replacement for the vacuum tube output stages then
used in military transmitting equipment.

    The first commercial overlay, the 2N3375, produced 10 watts of
output power at 100 Mhz and could handle 4 watts at 400 Mhz. Comparable
interdigitated structures of that day were capable of 5 watts at 100 Mhz and
0.5 Mhz and 0.5 watts at 400 Mhz.
SOURCE: "Solid State - a worthy challenger for RF power honors" by
D.R.Carley. Electronics. (February 19,1968) p.100.

SEE ALSO: "The overlay - a new UHF power transistor" by D.R.Carley,
P.L.McGeough and J.F.O'Brian. Electronics (August 23,1965) p.70.

1964   MICROELECTRONICS (Beam Lead)   M.Lepselter   (U.S.A.)

    In 1964, Martin Lepselter of Bell Telephone Laboratories invented
the beam lead as a mechanical and electrical interconnection between the IC
and its case.

SOURCE: "Solid State Devices - Packaging and Materials" R.L.Goldberg
"Electronic Design" 24. (November 23,1972) p.127.

SEE ALSO: "Beam lead technology" by M.P.Lepselter. The Bell System
Tech.Journal. Vol.XVL No:2 (February 1966) p.233.

1964   TELEMEDICINE   Various (U.S.A. et al)

The development of telemedicine in the United States can be divided into three stages: 1964-1969, 1969-1973 and 1973-present.

The first stage involved experimentation by medical practitioners on the clinical applications of telecommunications technology. The primary concern was the feasibility of two-way transmission of diagnostic information and clinical encounters via microwave links and video equipment.

Starting in 1964, the first interactive TV telemedicine project for the delivery of health care was carried out - a closed-circuit TV link between Nebraska Psychiatric Institute, Omaha, Neb., and Norfolk State Hospital, 112 miles away - under financing by the National Institute of Mental Health.

In 1967, an interactive TV link was installed between Massachusetts General Hospital and Logan International Airport, Boston, Mass., with financial support from the United States Public Health Service (later expanded to a Massachusetts General-Bedford VA Hospital link with Veterans Administration funds).

While the Nebraska telemedicine system was used primarily for psychiatric consultation and administrative purposes, the Massachusetts General Hospital-Logan Airport system was the first programme to use telemedicine in physical diagnosis and general patient care. The medical procedures used in physical diagnosis that were found to provide effective treatment over the interactive TV link were teleradiology, telestethoscopy and teleauscultation, speech therapy, teledermatology and telepsychiatry. The successful demonstration of physical diagnosis procedures provided additional incentives for Federal agencies to encourage further developments in the field.

The second telemedicine stage was characterized by a trend toward the exchange of knowledge and experience among the participants, and by Government support and sponsorship of research and demonstration programmes. The major supporter was the Health Care Technology Division in the Department of Health, Education and Welfare, which funded seven research and demonstration projects during 1972: Illinois Mental Health Institute, Chicago, Ill.; Case Western Reserve University, Cleveland, Ohio; Cambridge Hospital, Cambridge, Mass; Bethany/Garfield, Chicago, Ill.; Lakeview Clinic, Waconia, Minn; Dartmouth Medical School's INTERACT, Janover N.H.; and Mount Sinai School of Medicine New York, N.Y. In addition, the National Science Foundation funded two telemedicine projects in 1973: the Boston Nursing Home project for geriatric patients in nursing home that usually refer patients to Boston City Hospital, and the Miami-Dade project between Dade County and Jackson Memorial Hospital, Miami Fla.

During this stage, issues other than technical ones received some attention. These included consideration of the appropriate organisational and environmental settings for telemedicine implementation, manpower mixes and the role of non-M.D. providers, and rudimentary approaches to evaluation of telemedicine's impact on health-care delivery. The contributions of telemedicine to society as a whole were variously presented, but although some evaluation projects were started during this period, there were no significant efforts to investigate or document those benefits.

The initial evaluation efforts did not reveal conclusive results, but a comparison between the telephone and interactive TV encounters showed the former to be of shorter duration and more efficient for some aspects of patient care.

The third, and present, stage started in 1973 and its characteristic feature is the idea of telemedicine as an innovative mode of medical-care delivery. Two factors must be dealt with during this stage; sooner or later, telemedicine has to become self-supporting, or at least economically viable on its own; and the evaluation of telemedicine has to follow the concepts and method of evaluation in the medical-care field - i.e. evaluation in terms of structure, process and outcome variables. The major new challenge for telemedicine has become its economic viability - how to make it pay for itself. To date, for telemedicine programmes to survive, they've had to be heavily subsidized.

It has been recognised that various problems in medical care may be redressed by telemedicine, but these depend on the vantage point of the user. For those persons where time and distance barriers make it difficult to obtain access to medical care, telemedicine is obviously very useful. The providers recognize potential benefits, including greater opportunities to interact with other physicians, to consult with specialists without worrying about the possibility of "losing" their patients, to have more free time, and to supervise the work of a nurse practioner or physician assistant in a remote clinic. The benefits to the system of medical care lie in the greater ability of the system to co-ordinate the activities undertaken by the various health actors or providers in their respective roles.

SOURCE: "Coming - the era of telemedicine" by R.Allen. IEEE Spectrum. (Dec. 1976) p.33.

1964   MICROELECTRONICS (DIP)   B.Rogers (U.S.A.)
        (Dual-in-line package)

Bryant (Buck) Rogers fostered the invention of the DIP while at Fairchild Semiconductors in 1964. It originally had 14 leads and looked just as it does today.

SOURCE: "Solid State Devices - Packaging and Materials" R.L.Goldberg "Electronic Design" 24,(November 23,1972) p.127.

1964   "ETCH-BACK" TECHNIQUE IN PRINTED WIRING   Autonetics (U.S.A.)
        PLATED THROUGH HOLES

The interconnection of the internal layers of circuitry is made at the area where the drill penetrates through the copper pad exposing a cylinder of copper equivalent to the diameter of the drilled hole times 0.0044 (π times thickness of one ounce of copper). This small area of exposed copper can also be contaminated with epoxy smeared onto it during the drilling operation which can affect the resultant adhesion of the copper to the electroless copper deposit. Therefore, a smoothing process was developed at Autonetics which would expose a greater amount of copper at the interconnection areas to provide a more reliable bond. This, coupled with the fact that the smoothing operation also removes from the copper any smeared epoxy, provides for a more reliable interconnection than the standard T-joint.

SOURCE: "Electroplating of Plated Through-Hole Interconnection Circuit Board" by L.J. Quintana, AFS Proc. 1964, p. 175.

1964   WORD PROCESSOR   (I.B.M.)   (U.S.A.)

One specialized office application that attracted computers was word processing. IBM, already a dominant manufacturer of electric typewriters, is credited with creating the market in 1964 when it introduced a magnetic-tape typewriter. This unit could store information on magnetic media for later modification and automatic retyping.

SOURCE: Electronics (April 17, 1980), p.387.

1965   SNAPSHOT Satellite   (U.S.A.)

Launched 3rd April,1965. SNAP-10A nuclear reactor power supply operated at more than 500 W for 43 days.

SOURCE: Table of Artificial Satellites Launched Between 1957 and 1976. International Telecommunication Union. Geneva (1977)

1965   PEGASUS-1 Satellite   (U.S.A.)

Launched 16th February. Measured quantity, size and velocity of micro-meteoroids, 25 000 solar cells, batteries. Turned off in August 1968.

SOURCE: Table of Artificial Satellites Launched Between 1957 and 1976. International Telecommunication Union. Geneva (1977)

1965   GGSE-2(Gravity-Gradient-2) Satellite   (U.S.A.)

Launched 9th March, 1965. Gravity Gradient Stabilization Experiment satellite.
First satellite with two axis gravity gradient stabilization system. Performed a
"turnover manoeuvre" by the rod-retraction method. Solar cells.

SOURCE: Table of Artificial Satellites Launched Between 1957 and 1976.
International Telecommunication Union. Geneva (1977)

1965   LES-1 Satellite   (U.S.A.)

Launched 11th February, 1965. Tested devices and techniques for possible use in
United States communication satellites and satellite communication system.
Solid state X-band transmitter earth-sensing/antenna-switching system, mag-
netic spin-axis orientation system. 2376 solar cells (26 W minimum)

SOURCE: Table of Artificial Satellites Launched Between 1957 and 1976.
International Telecommunication Union. Geneva (1977)

1965   WIEGAND WIRE   J. Wiegand (U.S.A.)

About 10 years ago, John Wiegand discovered that by properly work-
hardening a magnetic wire, it is possible, along the exterior "shell" of the wire,
to produce a coercive force significantly greater than the coercive force in the
wire's core. By virtue of this magnetic differential, and depending on certain
external conditions, the direction of magnetization in the core of the wire can be
the same or opposite to that in the shell. And switching from one state to the
other is easily and repeatedly induced at well-defined magnetic-field levels.

Short lengths of wire exhibiting the Wiegand effect can serve as the
heart of magnetic pulse generators that have distinct advantages over similar
devices, including non-contact operation and a facility for being "read" by
detection devices having virtually no input power. Other important advantages
are that pulse signals are not rate sensitive, meaning the amplitude of the pulse
signal remains the same regardless of speed of operation; they offer any combination of pulse-generation direction and polarity, that is, uni-directional or
bi-directional, unipolar or bipolar. Thus any combination of direction and
polarity are available for pulse generation. And such devices are capable of
withstanding severe environments, including temperatures from -95°F to
+300°F.

Over the years, Wiegand has developed material composition and work
hardening procedures to a point where brief pulses (10^{-4} duration) at levels of
2 milliwatts can be produced. With properly-designed detectors, peak voltages
of 500 millivolts in the 50-ohm load have been observed.

SOURCE: "Wiegand Wire: new material for magnetic-based devices" by P.E.
Wigen. Electronics (July 10, 1975) p. 100.

SEE ALSO: "Wiegand effect pushing its way into new products"
Electronics. (April 14, 1977) p. 39.

1965   FR-1 Satellite   (France)

Launched 6th December, 1965. French satellite launched by a United States
rocket. Study of VLF radio wave propagation in the ionosphere and measurement of the electron densities in the vicinity of the orbit. Transmitted until
28th February 1969. 3840 solar cells (17 W) batteries.

SOURCE: Table of Artificial Satellites Launched Between 1957 and 1976.
International Telecommunication Union. Geneva (1977)

1965   OVI-2 Satellite   (U.S.A.)

Launched 5th October, 1965. Radiation measurements. Proton and electron
dosimeters and spectrometers. X-ray detector, magneto-meter and two-tissue-
equivalent ion chambers. 5000 solar cells (22 W), battery.

SOURCE: Table of Artificial Satellites Launched Between 1957 and 1976.
International Telecommunication Union. Geneva (1977)

1965   SMOOTH-SURFACED WIRE DRAWING   K.M.Olsen
                                                            R.F.Jack   (U.S.A.)
                                                            E.O.Fuchs

        A technique for producing wire with a very smooth surface by drawing
it through dies submerged in an ultrasonically agitated liquid has been devised
at Bell Laboratories. The agitated liquid continuously cleans the wire and dies
so that the drawn wire is relatively free of embedded particles and surface
scratches.

        Reduction of surface imperfections in wire improves its properties in
some instances. For example, a smooth finish is desirable in those types of
magnetic memories that store information on a thin film of metal plated onto a
wire. The wire finish should be as smooth as possible so that the film can be
deposited evenly.

        In this technique, the ultrasonic energy forms extremely minute vapor
cavities in the liquid wherever it contacts a solid surface. The expansion and
collapse of these cavities - known as cavitation - "scrubs" the wire clean of
foreign particles before it enters the dies to be reduced. The ultrasonic
agitation keeps the particles suspended in the liquid and prevents them from
collecting in the entry area of the dies; thus they do not score the wire as it it
drawn through the dies.

SOURCE: "Very smooth-surfaced wire produced by new drawing technique"
Bell Laboratories Record (October 1965) p. 390.

Copyright 1965, Bell Telephone Laboratories, Inc. Reprinted by permission,
Editor, Bell Laboratories RECORD.

1965   SATELLITE - INTELSAT I   International

        The first internationally owned satellite, INTELSAT I, was put into
operation in 1965. It was placed in a geo-stationary equatorial orbit, that is at
an altitude of 22 400 miles, in a longitudinal position 30° West for transatlantic operation. It had a mass of 39 kg (85 lb) primary power 45 W from
solar cells and was capable of relaying 120 voice circuits or one television
channel. The INTELSAT I system was to some extent experimental for two
main reason. Firstly, it was to ascertain whether reliable communication
could be maintained in spite of the high path loss of 200 dB; however, the
Earth stations employed the now well-known parabolic reflector type aerials -
diameter 85-100 ft - with cryogenic-cooled low-noise amplifiers and in this
respect it was a great success. Secondly, it was to determine whether the
transmission delay, Earth - satellite - Earth, of 250 ms was operationally
acceptable.

        The decision was taken to continue with satellites in the geo-
stationary orbit and this is used for all internationally-owned satellites today.

SOURCE: "Fixed Communications" by A.S.Pudner. Radio and Electronic
Engineer. Vol.45. No:10 (October 1975) p. 547.

1965   SELF-SCANNED INTEGRATED   G.P.Weckler (U.S.A.)
            PHOTODIODE ARRAYS

        The possibility of forming image sensors from arrays of silicon photo-
diodes on a single silicon chip has been recognised for many years, probably
since the inception of integrated circuit technology some fifteen years ago. It
was quickly apparent that the array size was limited, not by the number of
diodes that could be included on the silicon, but by the number of output leads
necessary to form connections to these diodes. To circumvent this problem, it
was necessary to scan the diodes, that is, to multiplex them in to a single output lead by means of switching circuitry on the same integrated chip. A second
problem was that of detecting the minute photocurrents produced by the necessarily very small diodes. The technique of charge integration had been used in
the Vidicon for some years, and it was pointed out by G.P.Weckler in 1965 that
this technique could be used with photodiode arrays, the switching being achieved
by m.o.s. transistors. The one step necessary to complete the picture was now
to include a shift register on an integrated circuit with the diodes and m.o.s.t's
to perform the serial multiplexing function. The first fully self-scanned arrays
using this technique were announced in 1967.

SOURCE: "Applications of self-scanned integrated photodiode arrays" by
P. W. Fry. The Radio and Electronic Engineer. Vol. 46. No:4 pp.151-160
(April 1976)

SEE ALSO: "Operation of p-n junction photodetectors in a photon flux integrating mode" by G. P. Weckler. IEEE J. Solid State Circuits. S C-2 No:3. pp. 65-73
(September 1967)

"Development and potential of optoelectronic techniques" by P. J. W.
Noble. Component Technology 2, No:8 pp. 23-8 (December, 1967)

1965   PROTON-1 Satellite   (U. S. S. R.)

Launched 16th July, 1965. Investigation of solar cosmic rays. Investigation of
the energy spectrum and chemical composition of particles of primary cosmic
rays in the energy range up to 10^{14}eV. Investigation of nuclear interaction
of ultra-high energy cosmic rays up to 10^{12}eV. Determination of the absolute
intensity and energy spectrum of electrons of galactic origin. Determination of
the intensity and energy spectrum of gamma rays of the galaxy with energies
over 50 million eV. Solar cells, batteries. Decayed on 11th October, 1965.

SOURCE: Table of Artificial Satellites Launched Between 1957 and 1976.
International Telecommunication Union. Geneva (1977)

1966   FLIP-FLOP BONDING TECHNIQUE   M. Wiessenstern   (U. S. A.)
            (FLIP-CHIPS)   and G. A. S. Wingrove

        The flip-flop bonding structure and method was invented and subsequen-
tly patented in 1966 by Wiessenstern and Wingrove. Since that time nearly every
semiconductor manufacturer has experimented with various forms of flip-flop
bonding for the purpose of assembling integrated circuits, and possibly some
discrete components, into larger subsystems. To this day, no successful
method of flip-chip bonding has become generally utilized on the open market.

SOURCE: "A multichip package utilizing In-Cu flip-chip bonding" by
A. P. Youmans, R. E. Rose and W. F. Greenman. Proc. IEEE. Vol. 57. No:9
(September 1969) p. 1599.

SEE ALSO: "Semiconductor device assembly with true metallurgical bonds"
by M. Weissenstern and G. A. S. Wingrove. U. S. Patent No:3256465 -June 14, 1966.

"Joining semiconductor devices with ductile pads" by L. F. Miller.
3 rd Annual Hybrid Microelectronics Symposium (Oct. 29th, 1968)

1966   OAO-1 Satellite   (U. S. A.)

Launched 8th April, 1966. Orbiting Astronomical Observatory. To study ultra-
violet, X-ray and gamma ray regions of the spectrum. Batteries failed on
second day in orbit. Solar cells.

SOURCE: Table of Artificial Satellites Launched Between 1957 and 1976.
International Telecommunication Union. Geneva (1977)

1966   SURVEYOR-1   Satellite   (U. S. A.)

Launched 30th May 1966. Soft landed on the moon on 2nd June 1966. Transmitted
11 150 photos up to 13th July 1966. 3960 solar cells (77 W) batteries.

SOURCE: Table of Artificial Satellites Launched Between 1957 and 1976.
International Telecommunication Union. Geneva (1977)

1966   OPTICAL FIBRE COMMUNICATIONS   K. C. Kao and G. A. Hockham
                                                                        (U.S.A.)

            Another method of providing guidance, and of cunningly circumventing
the problem of light travelling in straight lines, is to use a fibre consisting of
a glass core having a high refractive index surrounded by a cladding of lower
index. As early as August 1964, in an address to the British Association for
the Advancement of Science, the author speculated on the use of light and glass
fibres in the telephone network, instead of electric currents and wires, but
developments did not start in earnest until publication of the classic article of
Kao and Hockham of STL in 1966. At the time the problem seemed a formidable
one; the attenuation of existing fibres was about 1000 dB/km, the band-width
was expected to be low and fibre bundles were fragile. Since then enormous
strides have been made resulting in fibre attenuations of 2 dB/km produced as
a matter of routine, bandwidths of 1 GHz in a 1 km length of fibre having a
diameter of 100 μm. and fibres coated with nylon which are too strong to be
broken by hand. Such fibres are flexible and capable of being incorporated into
simple but effective forms of cable. The bandwidth of a single fibre is much
greater, and the attenuation lower, than existing copper coaxial cables and the
diameter is considerably smaller. Thus the capacity of the present telephone
network could be very greatly increased, with little additional installation
expense, by the gradual introduction of optical fibre cables.

SOURCE: "Lasers and Optical Electronics" by W. A. Gambling. The Radio
and Electronics Engineer" Vol. 45. No:10 (October 1975) p. 541.

SEE ALSO: "Dielectric-fibre surface waveguides for optical frequencies"
by K. C. Kao and G. A. Hockham. Proc. Instn. Elect. Engrs. 113. pp. 1151--8
1966.

1966   BIOSATELLITE-1 Satellite   (U.S.A.)

Launched 14th December, 1966. To determine the effects of the space environment on various life processes and study the effect of weightlessness on the
life processes of certain organisms.and the effects of radiation on organisms
in weightlessness. Due to failure of the retro-rockets to fire, it was not
possible to recover the capsule. Batteries.Decayed on 15th February, 1967.

SOURCE: Table of Artificial Satellites Launched Between 1957 and 1976.
International Telecommunication Union. Geneva (1977)

1966   ATS-1 Satellite   (U.S.A.)

Launched 7th December, 1966. Experiments to advance the fields of spacecraft
communications (aircraft and ground) meterology (photos, transmission of
weather facsimile) and control technology. Number of scientific experiments to
measure the orbital environment of the satellite. 22 000 solar cells (185 W)
batteries.

SOURCE: Table of Artificial Satellites Launched Between 1957 and 1976.
International Telecommunication Union. Geneva (1977)

1966   LUNAR ORBITER-1   (U.S.A.)

Launched 10th August, 1966. Flying photographic laboratory. Obtained high
resolution photographs of various types of surface on the moon to assess their
suitability as landing sites for APOLLO and SURVEYOR spacecraft; monitored
the meteroids and radiation intensity in the vicinity of the moon; provided
precise trajectory information for use in improving the definition of the moon's
gravitational field. 10 856 solar cells, battery.

SOURCE: Table of Artificial Satellites Launched Between 1957 and 1976.
International Telecommunication Union. Geneva (1977)

1966   NITRIDE-OVER-OXIDE SEMICONDUCTORS   F. H. Horn (U. S. A.)

After more than three years of legal proceedings between GE and International Business Machines Corporation, the U. S. Patent and Trademark Office has upheld GE's claim to priority of the invention, thus reaffirming GE's right to the patent, number 3, 597, 667.

In one form of the GE invention, a thin film of silicon nitride is placed between the gate and silicon dioxide in metal-oxide-semiconductor field-effect transistors (MOSFETs). This structure virtually eliminates the contamination by alkali ions that previously caused widespread failure of the tiny devices.

In another application of the invention, these nitride-over-oxide layers are used in standard bipolar transistors as a surface and junction-sealing passivation layer. Both the manufacturing yield and reliability of modern integrated circuits and semiconductor devices are "substantially enhanced" by the GE invention.

The original GE patent application was filed on March, 1, 1966, following a discussion of semiconductor instability problems between GE scientists Dr. Dale M. Brown and Dr. Horn, during which Dr. Horn suggested silicon nitride over silicon dioxide as a passivation technique for overcoming these problems. The idea was tried and successfully demonstrated shortly thereafter.

SOURCE: GE Public Information Release (Oct. 11, 1976) (GE Research & Development Center, Schenectady NY 12301. p. 2.

1966   ESSA-1 Satellite   (U. S. A.)

Launched 3rd February 1966. Meteorogical satellite. Environmental Survey SAtellite. Part of the TIROS Operational System (TOS). Advanced vidicon camera system (AVCS) Switched off on 8th May 1967, 9100 solar cells, batteries.

SOURCE: Table of Artificial Satellites Launched Between 1957 and 1976. International Telecommunication Union. Geneva (1977)

1966   PAGEOS-1 Satellite   (U. S. A.)

Launched 24th June 1966. PAssive Geodetic Earth-orbiting Satellite to be used in obtaining information for precision mapping of the earth's surface. Carries no instrument.

SOURCE: Table of Artificial Satellites Launched Between 1957 and 1976. International Telecommunication Union. Geneva (1977)

1967   LASER TRIMMING OF THICK FILM RESISTORS   ---   (U. S. A.)

The trimming of electronic components started in 1967 when the first experiments were conducted on the trimming of thick film resistors with CO_2 lasers. Two years later "Q-switched" laser systems were installed in the General Motors Delco Electronics plant in Indiana for the manufacture of thick film voltage regulators for automobiles.

SOURCE: "Bright future for laser trimming" by W. B. Cozzens. Electronic Engineering. (February 1976) p. 58.

1967   SOYUZ-1 Satellite   (U. S. S. R.)

Launched 23rd April 1967. Manned spacecraft. Re-entered 24th April 1967 after 17 orbits. Failed to land. Pilot: V. Komarov killed. Solar cells.

SOURCE: Table of Artificial Satellites Launched Between 1957 and 1976. International Telecommunication Union. Geneva (1977)

1967   AUDIO NOISE REDUCTION SYSTEM   R.M. Dolby (U.S.A.)

        Utilizing the masking effect, together with signal compression and
expansion, the Dolby Laboratories A301 achieves noise reduction (a) by boosting
low-level signal components during recording whenever possible (compression),
followed by complementary attenuation during playback (expansion), and (b) by
the masking effect whenever the signal level is already so high that compression
and expansion are not possible.

        Since masking is less effective with noise frequencies somewhat
removed from the signal frequency, it is necessary to deal with the various
portions of the spectrum independently. The noise reduction system then yields
a lower - and apparently constant - noise level, the classical hush-hush or swish
of normal compression and expansion being absent.

        The A301 system splits the audio spectrum into four bands and com-
presses and expands each of these in an essentially independent manner. Separate bands are provided for the hum and rumble frequency range (80 Hz, low-
pass), for the mid-audio range (80 Hz - 3 kHz, band-pass), for medium high
frequencies (3 kHz, high-pass), and for high frequencies (9 kHz, high-pass).
A high-level signal in one band hence cannot prevent noise reduction in another
band in which the signal level may be low.

        From another point of view, the system effectively produces a record-
ing equalization characteristic which continuously conforms itself to the incoming signal in such a way as to improve the signal to noise ratio during playback.

SOURCE: Dolby Laboratories Technical Report A301.

SEE ALSO· "An audio noise reduction system" by R.M. Dolby. J. Audio Eng. Soc.
15, 383 (1967)

        "Audio noise reduction: some practical aspects" by Ray M. Dolby.
"Audio" magazine. (June & July 1968)

        "The Dolby noise-reduction system - its impact on recording" by
John Eargle, "Electronics World" (May 1969).

        "Wireless World Dolby Noise Reducer", Part 1 (An introduction to the
Dolby noise reduction system), by Geoffrey Shorter, "Wireless World" (May 1975)
pp. 200-205.

1967   DIADEME-1 (D1-C) Satellite   (France)

Launched 8th February, 1967. Geodetic satellite. Honeycomb laser reflectors.
Operational in spite of low apogee. 2304 solar cells, batteries. Transmitted
until 2nd January 1970.

SOURCE: Table of Artificial Satellites Launched Between 1957 and 1976.
International Telecommunication Union. Geneva (1977)

1967   ION BEAM COATING   K.L. Chopra
                                          M.R. Randlett   (U.S.A.)

        The history of Ion Beam Coating (IB C) covers a period of more than ten
years beginning with metallic coatings reported by Chopra and Randlett.
Carbon deposition with "diamond like" properties was initially reported by
Aisenberg and Chabot in 1970 and 1971. Rapid expansion of the field resulted
when Spencer Schmidt et al. showed that essentially any solid material can be
deposited when a target is bombarded with an energetic ion.

        Although ion beam milling has been accepted for several years as the
most desirable technique when compared with chemical or plasma etching for
the fabrication of high resolution micron or submicron circuitry in both research and production installations, ion beam deposition is just now being
accepted by research laboratories as the single economical process which
affords real flexibility in thin film fabrication. High resolution microfabrica-
tion process with electron beam or x-ray lithography require improved deposition technology. The answer may be ion beam coatings both for research and
production applications.

SOURCE: "Ion Beam Coating: A New Deposition Method", by George R.
Thompson, Jr., Solid State Technology, December 1978, p. 73.

SEE ALSO: "Duoplasmatron Ion Beam Source for Vacuum Sputtering of Thin Films" by K.L. Chopra and M.R. Randlett. Rev. Sci. Instrum., Vol. 38, No. 8 (1967), p.1147.

S. Aisenberg and R. Chabot, J. Vac. Sci. Technol., Vol. 6 (1970), p.112.

S. Aisenberg and R. Chabot, J. Appl. Phys., Vol. 42 (1971), p.2953.

E. G. Spencer and P.H. Schmidt, J. Vac. Sci. Tech., Vol. 8 (1971), p.368.

"Deposition and Evaluation of Thin Films by D.C. Ion Beam Sputtering", Schmidt, P.H. Castellano R.N., Spencer E.G., Solid State Technol., Vol. 15, No. 7,   (1972), p.39.

1967   "ROTATOR" CIRCUIT NETWORK   L.O. Chua   (U.S.A.)

        This paper presents a new linear, reciprocal, active two-port network element called a rotator, of which there are three types: an R-rotator, an L-rotator, and a C-rotator. They have the unique property that whenever a nonlinear resistor, inductor, or capacitor is connected to one port of an R-, L-, or C-rotator, respectively, the resulting two-terminal network behaves as a new resistor, inductor or capacitor whose characteristic curve is that of the original resistor, inductor or capacitor rotated by a prescribed angle about the origin.

        The rotator is realizable by either a π-network or a T-network of linear resistances, inductances or capacitances. It can also be realised by a balanced lattice network of linear elements. Operational laboratory models are reported, and experimental data agree remarkably well with theoretical predictions.

        The sensitivity, power rating, and stability performances of rotators are considered in detail in this paper and practical stability criteria are given. They are shown to be indispensable building blocks for realising multi-valued elements and some potential applications are described.

SOURCE: "The rotator - a new Network Component" by L.O. Chua. Proc. IEEE Vol. 55. No:9 (Sept.1967) p.1566.

1967   TRAPPATT Diode   H.J. Prager
           (Trapped plasma avalanche   K.K.N. Chang   (U.S.A.)
           transit time diode).   S. Weisbrod

        The trapatt mode was discovered in 1967 by Prager, Chang and Weisbrod. It has permitted the realisation of high-efficiency solid-state microwave oscillators and amplifiers. D.C.-r.f. conversion efficiencies as high as 60% are obtained at frequencies of 1 - 2 GHz, and several authors have reported efficiencies as high as 35% at X-band frequencies.

        Trapatt action occurs when a rapidly increasing reverse-bias voltage of magnitude greater than the breakdown voltage is applied across the depleted diode. An avalanche zone sweeps rapidly from the junction, through the depletion layer to the substrate, leaving in its wake a dense plasma of holes and electrons and collapsing the electric field. The diode drops into a low-voltage high-current state, and the carriers are said to be trapped, as their drift velocities fall well below their saturated values. As the carriers drift slowly out of the active region under low-field conditions, the electric field within the diode recovers. When the electric field has fully recovered and the current returns to essentially zero, the cycle is repeated. The current and voltage waveforms produced are favourable to the production of high efficiencies. The frequency of operation is much lower than in the impatt mode, since the carriers spend a long part of the cycle with drift velocities well below the saturated drift velocity.

SOURCE: "Design and performance of trapatt devices, oscillators and amplifiers" by C.H. Oxley, A.M. Howard and J.J. Purcell. IEE Solid-State and Electron Devices. Vol.1. No:1 (Sept.1976.) p.24.

SEE ALSO: "High power, high efficiency silicon avalanche diodes at ultra frequencies" by H.J. Prager, K.K.N. Chang and S. Weisbrod. Proc. IEEE Vol. 55 (1967) p.586.

1968   <u>AMORPHOUS SEMICONDUCTOR SWITCHES</u>   S.R.Ovshinsky (U.S.A)

Switching phenomenon have been noted for a decade but Ovshinsky first attracted international attention by producing a discrete component commercially in 1968. Figure 7(a) shows an early switch made by his company in which a thin film of undisclosed glassy amorphous material is sandwiched between massive carbon electrodes in a thermister type package. Another possible discrete geometry used in earlier studies is the cross-over sandwich using thin film electrodes (Fig. 7(b).

These devices and their more modern counterparts are so-called threshold-switches, and their circuit configuration and resulting I-V characteristics are shown in Fig. 8. These devices have the remarkable property of having an initial OFF resistance, corresponding to region (a), of order tens of megohms, which drops to a much lower ON resistance, region (c) of around 100 Ω when the voltage across it exceeds a certain threshold value V_{th} typically around 10^V. The switching, which occurs along a load line corresponding to the load resistor, region (b), happens in a time typically of order nanoseconds, but there is a delay time which can be as long as 10 μs, between the application of a switching voltage and the onset of switching. If the current in the On-state is allowed to fall below a minimum holding value, I_h, the device reverts to its high resistance state, along region (d). In this sense, the device behaves as a monostable element. Notice that the switch is insensitive to the polarity of the supply; the I-V curve for negative voltages is a mirror image of that for positive values.

<u>SOURCE:</u> "Amorphous semiconductor devices and components" by J.Allison and M.J.Thompson. The Radio and Electronic Engineer. Vol.46. No:1 (January 1976) p.12.

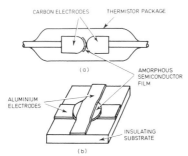

Fig. 7 Early discrete switches.

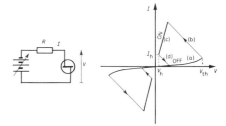

Fig. 8 Measuring circuit and voltage-current characteristic for a threshold device.

<u>SEE ALSO:</u> "Reversible electrical switching phenomena in disordered solids" by S.R.Ovshinsky. Phys.Rev.Letts.Vol.:21 (1968) p.1450.

1968   <u>IRIS (ESRO-1) Satellite</u>   (Europe ESRO)

Launched 17th May 1968. First European Space Research Organisation satellite. Measurements of the solar and cosmic radiations. 3456 solar cells, batteries. Decayed on 8th May 1971.

<u>SOURCE:</u> Table of Artificial Satellites Launched Between 1957 and 1976. International Telecommunication Union. Geneva (1977)

1968   <u>C-MOS (Complementary Metal-oxide-</u>   Westinghouse
            <u>semiconductor) Integrated Circuit</u>   GT & E Labs.   (U.S.A.)
                                                              RCA
                                                              Sylvania

At the October 1968 International Electron Devices Meeting in
Washington, the Westinghouse Molecular Electronics division announced two
experimental MOS field effect transistor schemes that may show the way to
more successful solid state memory systems. One, the MONOS (metal-oxide-
nitride-oxide-semiconductor) element for read-only memory applications, is a
non-volatile sandwich of silicon nitride and silicon dioxide. Turned on by a
negative voltage, MONOS could lead to circuits having all their active, passive
and storage elements on the same chip, according to Hung C. Lin, manager of
advanced techniques development.

The other development, a complementary MOS-bipolar structure, gets
around the fact that MOS FET's are usually limited to driving only low-
capacitance loads by placing vertical and lateral non bipolar transistors on the
same chip to drive higher loads. The resulting devices are 10 times faster
than most MOS FET's, Lin says, and are made with standard MOS manufacturing techniques. Potentially, he adds, the structures could be used not only as
large random-access memories but as logic and shift-register elements. And
the use of bipolar transistors at the output end could reduce the interface problems between MOS devices and others such as transistor-transistor logic or
diode-transistor logic.

The performance of complementary metal oxide semiconductor logic
is so much better than p-channel logic that there must be a good reason why its
not used more often. There is. It has been very difficult to fabricate n and p
channels in the same substrate.

Paul Richman, who with Walter Zloczower described the complementary MOS circuit at the International Electron Devices Meeting in 1968, says
that their difficulties with the extremely resistive substrate have been economic
rather than technological. The material is easy enough to make, but the only
source is a chemical firm in Germany.

Richman feels that the GT & E approach is the first simplified method
of making complementary MOS IC's. The RCA and Westinghouse fabrication
techniques have disadvantages, he says.

RCA's method of forming conventional n and p channels in the same
substrate requires extremely careful control of the diffusion process and
results in a rather high threshold voltage.

Westinghouse uses an elaborate procedure of etching pits, filling them
with epitaxial p-type material, then etching back to form the p channels. This
process involves critical mechanical operations and results in relatively slow
circuits.

Neither GT & E nor Sylvania has immediate plans for marketing
complementary MOS IC's. But Richman hopes that the new fabrication method
will open up the memory applications for which complementary MOS integrated
circuits are so well suited.

SOURCE: "Electronics Review - integrated electronics" Electronics.
(October 28th, 1968) p. 49.

1968   <u>SURVEYOR-7 Satellite</u>   (U.S.A.)

Launched 7th January 1968. Landed on moon on 10th January 1968. Took
21 000 photographs of lunar surface. Studied chemical characteristics of lunar
soil and dug seven trenches. 3626 solar cells (85 W) batteries.

SOURCE: Table of Artificial Satellites Launched Between 1957 and 1976.
International Telecommunication Union. Geneva (1977)

1968   THE "TRINITRON": Colour Cathode Ray Tube   Sony (Japan)

              A new colour CRT employing a single lens in-line gun was developed
        by Sony Corporation.   The single large diameter lens minimizes electron beam
        aberration, resulting in a high quality image.   Electrostatic plates attached to
        the top of the gun effectively converge the side beams at the phosphor screen.
        The unique arrangement of the electron optics of the gun permits modulating
        the electron scanning velocity, giving rise to further improvement of the picture
        image.   A new colour selection mechanism called the Aperture-Grill, which
        has a great number of slits instead of the holes or slots found in a conventional
        shadow mask, is incorporated in the CRT.   Since the Aperture-Grill has a
        greater beam transparency than a shadow mask, the CRT yields a brighter
        picture.   The vertically continuous phosphor stripes produces a high resolution
        image limited only by the electron beam diameter.   Additional advantages are
        that the Aperture Grill is less sensitive to terrestrial magnetism and that it is
        free from Moiré patterns.   The cylindrical face plate whose vertical curvature
        is almost infinite, reduces the ambient light problem.

              Sony Corporation has manufactured more than 20 million of this unique
        high-quality CRT, ranging from 5V" to 30V", the world's largest, with a
        variety of deflection angles.

        SOURCE :   "The 'TRINITRON' --- A New Colour Tube" by S. Yoshida,
        A. Ohkoshi & S. Miyaoka.   IEEE Trans., BTR-14, p. 19 - 27 (July 1968).

        SEE ALSO :   "The 'TRINITRON' --- A New Colour Tube" by S. Yoshida,
        A. Ohkoshi & S. Miyaoka.   Electronics & Radio Technician, Vol. 3, No. 4
        (December 1969).

        "A Wide-Deflection Angle (114°) TRINITRON Colour Picture Tube" by S. Yoshida,
        A. Ohkoshi & S. Miyaoka.   IEEE Chicago Spring Conference on BTR, June 12,
        1973.

        "25V Inch 114 degree TRINITRON Colour Picture Tube and Associated New
        Developments" by S. Yoshida, A. Ohkoshi & S. Miyaoka.   IEEE Chicago Spring
        Conference on BTR, June 10, 1974.

1968   "BARITT" DIODE   G.T.Wright (U.K.)

              In 1968 Wright described a new negative resistance microwave device
        based on the principle of barrier controlled injection and transit time delay -
        the BARITT diode.   His simple analysis suggested that the device should oper-
        ate at moderate power and low noise level.   In the same year, independently,
        Ruegg presented a paper on the simplified large-signal theory of a similar
        punch-through structure giving considerably optimistic prospects - an estimated
        efficiency of the order of 20% and power output of 10-100 W at 10 GHz.   These
        theoretical works were confirmed experimentally in 1970 when Sultan and Wright
        achieved negative resistance in npn silicon structures, and subsequently oscil-
        lations in pnp structures and in 1971 when Coleman and Sze reported oscillations
        in metal-semiconductor-metal structures.   Several experimental papers have
        since been presented, comparing the properties of different BARITT diode
        structures and pointing out the reliable and low-noise operation of the device at
        moderate power levels.

        SOURCE: "Large-signal analysis of the silicon pnp-BARRIT diode" by
        M.Karasek. Solid State Electronics Vol.19. (1976) p.625.

        SEE ALSO: G.T.Wright.   Elect.Lett.4. 543 (1968)

                    H.W.Ruegg.   I.E.E.E. Trans. ED-15, 577 (1968)

1968   APOLLO-7 Satellite   (U.S.A.)

        Launched 11th October 1968. First APOLLO manned spacecraft flight. Three
        astronauts on board: W.Schirra, D.Eisele, W.Cunningham. Tests for prep-
        aration of the manned lunar landing programme. Spacecraft landed on 22nd
        October,1968, 200 nautical miles south-south-west of Bermuda after 163 orbits
        (260.2 hours). Fuel cell power plant, storage batteries.

        SOURCE: Table of Artificial Satellites Launched Between 1957 and 1976.
        International Telecommunication Union. Geneva (1977)

1968   INTEGRATED CIRCUIT
        ALUMINIUM METALLISATION   R. Noyce   (U.S.A.)

        An insight into Noyce's style of technical leadership is provided by
Gordon Moore, a chemical physicist who was one of the eight founders of
Fairchild Semiconductors, and who quit in 1968 to join Noyce in starting Intel,
where he is now president and chief executive officer. "Bob was certainly the
idea man in the group......I can think of two things that at the time impressed
me even more than what he did for the integrated circuit. One was the use of
aluminium for transistor contacts. I remember struggling with all kinds of
complex alloys to find one metal that would make contact with both the emitter
and the base. One day Bob said, "Why don't you try aluminium".

        "So I tried aluminium and it worked beautifully. That really got the
double-diffused transistor out of the laboratory as a practical device. Then
there was the use of nickel to fabricate junctions with good electrical contacts.
Bob suggested it one day and it worked. These were both cases where he
proposed something I really thought wouldn't work and which then got us past
significant barriers."

SOURCE: "The genesis of the integrated circuit" by M.F. Wolff. IEEE
Spectrum (August 1976) p. 49.

1968   "MUTATOR" CIRCUIT NETWORK   L.O. Chua   (U.S.A.)

        The basic problem of synthesizing a nonlinear resistor, inductor, or
capacitor with a prescribed i-v, ϕ-i, or q-v curve is solved by introducing
three new linear two-port network elements, namely the mutator, the reflector,
and the scalor. The mutator has the property that a nonlinear resistor is
transformed into a nonlinear inductor, or a nonlinear capacitor, upon connecting
this resistor across port two of an appropriate mutator. The reflector has the
property that a given i-v, ϕ-i, or q-v curve can be reflected about an arbitrary
straight line through the origin. The scalor is characterized by the property
that any i-v, ϕ-i, or q-v curve can be compressed or expanded along a horizontal direction, or along a vertical direction. Using these new elements as
building blocks, it is shown that any prescribed single-valued (which need not be
monotonic) i-v, ϕ-i, or q-v curve can be synthesized.

        Active circuit realizations for each of these new elements are given.
Laboratory models of mutators, reflectors, and scalors have been built using
discrete components. Oscilloscope tracings of typical mutated, reflected and
scaled, i-v, ϕ-i, and q-v curves are given. The experimental results are in
good agreement with theory at relatively low operating frequencies. The practical problems that remain to be solved are the stability and frequency limitation of the present circuits.

SOURCE: "Synthesis of new nonlinear network elements" by L.O. Chua. Proc.
IEEE Vol. 36. No:8 (Aug. 1968) p. 1325.

SEE ALSO: "Additional types of mutators and active RC synthesis using
mutators" by T. Murata. Int. J. Electronics. Vol. 42. No:1 (1977) p. 33.

1969   TACSAT-1 Satellite   (U.S.A.)

Launched 9th February, 1969. Governmental telecommunications satellite
(TACtical communications SATellite). Capacity comparable to 10 000 two-way
telephone channels; upper de-spun portion contains a biconical horn for telemetry
and command, two microwave horns and five helical antennae for experimental
reception by mobile light-weight surface and airborne terminals having antennae
as small as 0.30 m in diameter, 60 000 solar cells, batteries.

SOURCE: Table of Artificial Satellites Launched Between 1957 and 1976.
International Telecommunication Union. Geneva (1977)

1969   AZUR Satellite   (Fed. Rep. of Germany)

Launched 8th November, 1969. Carries seven experiments designed to study
the earth's radiation belt, the aurorae, and solar particle events. More than
5000 solar cells, battery. Launched by NASA.

SOURCE: Table of Artificial Satellites Launched Between 1957 and 1976.
International Telecommunication Union. Geneva (1977)

1969   MAGNETIC BUBBLES   A.H.Bobeck, R.F.Fischer
                                            A.J.Perneski, J.P.Remeika, (U.S.A.)
                                            and L.G.Van Uitert

        A magnetic material usually consists of arrays of discrete localized
volumes of material, defined as domains. Each domain separated from its
neighbours by domain walls has a preferential orientation of the magnetization
vectors of all of the atomic magnetic moments within its volume. Domains may
have different orientations with respect to each other depending on their net
energy content and the force vectors acting upon them. In some cases, domains
can be produced and moved about in a thin plate or layer in a reproducible way.
This is the case for a number of ferro-magnetic and ferrimagnetic materials.
Figure 9 shows a typical domain structyre in a magnetic garnet at zero magnetic field with the random, worm-like domain patterns maintained, in the steady
state, by the inherent uniaxial magnetic anisotrophy of this material. If an
increasing magnetic field is applied perpendicular to the plane of the plate in
Fig.9. then the unfavourably oriented domains (with respect to the applied field)
may be made to shrink and then finally to collapse into cylindrical domains which
look like "magnetic bubbles" if they are observed in polarized light under a
microscope; they can be displaced in the direction of an applied magnetic gradient and their presence or absenbe at a certain position of the plate constitutes
the binary-coded information stored in the memory.

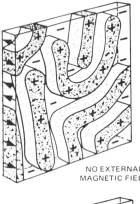

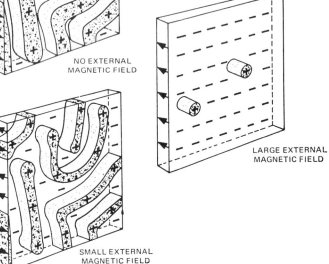

NO EXTERNAL
MAGNETIC FIELD

SMALL EXTERNAL
MAGNETIC FIELD

LARGE EXTERNAL
MAGNETIC FIELD

Fig. 9. Oppositely oriented magnetic domains shown with and
without an applied external magnetic field having indicated polarity
and direction. (After Bobeck and Scovil.[71])

SOURCE: "Magnetic domain bubble memories" by J.L.Tomlinson and H.H.
Weider. The Radio & Electronic Engineer. Vol.45. No:12.(Dec.1975) p.727.

SEE ALSO: "Application of orthoferrites to domain-wall devices" by A.H.
Bobeck, R.F.Fischer, A.J.Perneski, J.P.Remeika and L.G.Van Uitert.
IEEE Trans. on Magnetics. MAG-5, pp.544-54 (1969)

1969   SKYNET-A Satellite   (U.K.)

Launched 22nd November,1969. Government communication satellite to be
placed in synchronous orbit at 45°E longitude over Indian Ocean. Spin stabilized
7236 solar cells, batteries. Launched by NASA.

SOURCE: Table of Artificial Satellites Launched Between 1957 and 1976.
International Telecommunication Union. Geneva (1977)

1969   PARCOR   N.T.T. (Japan)

    A new method of speech analysis and synthesis, in which the speech spectra are expressed with the use of the partial autocorrelation (PARCOR) coefficients was proposed and developed by Nippon Telegraph and Telephone Public Corporation in 1969.

    In order to obtain a speech of high quality, it is necessary to extract the spectral envelope parameters and the driving source parameters and to reproduce these features of the original speech as accurately as possible. In extracting these parameters efficiently, PARCAR coefficients were introduced.

    Speech signals are sampled at every 125 μsec through a PARCAR analyzer. The signal amlitude just before and just after the sampling time is predicted with a set of n samples by the least square method. Then the deviation between the predicted value and the real value is measured. The extracted eight informations obtained during a certain time (one frame 15 msec) are the PARCOR coefficients. Speech informations are compressed to 57 bits which consist of the PARCOR coefficients of 8 x 5 = 40 bits (5 bits to a PARCOR coefficient) and driving source parameters of 17 bits. This corresponds to 3, 800 bits/sec. The speech compression by the PARCOR method is about one fifteenth of the 56, 000 bits/sec of the PCM method and is very efficient. Speech synthesis is just an inverse process of the speech analysis mentioned above. The PARCOR type speech analysis and synthesis system is superior to the conventional parameter editing and synthesis method. When the PARCOR synthesizer is composed of the recent high-speed logic elements (LSI IC), it is possible to respond simultaneously to many telephone circuits.

    This speech analysis and synthesis system based on PARCOR coefficients has many response words compared with that of the conventional recording and editing system and various flexible speech response services will be possible. Moreover, this analysis method is expected to produce new services such as the automatic speech recognition and the perception of speaking voices, etc.,

SOURCE : "Speech Analysis and Synthesis System based on Partial Auto Correlation Coefficients" by F. Itakura and S. Saito. Meeting Record of the Acoustical Society of Japan, 2-2-6, (October 1969) (In Japanese).

SEE ALSO : "Digital Filtering Techniques for Speech Analysis and Synthesis" by F. Itakura and S. Saito. Conference Record 7th Int. Cong. Acoust., 25Cl, Budapest, 1971.

"New Speech Analysis and Synthesis System PARCOR" by Fumitada Itakura. Nikkei Electronics, Vol. 2, No. 12, pp. 58 - 78, 1973 (In Japanese).

1969   "BUCKET-BRIGADE" DELAY CIRCUIT   F. L. G. Sangster
                                                   K. Teer   (U.S.A.)

    The general principle is that the signal to be delayed is sampled and stored in a cascade of capacitors inte rconnected by switches operated at the same frequency as the signal sampler.

    As a new signal sample can evidently not be stored in a capacitor before the signal sample present is completely removed, only half the number of capacitors actually do store information at any moment, the others being empty.

    In the past only rather complicated circuitry has been proposed for this function, so that even in integrated form there was no chance for an inexpensive compact design. A much simpler solution presents itself when signal sample transfer is not established by a charge transfer in the direction of signal travel but in the opposite direction, by what is essentially a charge deficit transfer. This principle leads to a much simpler resistorless circuit suitable for realization in integrated -circuit form.

SOURCE: "Bucket-Brigade electronics - possibilities for delay, time-axis conversion and scanning." by F. L. J. Sangster and K. Teer. IEEE Journal of Solid State Circuits. Vol. SC-4 No:3 (June 1969) p. 131.

1969   MICROELECTRONICS (Bipolar)   Bell Laboratories   (U.S.A.)
          C D I (Collector Diffusion Isolation)   Ferranti   (U.K.)

                    In 1970 manufacturers began to investigate bipolar processes which
          seemed to offer prospect of being competitive with m.o.s. For example there
          was the c.d.i. process (collector diffusion isolation) developed first at Bell
          Labs and then by Ferranti, the Isoplanar process of Fairchild, the Process
          IV which was suggested at Plessey's research centre at Caswell, and the
          Dutch Locos process developed by Philips. All of these were compatible with
          circuits which could operate in excess of 1.5GHz and all of then had the advan-
          tage of using less surface area than earlier processes. The c.d.i. system for
          example, started with a slice of 10 to 20Ω. cm p-type silicon into which n^+-layers
          were diffused. These were later to be the collectors of transistors formed in a
          1Ω. cm p-type epitaxial layer put down on top of them. The n^+ diffusions were
          made through the epitaxial layer to make contact with the now buried n^+ layers
          laid down at first. These not only acted as collector contacts but isolated the
          area within. In this base area the n^+ emitter diffusion is made, as well as any
          second emitter for a Shottky diode. After the oxide has been deposited and holes
          cut in it to gain access to the electrodes, silicon is grown in the holes to the s
          same level as the oxide, thus giving a flat surface.

          SOURCE: "The semiconductor story" by K.J.Dean. Wireless World.
          (April 1973) p.170.

          SEE ALSO: "Collector kdiffusion isolated integrated circuits" by B.T.Murphy
          V.J.Glinski, P.A.Gary and R.A.Pedersen. Proc.IEEE Vol.57. No:9
          (Sept.1969) p.1523.

1969   "MAGISTOR" MAGNETIC SENSOR   E.C. Hudson
                                                   I.B.M. (U.S.A.)

                    The Magistor, invented by E.C. Hudson, Jr., is a dual-collector planar
          transistor operating below avalanche breakdown with beta values in the range of
          30 to 100. In effect, this beta appears to amplify a typical Hall voltage. The
          sensitive axes are orthogonal to the substrate surface.

          SOURCE: "A magnetic sensor utilizing an avalanching semiconductor device"
          by A.W. Vinal. IBM J. Res. Dev. Volume 25, No. 3 (May 1981). p.196.

          SEE ALSO: Robert H. Cushman, Ed., "Transistor Responds to Magnetic
          Fields,". Electronic Design News (February 15, 1969) pp.73 - 78.

1969   SEMICONDUCTOR MEMORY SYSTEM   B.Agusta
                                          J.K.Ayling
                                          R.D.Moore   (U.S.A.)
                                          G.K.Tu

                    Since the first disclosure, by Agusta and Ayling et al, of the actual
          application of semiconductor memory in a computer system, the rapid develop-
          ment of silicon technology has led to the gradual (and probably eventually com-
          plete) replacement of magnetics by semiconductors in the memory field due to
          the speed, density, cost and power advantages of the latter.

          SOURCE: "Nonvolatile semiconductor memory devices" by J.J.Chang. Proc.
          IEEE. Vol.64 No:7, (July 1976) p.1039.

          SEE ALSO: "A 64 bit planar double-diffused monolithic memory chip" by
          B.Agusta ISCC Digest of Tech.Papers (February 1969) p.38.

                    "A high-performance monolithic store" by J.K.Ayling, R.D.Moore and
          G.K.Tu. ISCC Digest of Tech.Papers (February 1969) p.36.

1970   TUNG-FANG-HUNG (CHINA 1) Satellite   (People's Republic of China)

          Launched 24th April,1970. First satellite of the People's Republic of China.

          SOURCE: Table of Artificial Satellites Launched Between 1957 and 1976.
          International Telecommunication Union. Geneva (1977)

1970   CHARGE COUPLED DEVICES   W.S.Boyle & G.E.Smith   (U.S.A.)

        Storing charge in potential wells created at the surface of a semi-
conductor and moving the charge (representing information) over the surface
by moving the potential   minima.

Principle of Operation
Creation of Potential Wells

Consider the application of an increasingly positive voltage to the gate of the m.o.s. structure shown in Fig. 1(a). As is to be expected, the influence of the gate on the underlying semiconductor closely resembles that of the gate of an m.o.s. transistor, i.e. a small positive gate. bias causes the repulsion of majority carriers (i.e. holes) from the semiconductor immediately beneath the gate (Fig. 1(b)), whilst a gate bias in excess of the threshold voltage, V_{th}, makes it possible for an inversion layer to form at the oxide-semiconductor interface (Fig. 1(c)). Unlike the situation in an m.o.s. transistor, however, the existence of a gate voltage in excess of V_{th} does not necessarily mean that an inversion layer will form immediately in the structure of Fig. 1(a). This is because, whereas in an m.o.s.t. there is a source diffusion capable of supplying, almost instantaneously, a large number of minority carriers to form the channel, no such source exists adjacent to each c.c.d. electrode.

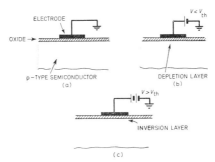

Fig. 1. Single c.c.d. electrode showing the creation of depletion and inversion layers under the influence of an increasingly positive electrode voltage.

SOURCE: "Charge coupled devices - concepts, technologies and applications" by J.D.A.Benyon. The Radio & Electronic Engineer. Vol.45. No:11 (Nov.1975) p.647.

SEE ALSO: "Charge coupled semiconductor devices" by W.S.Boyle and G.E.Smith. Bell System Tech.J. Vol.49 (1970) p.583.

1970   X-RAY LITHOGRAPHY FOR   E.Spiller   E.Castellani
        BUBBLE DEVICES   R.Feder   L.Romankiw   (U.S.A.)
                                       J.Topalian   M.Heritage

        X-rays have been used for several decades to obtain images of objects. First proposals to use x-rays for the fabrication of microelectronic devices were made in 1970. X-ray lithographic systems are very simple, they have a high throughput because many wafers can be exposed simultaneously and they have a resolution which is at least as good as that of electron beam systems. In 1972 Spears and Smith produced the first devices using proximity printing with x-rays and demonstrated the high resolution capability of x--ray lithography. In particular, x-ray lithography lends itself to the fabrication of high resolution devices requiring no alignment capability, such as magnetic bubble devices.

SOURCE: "X-ray lithography for bubble devices" by E.Spiller, R.Feder, J.Topalian. E.Castellani, L.Romankiw and M.Heritage. Solid State Technology (April 1976) p.62.

SEE ALSO: "X-ray projection printing of electrical circuit patterns" by R.Feder. IBM Report TR22.1065. (August 1970)

        "High resolution pattern replication using soft x-rays" by D.L.Spears and H.I.Smith. Electron. Lett. Vol.8. 102 (1972)

        "Evolution of bubble circuits processed by a single mask level" by A.H.Bobeck, I.Danylchuck, F.C.Rossol and W.Strauss. IEEE Trans. on Magnetics MAG-9, 474 (1973).

1970   "FLOPPY DISC" RECORDER   IBM (U.S.A.)

        The concept of floppy discs is rapidly becoming accepted in many
areas of data processing. Indeed, it has become a very much in-device when
talking about data entry and data communications systems. And yet the very
concept of floppy discs appears to have started almost by accident.

        The first, designed and developed by IBM, was simply a component
of the IBM 370 when introduced in 1970 - and a pretty obscure one at that. In
its original form it was part of the diagnostic system for the 370/155 and 165
and formed part of the controller for the IBM 3300 disk drive. At that stage
it was a read-only device and enabled diagnostic programs to be introduced
quickly, to identify faults and help to reduce maintenance time.

        Indeed, it was not until 1973 that it van be properly said to have formed
an integral part of an IBM system. That was the 3740 data entry system.....
which promptly established an industry standard and, because it provided three
times the storage and much faster access, rendered what competition there
was virtually obsolete overnight.

SOURCE: " 'Accident' became the floppy disc" by W. Boffin. Electronics
Weekly (May 12, 1976) p. 11.

1970   NATO-1 Satellite   (International NATO)

Launched 20th March, 1970. Telecommunication satellite. Geostationary spacecraft stationed over the equator at approximately 18°W longitude. Hundreds of
communications of various types (voice, wide-band data, telegraph and facsimile
data). More than 7000 solar cells, batteries.

SOURCE: Table of Artificial Satellites Launched Between 1957 and 1976.
International Telecommunication Union. Geneva (1977)

1970   OFO-1 Satellite   (U.S.A.)

Launched 9th November, 1970. Orbital Frog Otolith satellite. Carried two frogs
which were monitored for about five days alternately in the weightlessness of
space and during periods of simulated gravity created by spinning. Battery packs.
Decayed on 9th May 1971.

SOURCE: Table of Artificial Satellites Launched Between 1957 and 1976.
International Telecommunication Union. Geneva (1977)

1971   LIQUID CRYSTAL STUDY OF OXIDE DEFECTS   J.M.Keen (U.K.)

        One of the most convenient techniques was first reported by Keen.
It consists of introducing, between the oxidized silicon and a tin oxide coated
glass slide, a thin film of negative nematic liquid crystal. On applying a
voltage across this "capacitor" structure, defects can be seen as highly
turbulent regions of liquid crystal. For plane electrodes without an oxide layer
the same turbulence is present everywhere. In the case of oxides containing
defects, however, the turbulence is particularly violent making location of a
defect easy.

SOURCE: "Polarity dependent oxide defects located using liquid crystals" by
A.K.Zakzouk, W.Eccleston and R.A.Stuart. Solid State Electronics (1976)
Vol.19. p.133.

SEE ALSO: J.M.Keen Electron Lett.7. 15. 432 (1971)

1971   SHINSEI Satellite   (Japan Space & Aeronautics Research Institute of
                                        Tokyo University.)

Launched 28th September, 1971. Scientific observation satellite. Objective: to
measure solar and cosmic radiation.

SOURCE: Table of Artificial Satellites Launched Between 1957 and 1976.
International Telecommunication Union. Geneva (1977)

1971   HOLOGRAM MATRIX RADAR   K. Iizuka
                                       V. K. Nguyen   (Canada)
                                       H. Ogura

        The concept of hologram matrix is proposed. This concept was incor-
porated into the design of a novel radar which, unlike conventional radars,
determines the distance by the spatial distribution of the scattered wave rather
than by the lapse of time. The radar based upon this principle was developed
and built for the purpose of mapping ice thickness in the range of 0.5~5m. but
it has potential applications in other fields.

        Such a radar has real-time processing capability, resulting from an
amalgamation of the antenna and computer subsystems. The programability of
the radiation pattern by software of the processing simplifies the construction
of the radar. Capability of dual focussing of the transmitter and receiver
eliminates the necessity of either pulsing, or frequency modulation of the
transmitting signal. Superior performance in the short range, with high resolution, is particularly advantageous for measuring lossy ice.

        These features were substantiated by experimental results obtained
from the field operation of the system.

SOURCE: "A hologram matrix radar" by K. Iizuka, H. Ogura, J. L. Yen,
Van-Khai Nguyen and John R. Weedmark. Proc. IEEE Vol. 64. No:10
(Oct. 1976) p. 1495.

SEE ALSO: "Review of the electrical properties of ice and HISS down-looking
radar for measuring ice thickness" by K. Iizuka, V. K. Nguyen and H. Ogura.
Presented at the Aerospace Electronic Symposium in Toronto, Canada on March
16, 1971. Text is published in Can. Aeronaut. Space J. Vol. 17, No:10, pp. 429-
430, and pp. M28-M33. Dec. 1971.

        'Hologram matrix and its application to a novel radar" by H. Ogura and
K. Iizuka. Proc. IEEE (Lett.) Vol. 61, pp. 1040-1041, July 1973.

1971   CARRIER-DOMAIN MAGNETOMETER   B. Gilbert   (U.S.A.)

        A carrier-domain device, as conceived by Gilbert, consists of an
elongated bipolar transistor, within which emitter-current flow is restricted
to a small region known as a domain. The domain can be moved within the
device, subject to an external signal. Using this concept novel devices,
whose functions arise directly from their geometry, can be designed. One
form of carrier-domain device proposed by Gilbert is a magnetic-field sensor
in which two domains are caused to rotate together around a circular device
by the application of a magnetic field normal to the silicon surface. Output
current pulses are produced at a rate proportional to the magnetic-flux density.
The first successful carrier-domain magnetometer (c.d.m.) based on this
design has been fabricated and operated by the authors, and brief details of its
operation have been published.

SOURCE: "The carrier-domain magnetometer: a novel silicon magnetic
field sensor". M.N. Manley and G.G. Bloodworth. Solid-State and Electron
Devices, November 1978, Vol. 2, No. 6, p. 176.

SEE ALSO: GILBERT, B.: "New planar distributed devices based on a
domain principle". IEEE ISSCC Technical Digest, 1971, p. 166. (Ref. 1), p.183

GILBERT, B.: "Novel magnetic field sensor using carrier domain rotation:
proposed device design". Electron. Lett., 1976, 12, pp. 608-610. (Ref. 2), p.183

MANLEY, M.H., BLOODWORTH, G.G., and BAHNAS, Y.Z.: "Novel magnetic
field sensor using carrier domain rotation: operation and practical performance"
ibid., 1976, 12, pp. 610-611. (Ref. 3), p. 183.

1971   ELECTRONIC DIGITAL WATCH   Time Computer Corporation (U.S.A.)

        The first electronic digital watch was introduced in the fall of 1971:
the Pulsar, retailing for $ 2,000 with an 18-carat gold case bracelet. Touch a
a button and the light-emitting diodes showed the time. (It took about a year
for Pulsar to add the day and date.)

SOURCE: Electronics (April 17, 1980), p. 401.

1971   FAMOS (Floating-gate avalanche-injection
            metal-oxide-semiconductor) Integrated   D. Frohman-Bentchkowsky
            Circuit.   (U.S.A.)

            Famos describes the floating-gate avalanche-injection metal-oxide
semiconductor transistor that Dov Frohman-Bentchkowsky developed at Intel
Corp. in 1971.

            The Famos device is essentially a silicon-gate MOS field-effect
transistor in which no connection is made to the floating silicon gate. Instead,
charge is injected into the gate by avalanches of high-energy electrons from
either the source or the drain. A voltage of -28 volts applied to the pn junction
releases the electrons.

            Data is stored in a Famos memory by charging the floating-gate
insulator above the channel region. The threshold voltage then changes, and
the presence or absence of conduction is the basis for readout.

            The Famos cell has generally been considered more reliable than
nitride storage mechanism used in reprogrammable metal-nitride-oxide-semi-
conductor memories. In MNOS memories, carriers tunnel through a thin oxide
layer into traps at the oxide-nitride interface. But a partial loss of stored
charge during readout limits the number of readout cycles to approximately 10^{11}.

            In Famos memories, on the other hand, there is no loss of charge due
to reading. Moreover, over time, the loss of stored electrons is negligible,
less than one per cell per year, and information retention is excellent.

            SOURCE: "The Famos principle". Electronics. (March 3rd, 1977) page 109.

1971   DSCS-1 Satellite   (U.S.A.)

Launched 3rd November, 1971. Defense Satellite Communication System. Synchronous satellite carrying multichannel communications payload. Four
antennae, two for wide earth coverage and two with narrow beams for ground
controlled direction beaming for high-volume communications. Capacity: 1300
circuits.

            SOURCE: Table of Artificial Satellites Launched Between 1957 and 1976.
International Telecommunication Union. Geneva (1977)

1971   CERAMIC CHIP CARRIER   3M Co.,   (U.S.A.)

            A popular IC package was the ceramic chip-carrier. About one third
the size of a comparable DIP, it originated in 1971 at the 3M Co., in St. Paul,
Minn. It was a square, multilayered ceramic package whose bottom periphery contained a pattern of gold bumps on 40- or 50-mil centers. The chip
was bonded to a gold base pad inside a cavity within the ceramic. The small
hermetically sealed package could be easily attached or removed from pc
boards and hybrids.

            SOURCE: Electronics (April 17, 1980), p. 389.

1971   OREOL-1 Satellite   (France/U.S.S.R.)

Launched 27th December, 1971. Objective: to study the polar aurora, including
proton intensity and ion composition. Arcade Franco-Soviet equipment.

            SOURCE: Table of Artificial Satellites Launched Between 1957 and 1976.
International Telecommunication Union. Geneva (1977)

1971   SALYUT-1 Satellite   (U.S.S.R)

Launched 19th April, 1971. Objectives: scientific research and testing of on-
board systems and units. Control by remote command or by crew. Visited
by crews of Soyuz-10 and Soyuz-11. The latter spent 23 days in Salyut.
Decayed on 11th October, 1971.

            SOURCE: Table of Artificial Satellites Launched Between 1957 and 1976.
International Telecommunication Union. Geneva (1977)

1972   <u>MICROCOMPUTER</u>   Intel (U.S.A.)

For three years a great revolution has been taking place in digital
electronics. Since 1972, when the first microcomputer was introduced by Intel,
these devices have been very successfully used in a wide range of applications,
including process control, data communications, instrumentation and commercial systems. The key to this success is due to the price/performance ratio
enhancement that occurs when microcomputers are used in a system when
compared with more traditional approaches.

<u>SOURCE</u>: "The microcomputer comes of age" by H. Kornstein.
"Microelectronics" Vol. 8. No: 1 (1976) p. 17.

1972   <u>VIDEO GAMES</u>   Magnavox (U.S.A.)

First to market a video game consumers could buy and take home was
Magnavox, in 1972. However, their original Odyssey game was not an immediate sensation, perhaps because it had no automatic score-keeping feature,
lacked sounds, and required a plastic overlay on the TV screen to simulate the
net, goals and boundaries of a playing field (static electricity held the overlay
in place). Since then , Odyssey has evolved through four model changes and is
now offered with automatic serve, digital scoring and sound for $89.95. Also,
all stationary playing-court features are electronically generated. Magnavox's
game circuits are produced by Texas Instruments and the General Instrument
Corp.

The second major milestone in the evolution of home video games was
established just over a year ago when Atari and Sears teamed up to produce and
market Hockey Pong for the 1975 Christmas season. The product had many of
the important features associated with Atari's successful line of coin games and
sold briskly. By working closely with one of its integrated circuit suppliers
(American Microsystems, Inc.) Atari was assured of a supply of dedicated, prop-
rietary game chips. These chips greatly reduced the parts count, and costs,
associated with Atari's original coin-game products that were constructed with
hundreds of standard logic circuits. The coin games normally sell for $1000-
$3000 each - but consumer acceptance of the add-on TV version was expected to
peak somewhere under $100 retail.

Meanwhile, with Odyssey more streamlined and the Atari/Sears
venture a proven success, the lucrative aspects of consumer TV games became
obvious to other entrepreneurs. For many, the first opportunity to enter the
market came in March 1976 when the General Instrument Corp. (GI) Hicksville,
N.Y., announced production of its AY-3-8500 TV-game integrated circuits.
Product acceptance was swift and favourable. New companies anxious for an
early shot at the home video game sweepstakes quickly snapped up GI's projected 1976 production capacity of game chips.

<u>SOURCE</u>: "Electronic Gamesmanship" by D. Mennie. IEEE Spectrum (Dec. 1976)
p. 27.

<u>SEE ALSO</u>: "Video games : Perishable or Durable" Journal of Electronics
Industry. Japan. Vol. 23. No: 10 (Oct. 1976) p. 38.

1972   <u>LANDSAT-1 (ERTS-1) Satellite</u>   (U.S.A.)

Launched 23rd July 1972. Earth Resources Technology Satellite. Objectives: to
obtain coverage of the United States and other major land masses with multi-
spectral, high spatial resolution (60 m) images of solar radiation reflected from
the earth's surface. These images will be used in agricultural, geological, geo-
graphical, hydrological and oceanographical research.

<u>SOURCE</u>: Table of Artificial Satellites Launched Between 1957 and 1976.
International Telecommunication Union. Geneva (1977)

1972   <u>MICROELECTRONICS (V-MOS technique)</u>   T.J.Rodgers   (U.S.A.)

Because of the work of a 27-year-old research engineer, American Microsystems Inc., in Santa Clara, California, is on the verge of committing itself in a big way to V-MOS - an n-channel metal-oxide semiconductor technology that will compete with the new, faster and denser bipolar static designs and processes.

The engineer is T.J.Rodgers, who, as a doctoral candidate in electrical engineering at Stanford University in nearby Palo Alto, invented the V-groove MOS process (Electronics, September 18. p. 65). His goal was to push MOS technology to its limits so it would achieve bipolar speeds as well as high speed-power products and high packing densities in read-only and static random-access memories, random logic and microprocessor designs.

<u>SOURCE</u>: "Young EE's ideas to alter AMI's direction" Electronics. (January 22nd,1976) p.14.

1972   <u>NITROGEN-FIRED COPPER WIRING</u>   J.D.Grier   (U.S.A.)

In late 1971, market researchers at Owens-Illinois Inc., Toledo, Ohio, decided that the rising cost of precious metals made the time right for research on non-noble conductors for thick-film microelectronics. A young chemist, John D.Grier, who had been with the firm for five years, was assigned as program manager.

Grier decided to concentrate on creating a workable nitrogen-fired copper paste. There had been earlier research on copper pastes, but these compositions used 100-micrometer copper particles to produce conductors with poor peel strength and low conductivity.

He went to 3-to-5-μm copper particles for the functional phase of the ink and found both a glass binder and vehicle that could survive firing at about 800°C. The late 1972 result was a patented, practical, screenable copper paste that had good peel strength and conductivity. At that point Grier correctly predicted that the new copper paste would be suitable for microstrip and thick-film hybrid applications.

<u>SOURCE</u>: "Electronics" (Oct.28, 1976) p.121.

1972   <u>X-RAY SCANNER</u>   E.M.I.   (U.K.)

The skull surrounds the brain and provides a very good protection for this most delicate and vital organ; it also heavily attenuates diagnostic X-rays. The brain is a relatively homogeneous organ, when imaged by X-rays, which does not have much contrast to show up its structure. These two problems make imaging of the brain by conventional X-radiography of very limited diagnostic value. Contrast techniques can be used to improve the imaging but they do involve some risk to the patients and the need for hospitalisation. They are expensive.

In 1972, EMI Limited introduced computerised axial tomography to overcome these limitations. This radical new technique was developed at the Central Research Laboratories of EMI. Clinical trials rapidly showed that this was a major advance in diagnostic imaging.

In computerized axial tomography the patient is scanned by a tightly collimated narrow beam of X-rays. The transmitted beam is detected and converted to an electric signal after passing through the patient. Another detector is used in the reference mode to measure the primary X-ray beam.

The frame, carrying the X-ray source and detectors traverses linearly across the patient, a large number of readings of X--ray intensity are taken and stored as it traverses, the gantry is then indexed round by a small angle and the process is repeated. This series of traverses and angular movements is repeated until a large matrix of data has been acquired.

The computer then uses this data to calculate the X-ray absorption coefficient map of this cross section of the anatomy. This can then be displayed as a brightness modulation map on a cathode ray tube or printed as a map of X-ray absorption numbers by a line printer. The computation cancels out the effects of absorption in other parts of the anatomy so that the problem of shadowing by the skull or bone structure is overcome.

<u>SOURCE</u>: "Section by Section" by Shelley Stuart. "Electronics Weekly" (April 7th,1976) p.16.

1972   AUTOMATIC CONTROL OF   W. Bardsley
        CRYSTAL GROWTH   G. W. Green
                                      C. H. Holliday   (U.K.)
                                      D. T. J. Hurle

        In this note, we describe a novel, alternative method of automatic
diameter control (or, more strictly, control of cross-sectional area, since the
crystals may be of non-circular section) for which certain advantages can be
claimed. Put simply, the method comprises "weighing" the growing crystal by
means of an industrial weighing cell from which the pull rod is hung. The
method requires that there are no constraints to the vertical motion of the pull
rod, and this is achieved by a gas bearing where the rod enters the growth
chamber. Normally, some of the ambient gas escapes through the gas bearing,
but for the initial evacuation and flushing before growth, the rod is sealed by a
constrictable rubber sleeve. The pull rod has a self-aligning bearing at its
upper end to provide a connection to the weighing cell, and is rotated by a low
friction pin and fork arrangement.

        The electrical signal from the weighing cell is compared with a signal
from a rectilinear potentiometer driven from the leadscrew nut and any difference is amplified and used to adjust the crucible heating power in that direction
which minimises the difference signal. The desired diameter is predetermined
by setting electrically the magnitude of the potentiometer output voltage per
unit distance of pull rod travel. The initial growth out from the diameter of the
seed crystal to the final diameter has also been automatically controlled by
introducing a non-linear element in series with the potentiometer output circuit.

SOURCE: "Automatic control of Czochralski crystal growth" by W. Bardsley,
G. W. Green, C. H. Holliday and D. T. J. Hurle. Journal of Crystal Growth
Vol. 16 (1972) p. 277.

SEE ALSO: "Developments in the weighing method of automatic crystal pulling"
by W. Bardsley, B. Cockayne, G. W. Green, D. T. J. Hurle, G. C. Joyce, J. M.
Roslington, P. J. Tufton and H. C. Webber. Journal of Crystal Growth. Vol. 24/
25 (1974) p. 369.

1972   1024 BIT RANDOM ACCESS MEMORY   Intel (U.S.A.)

        The RAM - father of them all, the 1103 from Intel started the stam-
pede to semiconductor memories. It was the first time that more than 1000
bits of read/write memory could be supplied on a single semiconductor chip
in a low-cost MOS configuration.

SOURCE: "Special report - semiconductor RAM's land computer mainframe
jobs" by L. Altman. Electronics (August 28, 1972) p. 64.

1972   INTEGRATED INJECTION LOGIC   K. Hart and A. Slob (Holland)

        Logic gates suitable for large-scale integration (LSI) should satisfy
three important requirements. Processing has to be simple and under good
control to obtain an acceptable yield of reliable IC's containing about 1000 gates.
The basic gate must be as simple and compact as possible to avoid extreme chip
dimensions. Finally, the power-delay time product must be so high that operation at a reasonable speed does not cause excessive chip dissipation.

        Multicollector transistors fed by carrier injection proved to be a novel
and attractive solution. A simplified (five masks) standard bipolar process is
used resulting in a packing density of 400 gates/mm^2 with interconnection widths
and spacings of 5 μm. The power-delay time product is 0.4 pJ per gate. An
additional advantage is a very low supply voltage (less than 1 V). This, combined
with the possibility of choosing the current level within several decades enables
use in very low-power applications. With a normal seven-mask technology,
analog circuitry has been combined with integrated injection logic (I^2L).

SOURCE: "Integrated Injection Logic - a new approach to LSI" by Kees Hart
and Arie Slob. IEEE Journal of Solid State Circuits. Vol. SC-7 No. 5. (Oct. 1972)
p. 346.

SEE ALSO: "Super integrated bipolar memory devices" by S. K. Wiedman and
H. H. Berger. Presented at the IEEE Int. Electron Devices Conference.
October 11-13, 1971.

1972   DEEP PROTON-ISOLATED LASER   J.P. Dyment
                                                    L.A. D'Asaro
                                                    J.C. North   (U.S.A.)
                                                    B.I. Miller, and
                                                    J.E. Ripper

Proton bombardment as a means of isolation is now widely used for a number of semiconductor devices. It was first demonstrated for (GaAl) As/GaAs heterostructure lasers in 1972.

High peak-power lasers with junctions at a depth of 40 μm from the surface are required for applications where fibreoptics, with 25 μm square fibres, are used to couple the output of several lasers to form a high brightness source.

SOURCE: "Deep proton-isolated lasers and proton range data for InP and GaSb . Solid-State and Electron Devices, January 1979, Volume 3, No. 1, p.1.

SEE ALSO: "Proton-bombardment formation of stripe-geometry hetero-structure lasers for 300K c.w. operation".
J.C. Dyment, L.A. D'Asaro, J.C. North, B.I. Miller and J.E. Ripper. Proc. IEEE, 1972, 60, pp. 726-728.

1972   VIDEO DISCS   Philips   (Holland)

Philips Gloeilampenfabrieken demonstrated a long-playing video disk in September 1972. It was a dramatic improvement over an AEG-Telefunken/ Decca black-and-white video disk that had been demonstrated in 1970. Because the Teldec disks had grooves and mechanical tracking, they suffered from short playing time (an 8-inch disk played for only 5 minutes) and high record wear. The Philips disks held 30 to 45 minutes of colour material and instead of grooves had submicrometer pits molded into a spiral track; a laser-generated light spot read the patterns.

SOURCE: Electronics (April 17, 1980), p.409.

1973   DRY ETCHING   Mitsubishi Electric Co., (Japan)

The application of dry etching techniques using plasma chemistry to semi-conductor processing was introduced by Mitsubishi Electric Co., Japan. It shows that gas plasma containing fluorine species are able to etch silicon and its compounds (SiO_2, Si_3N_4). In the experiments, the gas used was CF_4. A plasma was produced by rf (13.56MHz) discharge and a barrel type plasma reactor was used. The etching mechanism was principally considered to be a chemical reaction between silicon and the fluorine radicals in plasma. However, the details of the etching characteristics and the etching mechanisms were not known in those days.

This technique promised a number of advantages over wet etching methods in terms of improved precise pattern control, problems of etchant preparation and disposal, and cost. It was expected to play an important role in the fabri-cation of Si integrated circuits (SLI, VLSI).

SOURCE :   "Etching Characteristics of Silicon, and its Compounds by Gas Plasma" by H. Abe, Y. Sonobe and T. Enomoto, Jap. J. Appl. Phys., Vol. 12, No. 1 (1973).

1973   SCANNING ACOUSTIC MICROSCOPE   C. F. Quate   (U.S.A.)

In 1973 Professor Quate conceived an approach of elegant simplicity to produce a microscope that would use sound, rather than light, in order to form images. This achievement which had been the aim of applied scientists for more than fifty years, led to the extremely rapid development of a micro-scope which already exceeds the resolution of optical microscopes.

The key idea, which was the recognition of the fact that velocities of acoustic waves in some solids can be as much as seven times greater than the velocity in water, resulted in the production of a lens which could focus a beam of sound, on its axis, without significant aberrations. Whilst such a lens cannot image a complete field, Professor Quate recognized that the axial focus was enough for the realization of a mechanically-scanned microscope in which the image was reproduced point by point. The scanning acoustic microscope has opened up a completely new field of microscopy which permits the direct imaging of biological specimens and the examination of silicon integrated circuits and other solid objects.

SOURCE :   "Major Prizes for Opto-Electronics Inventions".   The Radio and Electronic Engineer, Volume 52, No. 3 (March 1982), p. 107.

SEE ALSO :   "Seeing Acoustically" by R. K. Mueller and R. L. Rylander. IEEE Spectrum (February 1982), p. 28.

"Thermal Imaging via Cooled Detectors", by D. B. Webb.   The Radio and Electronic Engineer, Volume 52, No. 1 (January 1982), p. 17.

"Recent Developments in Scanning Acoustic Microscopy" by D. A. Sinclair, I. R. Smith and H. K. Wickramasinghe.   The Radio and Electronic Engineer, Volume 52, No. 10 (October 1982), p. 479.

1973   **LOGIC-STATE ANALYSER**
(displaying binary notation in 1s and 0s)   C. H. House (U. S. A.)

LOGIC-TIMING ANALYSER
(for recording, displaying and analysing   B. J. Moore (U. S. A.)
 complex timing relationships)

One problem, two men, two solutions: yet both designers were right; both of their designs were needed. So Charles H. House of Hewlett-Packard Co's Colorado Springs (Colo.) division, and B. J. Moore of Biomation Corpn. in Cupertino, California, developed two markedly different diagnostic instruments that were the first such electronic tools for studying, designing and trouble-shooting complex digital logic circuits and systems.

The 160 1L is a plug-in unit for HP's 180 series oscilloscopes, giving a 12-channel, 16-word-memory logic-state analyser for 10-megahertz operation. The Biomation 10-MHz 810-D digital logic recorder stores 256 logic states on each of eight channels, displaying waveform-like timing diagrams on an oscilloscope.

SOURCE: "Logic-analyser originators cited for testing innovation 1977 Award for Achievement" - Electronics (Oct. 27th, 1977) page 83.

1973   **SKYLAB-1** Satellite   (U. S. A.)

Launched 14th May 1973. Manned orbital research laboratory. Objectives: to determine man's ability to live and work in space for extended periods; to extend the science of solar astronomy beyond the limits of earth-based observation; to develop improved techniques for surveying earth resources, to make various investigations requiring a constant zero gravity environment.

SOURCE: Table of Artificial Satellites Launched Between 1957 and 1976. International Telecommunication Union. Geneva (1977)

1973   **SAMOS** Satellite   (U. S. A.)

Launched 13th July 1973. Low Altitude System Platform. Reconnaissance satellite equipped with a horizontal scan radar. Recovered on 12 October, 1973.

SOURCE: Table of Artificial Satellites Launched Between 1957 and 1976. International Telecommunication Union. Geneva (1977)

1974   **ELECTRON BEAM LITHOGRAPHY**   Bell Laboratories   (U. S. A.)

Electron-beam lithography was the key to making the masks for the optical lithography units. Without its ability to make masks with micrometer-wide lines, no LSI lithography technique based on the use of either masks or reticles would have been possible.

One of the first electron-beam systems came from Bell Laboratories in 1974. Called the Electron-Beam Exposure System, it made masks by using a raster-scanned beam aimed at a continuously moving table. Wafer alignment with the beam was controlled by a laser interferometer.

SOURCE: Electronics (April 17, 1980), p. 388.

1974   C.A.T.T.(CONTROLLED AVALANCHE   S.P.Yu
       TRANSIT-TIME TRIODE)   W.R.Cady   (U.S.A.)
                                               W.,Tantraporn

The use of avalanche and transit-time effects in microwave transistor-like structures for increased gain and higher-frequency operation has been proposed by us and others in recent publications. We have previously described the basic principle and large-signal theory of the controlled-avalanche transit-time triode (c.a.t.t.) and have also reported some initial experimental results in the 1 - 3 GHz region. The purpose of the present paper is to discuss more fully certain aspects of c.a.t.t. design and operation resulting from the avalanche-multiplication process and those whose importance has become clearer through our further investigations.

SOURCE: "Avalanche multiplication in C.A.T.T.'s " by J.R.Eshbach, S.P.Yu and W.R.Cady. IEE Solid State and Electron Devices, Vol.1. No:1 (Sept.1976) p. 9.

SEE ALSO: "A new three-terminal microwave power oscillator" by S.P.Yu, W.R.Cady and W.Tantraporn. IEEE Trans.ED-21 (1974) p.736.

"The third terminal in microwave devices" by J.E.Carroll. Proc. European Solid-state device research conf. Nottingham (1974)

"Transistor improvements using an impatt collector" by A.M. Winstanley and J.E.Carroll. Electron.Lett.Vol.10 (1974) p.516.

1974   PRESTEL System   S. Fedida (United Kingdom)

Mr. Fedida invented the concept of viewdata whilst working at the Post Office Research Centre in the early 'seventies. It combines a modified television set, a telephone line and a computer: a push button control panel calls up a 'page' of the information required by a subscriber on to a television screen using a telephone line link routed into a computer data bank. The simplicity of operating the system provides the potential for the mass marketing of information on a wide range of general and technical subjects.

SOURCE: "1979 MacRobert Award for Software System Inventor." S. Fedida. The Radio and Electronic Engineer, Vol. 50, No. 1/2, p. 10.

1974   WESTAR-1 Satellite   (U.S.A.)

Launched 13th April,1974. First United States domestic communication satellite placed in synchronous orbit over equator at 99°W. Can transmit 12 colour television channels or up to 14 400 one-way telephone circuits through five earth stations located close to New York, Atlanta, Chicago, Dallas and Los Angeles.

SOURCE: Table of Artificial Satellites Launched Between 1957 and 1976. International Telecommunication Union. Geneva (1977)

1974   16-BIT SINGLE CHIP MICROPROCESSOR   National (U.S.A.)

The semiconductor industry's first 16-bit, single-chip microprocessor is soon to be introduced by National Semiconductor Corp. Called PACE (for processing and control element), the device will handle 16-bit instructions and addresses, and either 16-bit or 8-bit data. It is being built with p-channel silicon-gate MOS Technology because, the company says, p-MOS is a more predictable and better established technology than n-MOS and meets both of PACE's main requirements: 10-microsecond execution time for instructions, and enough density to fit the entire circuit on a single chip.

PACE requires only two power supplies, + 5 volts and -12 V, instead of the three required with n-channel fabrication.

SOURCE: "National to show 16-bit processor on single chip" Electronics. (November 28th, 1974) p. 35.

1975   THE GYROTRON   A.G. Gapanov et al. (USSR)

Although the foundations for high-power-gyrotron development were laid sometime previously, the first reported real breakthrough describing a working device was in 1975.   In this device, the electrons are made to rotate at a cyclotron frequency (which is also near the operating frequency or a sub-harmonic of it) of the static magnetic field.   Hence it is sometimes called a 'cyclotron resonance maser'.   All devices mentioned here are based on developments arising from the concept of the cyclotron resonance maser.

The impetus for the development of gyrotrons came from Russian workers in the nuclear-energy field, who hoped to be able to heat dense plasma, confined by a powerful magnetic field, by adsorbing microwave energy at the cyclotron resonance frequency.   If sufficient power could be absorbed in this way, then temperatures approaching those required for fusion could be reached.   The power level required is in the region of 10 to 20 MW for several seconds.   The frequency required for typical magnetic fields used in plasma containment machines (such as Tokamaks) is 50 to 100 GHz for magnetic fields of $2 \cdot 0$ to $4 \cdot 0$ T.

SOURCE: "The gyrotron" by M.J. Smith. Electronics and Power (May 1981), p. 389.

SEE ALSO: GAPANOV, A.V., et al.: Radiofizika, 1975, 18, p. 280.
               GAPANOV, A.V., et al.: 'A device for cm-mm and sub-mm
                  wave generation.' Copyright no. 223931 with priority of
                  March 24, 1967 (Official bulletin KD10 of SM USSR (11) p. 200,
                  1976).

1975   LOCMOS (Locally oxidised complementary-   Philips   (Netherlands)
            Metal-oxide-semiconductor) Integrated Circuit

LOCMOS is an acronym for locally-oxidized CMOS, a process invented by Philips Research Laboratories which produces a high performance, high density CMOS that costs no more than standard CMOS. The LOCMOS 4000 range which is pin-for-pin compatible with otherppopular 4000 ranges, needs less chip area per function and thus enables full buffered circuitry to be built into every device.

Features of LOCMOS 4000 include high noise immunity, standardized outputs and increased system speeds. The increased voltage gain, due to buffering, gives almost ideal transfer characteristics, and every device will give a guaranteed output of 400μA from a 5V power supply. Output impedance and propagation delay are independent of input pattern and reduced sidewall capacitance results in higher speed.

SOURCE: Mullard announce LOCMOS 4000" Mullard Press Information Sheet. (September, 1975) p. 1.

1975   VIKING-1 Satellite   (U.S.A.)

Launched 20th August 1975. Objectives: to explore the surface and atmosphere of the planet Mars. Includes an orbiter and a lander separating on approach to Mars.
Orbiter: spacecraft arrived at Mars in June 1976.
Lander: landed on Mars on 20th July 1976.

SOURCE: Table of Artificial Satellites Launched Between 1957 and 1976. International Telecommunication Union. Geneva (1977)

1975   MICROELECTRONICS (Integrated   F.K.Reinhart
            Optical Circuits)   R.A.Logan   (U.S.A.)

                For the first time, scientists have combined a laser with components
        such as modulators, filters, and lightguides in a single crystal microcircuit,
        just as multiple components are fabricated in an integrated electronic circuit.

                The devices, integrated optical circuits measuring usually about 6 by
        15 mils, operate within the structure of a semiconductor injection laser.

                This type of circuit represents an alternative to hybrid integrated
        optics where components - often fabricated from different material systems -
        are interconnected on a base. By contrast, the new monolithic optical circuit
        "contains" many of the required components within the same single crystal.

                Franz K.Reinhart and Ralph A.Logan of Bell Laboratories, Murray
        Hill, New Jersey, developed the new circuit.

        SOURCE: "Integrated optical circuits: another step forward" Bell Labs.
        Record (September 1975) p.349.

1975   SILICON ANODISATION   R.Cook (ITT)   (U.S.A.)

                The discovery that silicon itself can be anodized opens an unexpected
        path to cheaper, denser, faster integrated circuits. The low-temperature
        process produces in one step the dielectric needed to isolate the active elements
        on a chip, thus adding the advantages of dielectric isolation to any semiconductor
        technology, whether bipolar or metal-oxide-semiconductor.

                Direct silicon anodization was discovered quite by chance. An anodiz-
        ing voltage was accidentally increased beyond the point required to anodize
        aluminium. The aluminium was destroyed, but the silicon substrate beneath the
        aluminium, when examined under a microscope, was seen to have been trans-
        formed into a porous dielectric layer. Further experiment revealed that the
        dielectric on the silicon surface could be tailored to almost any desired thick-
        ness simply by adjusting the anodizing process.

        SOURCE: "Anodizing silicon is economical way to isolate IC elements" by
        B.Cook. Electronics (November 13th, 1975) p.109.

1975   4096-BIT RANDOM ACCESS MEMORY   Fairchild (U.S.A.)

                In a significant development, Fairchild Semiconductor has applied its
        oxide-isolated Isoplanar technology to an injection-logic configuration. The
        result: the industry's first 4,096-bit I^2L random-access memory. The part
        has a nominal access time of 100 nanoseconds, making it more than twice as
        fast as today's n-MOS 4-kilobit dynamic RAMs. The device will be ready for
        selective prototyping late this summer.

        SOURCE: "Fairchild develops first 4K RAM to use I^2L" Electronics.
        (June 26th, 1975) p.25.

1975   THIN FILMS - DIRECT BONDED   J.F.Burgess
            COPPER PROCESS   C.A.Neugebauer   (U.S.A.)
                                                 G.Flanagan & R.E.Moore

                In the direct copper to ceramic bonding process, bonding is accom-
        plished by heating Al_2O_3 or BeO substrates in contact with the Cu foil. Foil
        thicknesses from 250 to 1 mil can be used. The gas atmosphere consists
        principally of inert gas such as argon or nitrogen with a small addition of
        oxygen, typically of the order of a few hundredths of a percent. The length
        of time required for bonding is typically a few minutes.

                The temperature for bonding is critically important. Bonding does
        not take place unless the temperature exceeds 1065°C, but it must be below
        1083°C, which is the melting point of copper.

        SOURCE: "Hybrid packages by the direct bonded copper process" by
        J.F.Burgess, N.A.Neugabauer, G.Flanagan and R.E.Moore. Solid State
        Technology. (May 1975) p.42.

1975   STARLETTE Satellite   (France)

Launched 6th February 1975. Satellite for geodetic, geodynamic and earth-moon system studies.

SOURCE: Table of Artificial Satellites Launched Between 1957 and 1976. International Telecommunication Union. Geneva (1977)

1975   RADUGA-1 (STATSIONAR-1) Satellite   (U.S.S.R)

Launched 22nd December 1975. Geostationary communications satellite providing colour, black-and-white television, and telegraph and telephone channels. Is linked with the Orbita network. In orbit at 99°E.

SOURCE: Table of Artificial Satellites Launched Between 1957 and 1976. International Telecommunication Union. Geneva (1977)

1976   MICROELECTRONICS
            (16, 384 bit Random Access Memory)   Intel (U.S.A.)

A triumph of semiconductor device technology, the 16, 384-bit random access memory has arrived. Its bit density is unprecedented and springs from an enhanced n-channel silicon-gate technique, in which a double level of poly-silicon conductors shrinks the memory cell to 450 micrometers square. That is less than half the cell size in the densest 4, 096-bit RAM.

SOURCE: "Enter the 15, 384 bit RAM" by J.E. Coe and W.G. Oldham. Electronics. (February 19th, 1976) p. 114.

1976   AMORPHOUS SILICON SOLAR CELL   RCA   (U.S.A.)

A new type of solar cell has been developed at RCA Laboratories using amorphous silicon (a-Si) deposited from a glow discharge in silane (SiH_4). These solar cells utilize $\sim 1\mu m$ of a-Si and have been fabricated in hetero-junction, p-i-n, and Schottky-barrier structures on low-cost substrates such as glass and steel.

Discharge-produced a-Si has optical and electronic properties that are ideally suited for a solar cell material. The optical absorption coefficient is significantly larger than that of crystalline Si over the visible light range[1] and therefore most of the solar radiation with $\lambda < 0.7 \mu m$ is absorbed in a film $\sim 1 \mu m$ thick.

SOURCE: "Properties of amorphous silicon and a-Si Solar Cells" by D.E. Carlson, C.R. Wronski, J.I. Pankove, P.J. Zanzucchi and D.L. Staebler. RCA Review Vol. 38 (June 1977) p. 211.

SEE ALSO: "Amorphous silicon solar cell" by D.E. Carlson and C.R. Wronski. App. Phys. Lett. Vol. 28 (1976) p. 671.

1976   POLYSILICON RESISTOR LOADED RAM's.   Mostek (U.S.A.)

In 1976, Mostek introduced its Poly R process with the MK4104, a 4-K-by-1-bit static RAM. The part diverged from the usual static RAM designs in that it replaced the depletion-mode MOS transistor loads in its cell with ion-implanted polysilicon resistors. The design not only saved chip area but also greatly lowered power dissipation. Since the polysilicon resistors are actually laid over the four transistors, the cell of the 4104 shrank to 2.75 mil^2 - roughly half the size of conventional cells.

The power is reduced because the high resistivity of the polysilicon loads - typically, 5, 000 megohms, accurately controlled by ion implantation - squeezes the current flow down to less than 1 namoampere per bit. Another feature of the Poly R loads is their negative temperature coefficient which automatically compensates for increased leakages that normally occur at elevated temperatures. Moreover, the polysilicon loads allow data retention in the cells even at greatly reduced supply voltages.

SOURCE: "Concepts for a dense new RAM". Electronics (September 27, 1979), p. 132.

1976   COMPUTER (One board with
              programmeable I/ 0)   Intel Corp. (U.S.A.)

              A complete general-purpose computer subsystem that fits on a single
printed-circuit board has been a major goal all through the steady evolution of
LSI technology. Such a computer, consisting of a central-processing unit,
read/write and read-only memories, and parallel and serial input/output   -
interface components, could satisfy most processing and control applications
needed by original-equipment manufacturers. A single board computer could
greatly extend the range of computer applications by providing a single solution
to three problems that have often precluded the use of conventional computers.

              The primary reason for use of a single assembly of LSI devices rather
than a multiboard subsystem is economic. Extra board assemblies are costly
in themselves and need related equipment, such as backplanes and housing,that
also adds to cost.

              Compactness and low power consumption are often prerequisites for
products. Using LSI for all key computer functions reduces power consumption
and provides a higher functional density than conventional subsystem designs.
This new class of LSI devices - programmable input/output interface chips -
enables an 8-bit computer to be built as a subsystem on one printed-circuit
board.

SOURCE: "The 'super component': the one-board computer with programmable
I/0" by R.Garrow, J.Johnson and M.Maerz. Electronics (February 5th,1976)
page 77.

1976   MARISAT-1 Satellite   (U.S.A.)

Launched 19th February,1976. Maritime communications satellite positioned
at 15°W over the Atlantic Ocean.

SOURCE: Table of Artificial Satellites Launched Between 1957 and 1976.
International Telecommunication   Union. Geneva (1977)

1976   MICROELECTRONICS (Versatile Arrays)   Philips (Holland)

              A simple variation of standard silicon-gate technology has produced
extremely versatile arrays that make novel analog-to-digital converters, analog
type displays and light-pattern scanners. The arrays consist of devices similar to standard metal-oxide-semiconductor elements, except that a resistive
electrode structure replaces the normal metal insulated gate. This structure
permits a voltage gradient to be set up across the ends of the gate and then
manipulated to control the transistors either singly or in groups.

SOURCE: "Resistive insulated gates produce novel a-d converters, light
scanners" by M.V.Whelan, L.A.Daverveld and J.G.deGroot. Electronics.
(March 18th, 1976) p.111.

1977   CCD ANALOG-TO-DIGITAL CONVERTER   G.E.Corp. (U.S.A.)

              For the industrial marketplace, the Research and Development Center
of General Electric Corpn. in Schenectady, N.Y., has fabricated the first CCD
analog-to-digital converter as a p-channel MOS chip providing a resolution
equivalent to 10 to 12 binary bits.

              The GE converter chip is big, measuring 240 by 180 mils, but it
contains a comparator, voltage reference, clock, counter and all necessary
control logic - in fact, it even has decoder/drivers for gas-discharge displays.
Operating speed, though, is slow: about 20 milliseconds total for a 10-bit
conversion. The device, which runs at a clock frequency of up to 500 kilohertz,
resolves analog inputs to within better than 1 millivolt, digitizing them to an
accuracy of ±0.5 least significant bit.

              Chip operation relies on the transfer of fixed-size charge packets
from one site to another, with conventional digital circuitry controlling the
conversion process.

SOURCE: "CCD's edge towards high volume use" Electronics (March 17th,
1977) page 74.

1977   ANISOTROPIC PERMANENT   Matsushita Electric   (Japan)
            MAGNET

            Matsushita Electric has developed what they claim to be the world's
first anisotropic permanent magnet for practical use. The new magnet is made
of manganese, aluminium and carbon materials which are available in abundance
instead of cobalt and nickel, which are scarce and expensive. It has higher
magnetic energy than the ordinary 'alnico' type magnet which contains cobalt
and nickel and has good mechanical strength and machinability, allowing it to be
shaped and drilled. Samples should be available from June 1977.

            The new magnet consists of manganese, aluminium and carbon. The
basic composition of the magnet was developed by Matsushita in 1967; however,
the magnet produced was of 'isotropic' type, in which the direction of magnet-
isation was distributed at random. With the anisotropic version, the company
has succeeded in aligning the direction in which it can be magnetised, so
increasing the magnetic energy to more than five times that of the isotropic
magnet, giving a maximum energy product of 7 MG. Oe.

SOURCE: "First anisotropic permanent magnet" Electronic Equipment News.
(June 1977) p. 10.

1977   POCKET TV RECEIVER   Sinclair Radionics   (U. K.)

            What is claimed to be the world's first pocket television set was
launched in London early this year by Sinclair Radionics, the British company
which pioneered the revolution in miniature electronic calculators four years
ago.

            The result of a £500,000 12-year research and development programme
the Microvision, which has a 2 inch screen, is now in production at Sinclair's
new assembly plant in St.Ives, Huntingdon.

            The set is 4 inches wide, 6 inches from front to back, just $1\frac{1}{2}$ inches
deep and weighs $26\frac{1}{2}$ oz. Yet, operated by internal rechargable batteries or
direct from the mains, it produces a sharp black and white picture which, when
viewed at a distance of one foot, is of equivalent size and brilliance to that of
normal domestic compact portables at 6 feet and 24 inch models at 12 feet.

SOURCE: "Television with two inch screen" National Electronics Review.
Vol. 13. No:4 (July/August 1977) p. 74.

1977   TRIMOS (Triac + MOS)   Stanford University   (U. S. A.)

            Another innovation in power at ISSCC comes from California's
Stanford University, which has developed a way to put signal and power devices
on one and the same piece of silicon.

            Called Trimos, the new technology permits integrating an insulated-gate
triac with metal-oxide-semiconductor components, inviting a host of new
applications in crosspoint switching, output stages, and power control.

            Trimos is actually a merged device based on double-diffused MOS
technology - two high-voltage D-MOS transistors are merged around a common
drain. Contact is made to the source and diffused channel of each D-MOS
device, forming symmetrical anode and cathode contacts. The shared gate
metal forms the unit's control electrode.

            In its on state, the Trimos device exhibits a dynamic resistance of less
than 10 ohms and can pass currents on the order of amperes. A simple shunt
switch, in the form of a conventional MOS transistor, can be fabricated adjacent
to the Trimos unit for switching it out of its on state or inhibiting it from
triggering. Without such a bypass structure, the Trimos device typically has
turn-on and turn-off times on the order of 200 nanoseconds, and its single-
pulse dv/dt capability exceeds 1,000 volts per microsecond.

SOURCE: "Trimos combines triac, MOS devices". Electronics,
(March 2, 1978), p. 42.

1977   FLAD (Fluorescence-activated display)   Institute for Applied
                                                         Solid State Physics,   (Germany)
                                                         Freiburg.

        Display-system designers will soon have a new device to work with -
the fluorescence-activated display (FLAD).   Invented at the Institute for
Applied Solid State Physics in Freigburg, West Germany, the device uses a
layer of plastic material appropriately doped with fluorescent organic molecules.   In this layer, ambient light is collected, guided, and then emitted at
the segments of the display's digits.

        The FLAD dissipates the same power as a liquid-crystal display, but
its light intensity is said to be much higher.   Moreover, the light can be any
color in the spectrum between green and red, the inventors say.   West
Germany's Siemens AG will produce the FLAD display first, initially for use
in battery-operated digital tabletop and alarm clocks.   Later, FLADS will be
used in pocket calculators, portable instruments, and scales that indicate
price and weight.

        SOURCE: Electronics, March 17, 1977, p. 55.

1977   MICROELECTRONICS H - MOS   Intel   (U.S.A.)

        To achieve their new high-performance process called H-MOS, Intel
has chosen the direct device-scaling method for two reasons.   First, it evolves
directly out of standard silicon-gate processing and so requires neither new
device structures not complex circuit schemes (either requirement would make
yields and fabricating costs too unpredictable to guarantee their usefulness over
a wide range of semiconductor products).   Second, it fits in with the trend to
smaller and smaller circuit patterns, as photolithographic methods grow more
refined and electron-beam wafer-fabrication techniques stand ready to take
over.

        SOURCE: H-MOS scales traditional devices to higher performance level" by
R. Pashley, K. Kokonnen, E. Boleky, R. Jecmen, S. Liu and W. Owen.
Electronics (August 18thml977) p. 94.

1978   LASER-ANNEALED POLYSILICON   Texas Instruments   (U.S.A.)

        Researchers at Texas Instruments Inc.'s Central Research Laboratories
have succeeded in fabricating MOS devices in laser-annealed polysilicon on
silicon dioxide.   Not only will the devices have the speed and density of those
fabricated on sapphire, but the all-silicon construction could lead to true three-
dimensional circuitry.

        Unannealed polysilicon is made up of randomly oriented crystal grains
on the order of 500 angstroms across.   Carrier flow is impeded at each grain-
to-grain boundary and the resulting low mobility would yield a poor device at
best.

        Armed with a pulsed frequency-doubled neodymium-yttrium-aluminum-
garnet laser, TI scans the polysilicon surface to induce localized melting.   The
grains recrystallize with much larger dimensions so the number of interfaces
is reduced and mobility is enhanced.   With this set up, researchers have
observed grains as large as 10 micrometers across.

        To build what it calls silicon-on-insulator MOS FETS, TI begins with a
single-crystal p-type silicon substrate.   On this, it grows a 1-μm-thick oxide
layers and then deposits a 0.5-μm film of undoped polysilicon film.   The
samples then go onto an X-Y translation stage, which is heated to 350° C while
a stepping motor moves it in synchronization with the pulsed laser.

        Then the polysilicon is selectively etched down to the oxide level to
isolate islands for each transistor.   Boron ions are implanted to form the
channel, which is covered with a thermally grown 500-$\overset{\circ}{A}$ gate oxide.   A polysilicon gate is deposited on the gate oxide and implanted with phosphorus.   It
also is used for a self-aligned arsenic ion implantation of the source and drain.

        SOURCE: "All-silicon devices will match SOS in performance" by John G.
Posa, Electronics (November 22, 1979), p. 39.

1978   ISL (INTEGRATED SCHOTTKY LOGIC)   J. Lohstroh et al.
                                                          (Philips) (Holland)

Consider a pair of bipolar technologies: low-power Schottky transistor-transistor logic aiming at high speed for medium-scale parts like the 7400 family, and integrated injection logic, which merges transistors specifically for the high packing density needed in large-scale integration. What if the attributes of both could be found in a high-speed, low-power logic suitable for LSI. Apparently they can - in ISL, a newly developed technology that stands for integrated Schottky logic.

Developed by Jan Johstroh and colleagues at the Digital Circuitry and Memory Group of Philips Gloeilampenfabrieken in Eindhoven, the Netherlands, ISL has already performed admirably in "kit" parts - flip-flops, oscillators, and the like. Such devices have exhibited gate propagation delays of about 3.5 nanoseconds (half that of low-power Schottky and a quarter that of I^2L), with each gate drawing only about 400 microamperes. An ISL D-type flip-flop toggles comfortably at 60 megahertz, as compared with a limit of about 33 MHZ for a similar low-power Schottky device.

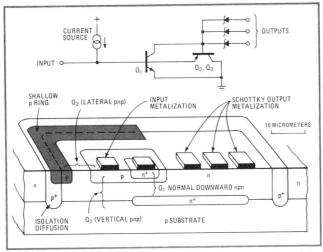

New logic. In ISL, the normal downward npn transistor inherently adds a vertical pnp device. A p ring parallels the vertical transistor with a lateral pnp one.

SOURCE: "Two popular bipolar technologies combine in Philips' device".
Electronics (June 8, 1978). p. 41.

1978   LIGHT BUBBLES   W. Ruehle
                                             V. Marello   IBM   (U.S.A.)
                                             A. Onton

Mobile "light bubbles", which appear to be electrical analogs to magnetic bubbles, have been generated in magnesium-doped zinc-sulfide thin films, IBM scientists reported at the Electronic Materials Conference in Santa Barbara, California. The 1-μm bubbles appear when 10-kHZ, 190-V ac current is applied to the film via sets of parallel 1-mm-wide metallic lines orthoganally placed on each side of the film. When voltage is applied to a pair of intersected electrodes, the intersected area will emit light bubbles that appear to move in discrete steps - and move faster when the frequency rises to 50 kHZ. The alternating-current thin-film electroluminescence, or actel, lasts as long as an hour and can be used to form images on the film by stimulating different areas with light or an electron beam or by applying voltages across the film. However, before the technique becomes practical, much work needs to be done to direct the bubbles.

SOURCE: IBM Scientists report observation of "light bubbles". Electronics,
July 6, 1978, p. 33.

SEE ALSO: "Electrical analogy to magnetic bubbles". Electronics Weekly,
(July 12, 1978), p. 7.

1978   <u>LIGHTWAVE POWERED TELEPHONE</u>   Bell Laboratories (U.S.A.)

     Bell Laboratories has developed a telephone powered by light carried
to it by a fiber-optic cable. Such a development brings the totally fiber-optic
phone system closer to reality, but it is more than a major technological
advance.

     The major problem was making the phone run off the power available
from the fiber cable. The biggest headache here, according to De Loach, was
the conventional telephone ringing mechanism.

     So a new ringer was designed that worked perfectly well at a couple of
volts. It uses an electroacoustic tone generator with a thin piezoelectric
active element. "This device has some exceptional specs," says De Loach.
"For example, its overall efficiency from the optical input to the acoustic
output is more than 33%".

     With the ringer problem solved, the power requirements of the rest of
the phone were readily satisfied. This power, as well as the ringer drive,
comes from a Bell-developed photodetector, which converts light to electrical
pulses. The same device can act as a photodiode, too.

     The photodetector's narrow-bandwidth conversion efficiency at 0.81
micrometer is 56% - the highest ever reported. It is a double-heterostructure
device with layers of gallium aluminum arsenide sandwiching one of gallium
arsenide and grown on a single crystal substrate. A hole etched in the substrate exposes some of either GaAlAs layer, and the glass fiber butted at that
point couples light to and from the photodetector.

     But the phone handset must send as well as receive, so Bell has designed
the photodetector to generate light as well, at a wavelength different from the
incoming light. With time-sharing and automatic switching, the phone sends
signals back to the central office at a 0.9-μm wavelength. It operates in a
duplex mode, taking further advantage of fiber's bandwidth capability.

     The laser light coming from the central office is on for about 95% of
the time. It is pulse-width-modulated by the voice or data. Since the modulation bandwidth is small compared with the carrier wavelength, the change in
pulse width is easily controllable.

<u>SOURCE :</u>   "Bell Labs develops telephone powered by lightwaves alone".
Electronics (November 23, 1978), p.39.

<u>SEE ALSO :</u>   "De Loach built the fiber-optic phone" by Harvey J. Hindin.
Electronics (October 25, 1979), p.231.

1978   <u>OMIST (Optical Metal Insulator Silicon Thyristor)</u>   A. G. Nassibian (Australia)
                                                                 R. B. Calligaro (Australia)

                                              J.G. Simmons (Canada)

     Recently a novel metal-tunnel oxide n/p^+ silicon device with I/V
characteristics similar to those of a silicon control rectifier (metal-insulator-
silicon-thyristor, the m.i.s.t.), but with added advantage of being of a much
simpler structure which is also compatible with l.s.i. techniques, has been
described. The device has been shown to have wide ranging digital and analogue
circuit applications, including oscillators r.a.m.s. and r.o.m.s. Furthermore,
the device has been shown to be light sensitive.

     This device has an advantage over the conventional analogue light-sens-
itive devices in that it is a true digital-optical switch. When the incident
light is above a certain threshold intensity the device switches between two well-
defined states. In optical systems, such a device can perform both optical
transduction and thresholding, thereby greatly simplifying the receiver system.

<u>SOURCE:</u>   "Digital optical metal insulator silicon thyristor (o.m.i.s.t.)"
by A.G. Nassibian, R.B. Calligaro and J.G. Simmons. Solid-State and
Electron Devices, September 1978, Volume 2, No. 5, p.149.

<u>SEE ALSO:</u>   "Bistable impedance states in m.i.s. structures through con-
trolled inversion" by H. Kroger and H.A.R. Wegener, Appl. Phys. Lett.,
1973, 23, pp.397-399.

1978   ANALOGUE ALL-ELECTRONIC CLOCK FACE   Hosiden Electronics
                                                  and NEC (Japan)

   Electronic parts manufacturer Hosiden Electronics and Nippon Electric
Company (NEC) have developed a 60-pole analog fluorescent display tube for
use in clocks and the drive circuitry to go with it.   The new inventions now
make it possible to introduce electronics on a full scale to analog clocks (which
use hands).

   Since the clock face has been turned into a fluorescent display tube, it
is now possible to achieve a range of colors from yellow to blue using filters.

   Ordinary fluorescent tubes are direct-head 3-electrode vacuum tubes
with anodes coated with phosphor, and they use the electro-luminescence
principle of light emission.   This kind of tube was first developed some ten
years ago in Japan since which time it has undergone improvements in performance.   It radiates a green light which is easily visible.   It has a low voltage,
low power consumption and high response speed making it suitable for use with
MOS LSI chips.   Other features include its dependability, long life and mass
production capability.

   Analog systems have been around for a long time in timepieces and
meters, and they have become an irreplaceable part of every-day life.   The
all-electronic analog clock has, therefore, a bright future now that this new
fluorescent display tube and accompanying drive circuitry have been developed.

SOURCE: "Analogue clocks can now go electronic".   JEI (Journal of
Electronics Industry), Japan.   April 1979.   p. 42.

1978   LASER OPTICAL RECORDING SYSTEM   Philips   (Netherlands)

   An ultra-compact diode laser optical recording system, the world's
first, has recently been introduced by Philips.   It allows high-density recording
and retrieval of up to 10^{10} bits of data, equivalent to about 500 000 typewritten
pages, on a pre-grooved 30 cm disk.   This capacity represents an improvement of ten times compared with the most advanced magnetic disk pack systems
currently available.   The system offers direct read-after-write with random
access; any address can be reached in a mean time of 250 ms, providing
virtually instant access to 5×10^{9} bits (the capacity of one side of the disk).

   The system uses similar techniques to those developed for VLP (Video
Long Play).   The real breakthrough, however, has come with the development
of a suitable miniature diode laser and matching recording material.   The
laser used is of the AlGaAs DH type and employs a 0.1 mm square semiconductor chip housed in a transistor-sized encapsulation.   Despite its small
size, the device develops a pulsed light output power equivalent to that of a
large gas laser and its associated modulator.

SOURCE :   "World's First Diode Laser Optical Recording System".
Electronic Components and Applications, Volume 1, No. 2 (February 1979),
p. 128.

SEE ALSO :   "An optical disk replaces 25 mag tapes" by G. C. Kenney,
D. Y. K. Lou, R. McFarlane, A. Y. Chan, J. S. Nadan, T. R. Kohler, J. G.
Wagner, F. Zernike.   IEEE Spectrum (February 1979), p. 33.

"Consumer electronics : personal and plentiful" by Don Mennie.   IEEE
Spectrum (January 1979), p. 62.

1978   TAMED FREQUENCY MODULATION   N. Wiedenhof   Philips
                                        J. M. Waalwijk   (Holland)

   Philips Research Laboratories in Eindhoven, have designed a different
system of frequency modulation for transmitting digital information.   Using
this method a very narrow spectrum is obtained while the quality of detection
is almost equal to the maximum that can be obtained with digital transmission.

   The new method has been given the name 'tamed frequency modulation',
TFM.   Because of its properties tamed FM is eminently suitable for digital
radio communication.

SOURCE :   "Tamed FM for Efficient Digital Transmission Via Radio".
Philips Research, Philips, Eindhoven.   (7810/0920/186E PR+PREL).

1978   LCP (LASER COLD PROCESSING) OF SEMICONDUCTORS

                                                    Quantronix Corporation (USA)

              A new and exciting technology has appeared in the last two years that
could be of major importance in producing tomorrow's very large-scale
integrated circuits as well as in raising the yields of today's LSI devices. It
is the use of a laser as a heat source for some of the many high-temperature
process steps in semiconductor manufacture - for instance, annealing a wafer
to eliminate crystal damage due to ion implantation, diffusing a wafer with
dopants, and growing crystalline material from amorphous or polycrystalline
material.

              In this new technique, an intense laser beam heats a semiconductor
surface to a temperature at which some desirable physical or chemical change
takes place in the material.   The main practical advantage of this process is
that the laser spot limits irradiation to specific areas and the short pulse
limits heating to a small depth while the rest of the material stays near
ambient temperatures.   Hence the name of the technique - laser cold processing, or LCP.

SOURCE :   "Laser cold processing takes the heat off semiconductors"
by R.A. Kaplan, M.G. Cohen, and K.C. Kiu.   Electronics (February 28, 1980),
p.137.

1978   ONE MEGABIT BUBBLE-MEMORY   Intel   (U.S.A.)

                                             Texas Instruments   (U.S.A.)

              There can no longer be any doubt that bubble-memory systems are a
reality - 1978 saw the introduction of not one but a pair of million-bit bubble
memory chips.   Intel Magnetics Inc., Santa Clara, Calif., was first with its
7110 chip, a 4-square-centimeter device organized into 256 4,096 bit loops,
which when operated at a 100-kilohertz field frequency provides an average
access of about 20 ms.   The organization of Texas Instruments' megabit chip
is 512 loops of 2,048 bits each; with shorter loops, it has an access of about
10 ns.

SOURCE: Electronics (October 25, 1979), p.133.

SEE ALSO: "Megabit bubble-memory chip gets support from LSI family"
by Don Bryson, Dick Clover and Dave Lee, Electronics (April 26, 1979), p.105.

1978   INTEGRATED OPTOELECTRONICS   A. Yariv et al.   (U.S.A.)

              Today a single optical fiber can transmit billions of bits of information
per second over many kilometers.   In fact, fibers can handle more bits of
information per second than conventional optical sources - lasers or light-
emitting diodes - can transmit and detectors can receive through them.
Combining lasers, detectors, and active electronic devices for modulating the
light on a single-crystal chip of GaAs - a means for reducing an electronic
bottleneck - was first suggested 10 years ago by Amnon Yariv of the California
Institute of Technology in Pasadena.   Research efforts since then indicate
that the time is ripe to put the idea into practice.

SOURCE :   "Integrated Optoelectronics" by N. Bar-Chaim, I. Ury and
A. Yariv.   IEEE Spectrum (May 1982), p.38.

SEE ALSO :   "Integration of an injection laser with a Gunn oscillator on a
semi-insulating GaAs substrate" by C.P. Lee, S. Margalit, I. Ury and
A. Yariv.   Applied Physics Letters, Volume 32, (1978), p. 806.

1979   TWO-LAYER RESIST TECHNIQUE FOR SUBMICROMETRE LINES

                                             Bell Laboratories/MIT (USA)

              In the two-level resist method, a very thin , 1,500-to-2,000-angstrom
amorphous upper layer of selenium and germanium is sputtered or evaporated
onto a polymer layer.   The latter, about 2 μm thick, may be made of any of
the standard resist polymers without the silver that renders them photosensitive.

This beginning layer is thick enough to compensate for silicon's microscopic roughness, which often makes fine geometries impossible because of depth-of-field or optical-interference effects. Thus the SeGe layer is an almost perfectly flat optical surface - which also cuts the reflections and refractions that make for uncontrolled line widths.

To make the thin upper layer photosensitive, the team soaks the wafer in a room-temperature potassium silver selenide solution for 30 seconds. Then it exposes patterns with ultraviolet light, usually at a 4,300-Å wavelength, but sometimes as short as 3,250 Å.

Kai notes that the upper layer has more than twice the contrast of conventional resists and that the amorphous material is finely grained, thus reducing the optical-dispersion effects that can impair resolution. In fact, ultimate resolution is finer than the minimum spot or line sizes possible with available UV sources.

SOURCE :   "Two-layer resist technique produces submicrometer lines with standard optics", by James B. Brinton. Electronics (February 14, 1980), p. 47.

1979   CCD COLOUR TV CAMERA   Sony (Japan)

For improved sensitivity and resolution, the new camera uses two imager chips, one to generate the green signal and the other to generate the red and blue signals . Sensitivity is high because an entire chip is used for green - a stripe filter is not required - and green light generally contains most of the energy in images.

The two chips are offset horizontally by one half the horizontal pixel pitch, and sophisticated signal-processing techniques are used to increase resolution to a value almost as high as what could be obtained with a single chip having twice as many pixels along each horizontal line. Thus a measured optical resolution of 280 test-pattern lines per picture height is obtained even though there are only 245 pixels across the width of each sensor. The measured vertical resolution is 350 lines from 492 pixels.

SOURCE: "Color TV camera using CCD imager chips gets first sale"..
Electronics (February 14, 1980). p. 79.

SEE ALSO: Electronics (July 5, 1979). p. 67.
                    "   (January 19, 1978). p. 33.
                    "   (July 20, 1978). p. 63.
                    "   (September 28, 1978). p. 68.

1979   AMORPHOUS SILICON LIQUID-CRYSTAL DISPLAY   RSRE (United Kingdom)
                                                        Dundee University

A small experimental liquid-crystal display that is addressed by a matrix of amorphous silicon thin-film transistors has been developed by researchers at Dundee University, Scotland, with funds from Britain's Royal Signals and Radar Establishment. The display panel is far from complex - it measures 1.6 by 2.2 cm (0.6 by 0.9 in.) and consists of a five-by-seven array of display elements each 2 mm square. But its development, which began four years ago, points to the potential of amorphous silicon thin-film transistors as a means of overcoming the addressing limitations of LCDs at low cost. The electrical performance of individual devices looks acceptable, with an on current of 5 µA, a 10^5 on- to off-current ratio, and a response time for each liquid-crystal element of less than 100 µs. The work is to be described in a paper at the Sixth European Solid State Device Research Conference at York, England, September 15 - 18, 1980, together with one from Plessey's Allen Clark Research Centre on device physics of amorphous silicon transistors - work also funded by the RSRE.

SOURCE :   "British address LCD with amorphous silicon thin-film transistors". Electronics (August 28, 1980), p. 67.

SEE ALSO :   Electronics (June 21, 1979, p. 69.

1979   LASER-ENHANCED PLATING AND ETCHING   I.B.M. (U.S.A.)

        Recently, an interesting and novel application of lasers was discovered
in which a laser beam impinging on an electrode is used to enhance local
electroplating or etching rates by several orders of magnitude. It has also
been discovered that with the aid of the laser it is possible to produce very
highly localized electroless plating at high deposition rates, to greatly enhance
and localize the typical metal-exchange (immersion) plating reactions, to
obtain thermo-battery-driven reactions with simple single-element aqueous
solutions, and to greatly enhance localized chemical etching. Since laser
beams can be readily focused to micron-sized dimensions and scanned over
sizeable areas, the enhancement scheme makes it possible to plate and etch
arbitrary patterns without the use of masks.

SOURCE: "Laser-Enhanced Plating and Etching: Mechanisms and Applications". IBM J. Res. Dev., Volume 26, No. 2 (March 1982), p.136.

SEE ALSO: R.J. von Gutfeld, E.E. Tynan, R.L. Melcher and S.E. Blum,
"Laser Enhanced Electroplating and Maskless Pattern Generation", Appl. Phys.
Lett. 35, 651 (1979).

R.J. von Gutfeld, E.E. Tynan, and L.T. Romankiw, "Laser Enhanced Electroplating and Etching for Maskless Pattern Generation", Extended Abstract No.
472, Electrochemical Soc. 79-2 (1979), 156th Meeting of the Electrochemical
Society, Los Angeles, CA.

1979   SATELLITE ECHO-CANCELLING CIRCUIT   Bell Laboratories (U.S.A.)

        By the beginning of 1980, telephone communications by satellite will
have undergone a drastic technical change. The voice-garbling echoes that
occur along the 45,000-mile-long satellite paths will be just about eliminated
by a new integrated circuit that was developed by researchers at Bell Laboratories.

        The chip - dubbed an echo canceler - could by itself double the number of
satellite circuits used by AT & T in its telephone network. So far, because of
echo problems, the company transmits transcontinental telephone calls only one
way by satellite.

        The echo for one-way transmission is controllable by the present echo
suppressors, which open the transmission paths when the echo's amplitude
becomes too high. But these devices do not meet high enough two-way transmission standards for AT & T. The new digital device removes an echo
signal from a circuit by sampling it electronically as it occurs, making a replica
of it, and adding the replica to the original signal so that the two cancel each
other. It is ideal for connection to Bell's all-digital telephone network, which
will be completed over the next 20 years.

SOURCE: "No more echoes". Electronics (October 25, 1979), p.228.

SEE ALSO: "Bell's Echo-Killer Chip". Donald L. Duttweiler. IEEE
Spectrum (October 1980), p.34.

"Silencing Echoes on the Telephone Network". Man Mohan Sondhi and
David A. Berkley. Proceedings of the IEEE, Vol. 68, No. 8 (August 1980),
p.948.

1979   FLOTOX (Floating-Gate Tunnel Oxide) PROCESS   Intel (U.S.A.)

        Flotox resembles the Famos structure except for the additional tunnel-
oxide region over the drain. With a voltage V_g applied to the top gate and with
the drain voltage V_d at 0 V, the floating gate is capacitively coupled to a
positive potential. Electrons are attracted through the tunnel oxide to charge
the floating gate. On the other hand, applying a positive potential to the drain
and grounding the gate reverses the process to discharge the floating gate.

        Flotox, then, provides a simple, reproducible means for both program-
ming and erasing a memory cell.

SOURCE: "16-K EE-PROM relies on tunneling for byte-erasable program
storage" by W.S. Johnson, G.L. Kuhn, A.L. Renninger, and G. Perlegos.
Electronics (February 28, 1980), p. 113.

1979   <u>SEVEN-COLOUR INK-JET PRINTER</u>   Siemens   (Germany)

Engineers of the Teletype and Data-Transmission group at Siemens AG in Munich have developed a prototype color ink-jet printer that can produce characters of seven different colors at the rate of 200 per second in both directions across normal paper. The unit consists of the company's older PT80 printer whose single-color ink-jet mechanism has been replaced by a multinozzle unit for red, green, and yellow ink. Proper mixing of these inks yields the colors blue, magenta, cyan, and black. Color and mixture data are stored in a floppy-disk memory whose program is derived from the output of a separate color scanner. One of the biggest problems involved, the Siemens engineers say, was developing nonsmearing inks that would stay liquid in the nozzles yet become dry right after they hit the paper. If customer reaction is favorable, the color printer will go into production. It will make possible multicolored graphic representations.

SOURCE: "Siemens develop seven-color ink-jet printer". Electronics, (September 13, 1979), p. 71.

1980   <u>OUTDOOR LARGE SCREEN COLOUR DISPLAY SYSTEM</u>   Mitsubishi Electric
                                                        Corporation
                                                        (Japan)

A large screen colour display system for use in outdoor video system has been developed and introduced by Mitsubishi Electric Corp. The display screen consists of a matrix array of small, high brightness light emitting tubes and can present sharp colour pictures even in full daylight. Many video display systems have been installed for various outdoor video services such as in sports stadiums, racetracks, or as an advertising media. However, these previous systems consist of an array of incandescent lamps, and have many problems especially in displaying colour pictures such as insufficient colour quality, high power consumption and short operating life.

To overcome these problems, Mitsubishi has developed the high brightness light emitting tube (LET). The LET is a small flood beam CRT (28mm in diameter) having a single phosphor of Red, Green, or Blue for each tube, and works for a single picture element in the screen. Brightness of the LET is 8,000 bits for a green tube, (over 20 times brighter than the usual tv picture tube).

SOURCE :   "AURORA VISION". A Large Screen Colour Display System" by K. Kurahashi, K. Yagishita, T. Tomimatsu and H. Kobayashi. (In Japanese). Technical Report of Inst. of TV Eng. (Japan) IPD49-3 (1980 3 19).

SEE ALSO :   "An Outdoor Large Screen Colour Display System" by K. Kurahashi, K. Yagishita, N. Fukushima and H. Kobayashi. 1981 SID Symp. Digest of Technical Papers. Vol. 12, 13, 1. (p.132).

1980   <u>MAT (MAGNETIC AVALANCHE TRANSISTOR)</u>   I.B.M. (U.S.A.)

A new semiconductor device for sensing uniaxial magnetic fields has been realized. The device is basically a dual-collector open-base lateral bipolar transistor operating in the avalanche region, and is referred to as a magnetic Avalanche Transistor. It exhibits high magnetic transduction sensitivity compared to traditional Hall-effect and conventional nonlinear magneto-resistive devices. Several hundred experimental devices have been designed, fabricated, and tested over the past two years. Many structural and some process parameters were varied. The magnetic sensitivity of a typical device was found to be proportional to substrate resistivity. A sensitivity of 30 volts per tesla was measured for devices which used 5-ohm-cm p-type substrates. The output signal measured between collectors is differential and responds linearly with field magnitude and polarity. A typical signal-to-noise ratio is 20 000 per tesla. The bandwidth is known to extend well beyond 5 MHz. The sensitive area is calculated to be on the order of 5 μm^2. This communication describes the basic structure, fabrication, and characteristics for the magnetic avalanche transistor.

SOURCE: "A magnetic sensor utilizing an avalanching semiconductor device" by A. W. Vinal. IBM J. Res. Dev., Volume 25, No. 3. (May 1981). p.196.

1980   256K DYNAMIC RAM   NEC-Toshiba   )
                                                          Musashino Electrical) Japan
                                                          Communication Labs.)

          NEC-Toshiba Information Systems Inc. and NTT-Musashino Electrical
Communication Laboratory, both of Tokyo, each present 256-K-by-1-bit
dynamic RAMs. The NEC-Toshiba chip has a 160-nanosecond access time
and a 350-ns cycle time, while the NTT device, which uses molybdenum to
speed signal propagation, accesses in just 100 ns and cycles in double that.
NEC-Toshiba's chip consumes 225 milliwatts of active power and 25 mW on
standby, while NTT's device uses 230 mW but only 15 mW on standby. Both
designs require a 256-cycle/4-millisecond refresh.

          NEC-Toshiba uses two levels of polysilicon, 1.5-μm direct-step-on-
wafer photolithography, and all-dry processing to build its RAM. The device
is already being shown in a 16-pin package, the pinout of which meets the
Joint Electron Device Engineering Council's standard with the eighth address
line on pin 1. The oblong die, measuring 191 by 338 mils, contains two 128-K
arrays. Each array is further split into two 128-by-512-bit sections, separated by 512 sense amplifiers that run the length of the chip. Various techniques,
including silicone coating, are used to keep the mean time between failure due
to alpha radiation below 30,000 device hours.

          NTT-Musashino's RAM is built with electron-beam direct writing, dry
processing, and three interconnection levels: molybdenum word lines,
aluminum bit lines, and polysilicon for storage capacitor electrodes and gates
of non-array devices. NTT's die is organized just like NEC-Toshiba's but
with one interesting difference: each 128-K array has attached to it a 2-K
block of redundant cells and a dummy sense circuit. The extra cells,
connected via four pairs of spare bit lines and two spare word lines, are replaced by electrically programming on-chip poly-silicon resistors during
wafer probing. The additional circuits take up no more than 10% of NTT's
nearly square 230-by-232-mil die.

SOURCE: "ISSCC: a gallery of gigantic memories". Electronics,
(February 14, 1980). p.138.

1980   FIBRE-OPTIC LASER DRIVEN SUPERHETERODYNE   S. Saito   )
                                                          Y. Yamamoto)Japan
                                                          T. Kimura   )

          Researchers at Nippon Telegraph & Telephone's Musashino Laboratories
have demonstrated for the first time the use of classical superheterodyne
detection in an optical-fibre transmission system. The technique represents a
significant step in the practical realisation of optical communication systems,
and could lead to the more efficient use of the low-loss frequency window
recently opened by advances in optical-fibre technology, since finer separation
of carrier frequencies will become possible. Improvements will also be made
in optical signal reception.

          The main signal was supplied by an AlGaAs laser, emitting at 820nm,
the drive current being directly frequency modulated by an r.f. signal. This
was matched to a local-oscillator signal from a similar laser by temperature
control and direct-current adjustment. No feed-back stabilisation system was
used. The mixed signal was aligned on a photodetector and the resulting f.m.
intermediate-frequency signal fed to a frequency discriminator for conversion
to an a.m. output. The spurious signal depth was very small and easily
filtered out.

          Digital (at 100Mbit/s) and analogue (at 300MHz) signals were transmitted
satisfactorily by the system, which is described in the 23rd October issue of
Electronics Letters. The system is claimed to have many advantages over
conventional amplitude-modulation systems, such as low power, simplicity
and better signal/noise ratio, and is expected to be improved in the future by
the use of recently developed frequency-stabilisation feedback systems.

SOURCE: "Classical steps in optical fibres". Electronics and Power,
(November/December 1980). p.855.

SEE ALSO: "Fibre optics adopts superheterodyne principles". Electronics,
(November 20, 1980). p.73.

1980   "MCZ (Magnetic Field Method of Silicon Crystal Growth)_ Sony (Japan)

            Sony has developed a new method for producing very high-quality single crystals of silicon with greater uniformity and lesser defect generation, which greatly reduces the wafer warpage and distortion, through the application of a high magnetic field in the silicon pulling process.

            The new silicon crystal growth method, called "MCZ (magnetic-field CZ) method," has been developed to meet the requirements of the coming age of ultra-high-density semiconductor devices, including CCDs and super-LSIs, which integrate tens of hundreds of thousands of elements into several-millimeter-square chips.

            Sony's MCZ method is the world's first of its kind, which applies a high magnetic field, instead of zero gravity, to mass produce very high-quality crystals of silicon for industrial applications.

            SOURCE: "Sony Develops Magnetic-Field Method for High-Quality Silicon Crystals". Journal of the Electronics Industry, Japan. August 1980. p. 42.

1980   FIBRE-OPTIC SUBMARINE CABLE_   Standard Telephones and Cables
                                                                          (United Kingdom)

            Tomorrow (Feb. 14th) weather permitting, Standard Telephones and Cables will start laying what is probably the first purpose-built fibre optic submarine cable in the world.

            The British Post Office cableship Monarch will lay the trial system, a five nautical mile loop of armoured cable made by STC, in Loch Fyne at Inveraray, Scotland.

            Results of the trial will be monitored by STC which will be looking at "all possible parameters", and the British Post Office. It is considered vitally important since STC is a major exporter of submarine cable and the Post Office and STC need experience of a realistic sea situation to face competition from foreign companies, especially those in the USA and Japan.

            The armoured cable is about two inches in diameter and has four fibres although it has been designed to take up to eight. In about a year's time, two regenerators will be inserted by Cableship Monarch. The housing for these regenerators has already been incorporated and the cable will be lifted and the regenerators placed within the housings.

            SOURCE :   "PO first with seabed optic link" by Denise Clark.
            Electronics Weekly (February 13, 1980), p. 1.

1981   HYDROPLANE POLISHING OF SEMICONDUCTORS_   J. V. Gormley
                                                                                   M. J. Manfra   (U.S.A.)
                                                                                   A. R. Calawa
                                                                                   (Massachusetts Institute
                                                                                   of Technology)

            A new polishing technique for semiconductor materials promises very smooth surfaces free of mechanical defects, faster polishing, and, by implication, improved yield and throughput. Called hydroplane polishing by its developers at the Massachusetts Institute of Technology's Lincoln Laboratory in Lexington, Mass., the system may beat today's mechanical and chemical polishing approaches. So far used on gallium arsenide and indium phosphide, the technique removes material at up to 30 μm per minute, as much as 60 times faster than other methods, and the surfaces produced are flat to within 0.3 μm and free of mechanical damage. In the new process, semiconductor wafers are mechanically suspended about 125 μm above the surface of a smooth, spinning disk coated with continually replenished etchant solution; the wafers thus hydroplane just above the disk's surface. The new method was developed to satisfy the stringent surface-quality requirements of molecular-beam epitaxy.

            SOURCE: "Hydroplaning could yield smoother IC wafers". Electronics,
            (December 15, 1981), p. 34.

            SEE ALSO: "Spinning etchant polishes flat, fast". Electronics (Jan. 13, 1982)
            p. 40.

1981   <u>PLANE-POLARIZED LIGHT OPTICAL FIBRE</u>   T. Suganuma
                                                   (Hitachi)   Japan

        The big difference in the new fiber is the presence of an 80-by-26-μm elliptical jacket surrounding the cladding and itself surrounded by the support. Boric oxide in the jacket material increases its temperature coefficient of expansion far above that of the support. Thus as the fiber drawn from the four-part preform (core, cladding, jacket, and support) cools down from the $2,000^{\circ}$C temperature at which it is fabricated, differences in the thickness of the jacket material exert anisotropic forces on the core along the major and minor axes of the ellipse. The direction with the higher compression, along the short axis of the ellipse, has the higher index of refraction.

        Because the index of refraction is highest along the shorter of the ellipse's perpendicular axes and lowest along the longer one, a single-polarized wave of light launched into the cable along either axis will be transmitted unchanged. Even after transmission over 1 km, the polarization, as measured by the extinction ratio, will be better than 30 decibels - that is, the conversion of energy from one plane into the orthogonal one is less than 0.1%. For a standard fibre under the best of conditions the ratio would probably not exceed 10 dB. And the slightest vibration of a standard fiber, which causes anisotropic mechanical pressure on the core, can reduce the ratio to about 3 dB.

<u>SOURCE:</u>   "Fibre transmits plane-polarized wave of light". Electronics, (July 28, 1981). p. 77.

<u>SEE ALSO:</u>   "Elliptically cross-sectioned fibre". Electronics, (August 30, 1979). p. 67.

1981   <u>HYDROGENATED AMORPHOUS SILICON FILMS</u>   V. Grasso
                                                   A. M. Mezzasalma (Italy)
                                                   F. Neri

        The preparation of hydrogenated amorphous silicon films was carried out by a new method consisting in mixing evaporated silicon from an electron-beam source with a stream of ionized hydrogen produced by an Ion Tech low energy source. This source is a cold cathode device which operates at lower pressure than conventional cold cathode sources.

<u>SOURCE:</u>   "A new evaporation method for preparing hydrogenated amorphous silicon films" by V. Grasso, A. M. Mezzasalma and F. Neri. Solid State Communications, Vol. 41, No. 9 (1982). p. 675.

1982   <u>FISSION TRACK AUTORADIOGRAPHY (A.E.R.E.)</u>   (United Kingdom)

        Since the 'soft error' was diagnosed in 1978, very-large-scale integrated (v.l.s.i.) circuit memory manufacturers have sought a method of detecting the minute amount of naturally-occurring radioactive impurities which, if present in v.l.s.i. circuit materials, can disrupt circuit performances.

        Now, however, Harwell has developed an extremely sensitive technique known as fission track autoradiography (FTA) which can detect the presence of uranium in concentrations as small as 2 parts in 10^9. This provides manu-facturers with a quality control enabling them to assess raw material, and components, thereby reducing the risk of component failure.

        The 'soft error' effect is produced by alpha-particle emissions from radioactive impurities present in any part of the v.l.s.i. circuit assembly. The energy possessed by an alpha-particle can produce an electric charge which may change the content of a single memory location, giving rise to computational errors. Because of this, semiconductor manufacturers are now specifying alpha-particle emission rates of less than 0.001 particles/cm^2/hour for their memory device materials.

        It is not possible to detect such emission levels directly. The Harwell FTA technique exploits uranium-235, the fissile isotope present as 0.72% of natural uranium; prepared specimens of semiconductor material are coated with a polyimide film solid state nuclear track detector (s.s.n.t.d.) and irradiated with thermal neutrons in Harwell's Materials Testing Reactor, DIDO. On irradiation, the U-235 undergoes fission and the resulting fission particles

are registered as tracks on the s.s.n.t.d.   Afterwards the polyimide film is
chemically etched to develop the fission tracks which can then be examined by
optical microscopy.   From the information gained it is possible to determine
precisely the amount of uranium present, down to 2 parts in 10^9 (or a surface
distribution of 3×10^{-6} µg/cm^2) and thus to calculate alpha-emission rates of
as little as 0.0002 particles/cm^2/hour.

SOURCE: "Fission Track Radiography for Checking V.L.S.I. Circuits".
The Radio and Electronic Engineer, Volume 52, No. 5 (May 1982), p.200.

1982   RECRYSTALLIZATION SILICON PROCESS   Texas Instruments (U.S.A.)

Using a moving graphite heater to create a thin layer of single-crystal
silicon atop an oxide insulator, Texas Instruments Inc., is developing what it
believes will be a practical alternative to expensive silicon-on-sapphire substrates for high-density, high-speed complementary-MOS integrated circuits.

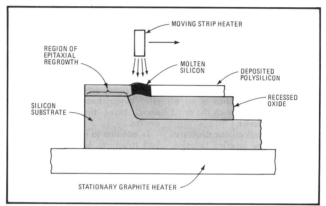

Hot spot. Strip heater that Texas Instruments moves across deposited polysilicon produces a
region of epitaxial regrowth above the single-crystal substrate. Recrystallization continues
above the insulator region, making a device-quality single-crystal layer.

SOURCE: "Oxide insulator looks the equal of sapphire for C-MOS ICs".
Electronics (June 2, 1982), p.45.

1982   AMORPHOUS PHOTOSENSORS   Sony (Japan)

Sanyo Electric recently succeeded in developing photosensors made of
amorphous semiconductor material for the first time in the world, and announced
that they would be released in spring.   Based upon the development of the
amorphous photosensors, Sanyo also developed one-chip full-colour sensors and
intends to expand their application to line sensors.   In the case of visible light
total spectra photosensors, the new amorphous photosensors cost only half of
conventional silicon monocrystal photosensors.

The amorphous sensor developed by Sanyo is made by forming a transparent conductive layer on a glass base, by forming PIN amorphous silicon on
it through the continuous separation forming method using silicon carbide in
the P layer, separating it into chips after attaching lead wires, and molding it
into a resin package after performing face-down-bonding/lead-frame bonding.
The mono-color and fullcolor sensors are then made by providing appropriate
filters; even in the case of the full-color sensor, it can be made as a single
chip, taking advantage of the large electric resistance between the lead wires.

SOURCE: "Sanyo Develops Amorphous Photosensors". JEI, Japan (April
1982), p.50.

Chapter 12
List of Inventors

(In alphabetical order, with dates of inventions)

Abraham H. 1918
Abramson 1949
Agusta B. 1969
Aitken H.H. 1939
Alderton R.H. 1957
Allen J.W. 1960
Anderson O.L. 1950s
Andreatch P. 1950s
Anson R. 1924
Appleton E. 1923, 1924
Ardenne Von 1935, 1937
Artom A. 1907
Armstrong E.H. 1912, 1922, 1933, 1951
Ashworth F. 1957
Ayling J.K. 1969

Babbage C. 1833
Bacon F.T. 1839
Bain A. 1843
Baird J.L. 1925
Baran P. 1964
Bardsley W. 1972
Basov N.G. 1953
Bauer M. 1874
Baxandall P.J. 1952
Bardeen J. 1948
Barkhausen H. 1919
Becquerel E. 1839
Bell A.G. 1876
Berg 1900
Berliner E. 1887
Bertout F. 1956
Betts A.G. 1905
Betulander G.A. 1912
Beutheret C. 1956
Beverage H.A. 1928
Black H.S. 1927
Blathy O.T. 1885
Bloch E. 1918
Blumlein A.D. 1931, 1932, 1935, 1936, 1941,
                    (& 1942.
Bobeck A.H. 1969
Bohr N. 1913
Boetzelaer L.W.J. van. 1925
Boole 1848
Boot H.A.H. 1939
Bowen E.G. 1940
Boyle W.S. 1970
Bradley C.S. 1885
Branley E. 1890
Brattain W.H. 1948
Braun F. 1897
Braunstein R. 1955
Breit G. 1924
Buck D. 1955
Buehler E. 1952
Bullen T.G. 1937
Burgess J.F. 1975
Bush V. 1931
Butement W.S. 1941

Cady W.G. 1921
Cady W.R. 1974
Cahill T. 1908
Calawa A.R. 1981
Calligaro R.B. 1978

Campbell G. 1915
Campbell-Swinton A.A. 1908
Carlson C. 1937
Carlson R.O. 1962
Carson J.R. 1915, 1936
Casimir Jonker 1955
Casstellani E. 1970
Chadwick J. 1932
Chang K.K.N. 1967
Chapin D.M. 1954
Christensen H. 1950s, 1960
Chopra K.L. 1967
Chua L.O. 1967, 1968
Clark L. 1870
Clarke A.C. 1945
Clarke C.L. 1884
Claude G. 1910
Clavier A.G. 1929
Cleeton C.E. 1934
Cockcroft J.D. 1932
Cook R. 1975
Cooke W.F. 1837
Coombs A.W.M. 1943
Cooper-Hewitt P. 1901
Crane H.D. 1960
Crookes W. 1878
Cross C.F. 1878
Cunaeus 1745
Curie J. 1880
Curie P. 1880
Czochralski J. 1917

Dalton J. 1808
Dancer 1839
Danielson E. 1902
Danko S.F. 1949
Darlington S. 1952
Dash W.C. 1958
Davenport T. 1837
Davy E. 1837
Dee P.I. 1941
De Loach B.C. 1964
Deri M. 1885, 1885
Dippy R.J. 1938
Dodge H.F. 1944
Dolby R.M. 1967
Dowling J.J. 1937
Dreyer J. 1934
Duane J.T. 1962
Dummer G.W.A. 1940, 1943, 1952.
Dunwoody H.C. 1906
Dyke W.P. 1960

Eccles 1919
Eccles W.H. 1912
Eckert P. 1947
Edison T.A. 1877, 1877, 1900
Einstein A. 1905
Eisler P. 1943
Emeis R. 1953
Esaki L. 1958

Faraday M. 1831, 1831, 1834, 1839, 1850.
Farnsworth P. 1919
Feder E. 1970
Fedida S. 1974

Fenner G.E.   1962
Fessenden R.   1906, 1912
Fermi E.   1934
Fischer R.F.   1969
Fischer Prof.   1939
Fitzgerald D.G.   1876
Flanagan G.   1975
Flechsig W.   1938
Fleming J.A.   1904
Fleming-Williams B.C.   1939
Fletcher N.H.   1954
Flowers T.H.   1943
Forest de   1906, 1912
Forrat F.   1956
Fox Movietone News   1927
Franklin E.   1940
Fritch   1938
Frohman-Bentchkowsky D.   1971
Frosch E.J.   1957
Fuchs E.O.   1965
Fuller C.S.   1954, 1956

Gabor D.   1948
Galvani L.   1780
Gambrell T.E.   1897
Gapanov A.G.   1975
Gaugain J.M.   1855
Geiger H.   1908
Giacoletto L.J.   1956
Gianole V.F.   1957
Gibbons P.E.   1960
Gilbert B.   1971
Geissler H.   1856
Gill E.W.B.   1922
Gormley J.V.   1981
Graaf van de   1930
Grainger P.   1930s
Grasso V.   1981
Green G.W.   1972
Grier J.D.   1972
Grove W.R.   1839, 1852
Gummel H.K.   1964
Gunn J.B.   1963

Haas W.J. de 1930, 1935
Haeff A.W.   1943
Hahn W.C.   1939
Hall E.H.   1879
Hall R.N.   1962
Hamming R.W.   1950
Harris A.F.   1897
Hart K.   1972
Hazeltine L.A.   1918
Heaviside O.   1901
Heil A.   1935
Heil O.   1935, 1935
Heiman F.P.   1962
Henry J.   1832
Herd J.F.   1923
Heritage M.   1970
Herschel   1800, 1840
Hertz H.R.   1887
Hilpert G.   1909
Hilsum C.   1961
Hockham G.A.   1966
Hoeni J.A.   1959
Hofstein S.R.   1962
Hollerith H.   1889
Holliday E.H.   1972
Holms J.H.   1884, 1887
Holonyak N.   1962
Holst   1928
Horn F.H.   1966

Horton J.W.   1928
House C.H.   1973
Houskeeper W.G.   1919
Hull A.W.   1918
Hunnings H.   1878
Hurle D.T.J.   1972

Iizuka K.   1971
Iverson K.   1950

Jack R.F.   1965
Jansky K.G.   1933
Johnson J.B.   1925
Johnston R.L.   1964
Jordan   1919
Josephson B.D.   1962
Joule J.   1847
Junger   1900
Juran J.M.   1951

Kao K.C.   1966
Kay A.   1952
Keck P.H.   1953
Keen J.M.   1971
Keith E.A.   1896
Kelvin Lord   1851
Kennelly A.   1901
Kerst D.W.   1941
Kilby J.S.   1959
Kimura T.   1980
Kingsley J.D.   1962
Kirchoff G.W.   1845
Kleist von   1745
Kleimock J.J.   1959
Knight G.   1772
Knoll M.   1932, 1935
Kompfner R.   1943
Kruger F.   1919
Kurtz K.   1919

Langevin P.   1915
Langmuir I.   1912, 1912, 1914
Lasher G.   1962
Laurence E.O.   1929
Lawson W.D.   1959
Leclanche G.   1868
Leibniz G.W.   1672
Lenz H.   1868
Lepselter M.   1964
Lewis W.B.   1940
Logan R.A.   1975
Lilienfeld J.   1930
Little J.B.   1948
Littleton J.T.   1931
Loewe S.   1926
Lohstroh J.   1978
Lombardi L.   1900
Loor H.H.   1959
Lovell A.C.B.   1941
Luc de   1800

Mallina R.F.   1953
Malter L.   1935
Manfra M.J.   1981
Marconi G.   1896
Marello V.   1978
Marrison W.A.   1928
Mattox D.H.   1963
Maunchly J.   1947
Maxwell J.C.   1865
Mclean D.A.   1956
McMullan D.   1935
Meade D.A.   1956

Meissner 1912
Meitner 1958
Miller J.M. 1919
Moore B.J. 1973
Moore J.B. 1928
Moore R.D. 1969
Moore R.E. 1975
Moore School 1943
Morrell J.H. 1922
Morse S.B. 1837
Morton G.A. 1935
Moscicki I. 1904
Mossbauer R.L. 1958
Mushenbrook von 1745

Nassibian A.G. 1978
Nathan M.I. 1962
Nelson H. 1961
Neugebauer C.A. 1975
Neumann von 1945
Neumann G. 1924
Newman M. 1943
Ngvyen V.K. 1971
Nicholson 1919
Nielsen S. 1959
Nipkow P. 1884
Nishizawa J. 1950
Nobilli C.L. 1828
Northrup E.F. 1918
Noyce R. 1968

Oatley C.W. 1935
O'Brien W. 1945
O'Connell J. 1956
Oersted H.C. 1820
Ogura H. 1971
Ohm G.S. 1826
Ohl R.S. 1949/50
Olsen K.M. 1965
Onnes K. 1911
Onton A. 1978
Ovshinsky S.R. 1968

Palmgren N.G. 1912
Pascal B. 1642
Pearson G.L. 1954
Perneski A.J. 1969
Peterson H.O. 1928
Pfann W.G. 1952
Pfleumer 1920
Picard G.W. 1907
Pierce J.R. 1943
Planck M. 1900
Plante 1860
Pleucker J.P. 1858
Pol van der 1926
Pollard P.E. 1937
Poulsen V. 1898
Power F.S. 1956
Prager H.J. 1967
Prokhorov A.M. 1953
Puckle O.S. 1933
Putley E.H. 1959, 1960

Quate C.F. 1973

Rabi I.I. 1939
Randall J.T. 1939
Reeves A.H. 1937, 1940
Reinhart F.K. 1975
Reis H. 1956
Reis J.P. 1860
Remeika J.P. 1969

Reynolds J.N. 1916
Ridley B.K. 1961
Rijlant P. 1921
Ritter J.W. 1801, 1803
Roberts J.G. 1916
Rodgers T.J. 1972
Rogers B. 1964
Romankiw L. 1970
Romig H.C. 1944
Rontgen W.K. 1895
Rosenthal A.H. 1940
Round H.J. 1926
Rowen J.H. 1963
Ruehle W. 1978
Ruska E. 1932
Rutherford E. 1902, 1908, 1911, 1918

Saito S. 1980
Sangster F.J.L. 1969
Sargrove J.A. 1947
Schottky W. 1918
Schweigger J. 1828
Seebeck T.J. 1821
Shannon C.E. 1938, 1948
Schelkunoff S.A. 1936
Schalow A.L. 1958
Scheutz P.G. 1854
Schilling 1812
Shewhart W.A. 1931
Shockley W. 1948, 1949/50
Siemens E.W. 1847, 1877, 1925
Simmons J.G. 1978
Simpson O. 1947
Sittig E.K. 1963
Slob A. 1972
Smart A.W. 1945
Smith G.E. 1970
Snoek J.L. 1909
Soddy F. 1902
Soltys T.J. 1962
Sommering 1812
Sosnowski J. 1947
Southworth G.C. 1936
Spiller E. 1970
Starkiewicz J. 1947
Stearn C.H. 1878
Stibitz G. 1939
Strowger A.B. 1889
Suganuma T. 1981
Swan J.W. 1878
Swann W.F.G. 1913
Swartz H. 1945
Sweet R.G. 1963

Tantraporn W. 1974
Tao Y. 1962
Teal G.K. 1948, 1952
Teer K. 1969
Telegen 1928
Tesla N. 1888
Teszner S. 1958
Theurer H.C. 1953, 1960
Thomson J.J. 1893, 1897
Topalian J. 1970
Topham F. 1878
Townes C.H. 1953, 1958
Tu G.K. 1969
Turing A. 1943, 1946

Uitert L.G. van 1969
Ulitovsky 1949
Uttley A.M. 1942

Varian S.F.   1939
Varian R.H.   1939
Vasalek J.   1921
Vogel P.   1961, 1962
Volta A.   1800
Voogd J.   1930

Wagner K.W.   1915
Wald A.   1943
Walker P.J.   1957
Walton E.D.S.   1932
Warner Bros   1926
Watkins T.B.   1961
Watson G.N.   1918
Watson-Watt R.A.   1923, 1924
Weber J.   1953
Weckler G.P.   1965
Weisbrod S.   1967
Wente E.C.   1917
Weston E.   1891
Wheatstone C.   1837, 1843, 1845
Wheeler H.A.   1926
Whiddington R.   1920

Weidenhof N.   1978
Wiegand J.   1965
Wiessenstern M.   1966
Williams F.C.   1942, 1942, 1946
Williams N.A.   1934
Williamson D.T.N.   1947
Wilkes M.V.   1948, 1951
Wilson C.T.R.   1912
Wimshurst J.   1882
Wingrove G.A.S.   1966
Wood   1915
Wray   J.T.   1857
Wright G.T.   1968

Yagi H.   1926
Yamamoto Y.   1980
Yariv A.   1978
Young A.S.   1959
Yu S.P.   1974

Zamboni   1800
Zipernowski C.   1885
Zworykin V.K.   1919, 1923, 1935.

Chapter 13
List of Books on Inventions and Inventors

Books on Inventions

(N.B. - Some of the older books are now out of print and difficult to obtain)

Abbott C.G. "Great Inventions" Smithsonian Institution, Washington (1932)

Aitken W. "Who invented the telephone" Blackie, London (1939)

Appleyard R. "Pioneers of electrical communications"MacMillan, London (1930)

Barnouw E. "A history of broadcasting in the United States" Oxford Univ. Press (1968)

Briggs Asa "The history of broadcasting in the United Kingdom" Vols.1 & 2. Oxford Univ. Press (1965)

Bryn E.W. "The progress of invention in the nineteenth century" Munn. New York (1900)

Carr L.H.A. & Wood J.C. "Patents for Engineers" Chapman & Hall, London (1959)

Carter E.F. "Dictionary of inventions & discoverers" Fdk. Muller, London (1966)

Chase C.T. "A history of experimental physics" Von Nostrand (1932)

Crowther J.G. "Discoveries & inventions of the 20th century" Routledge & Regan Paul Ltd. London (1966)

Crowther J.G. "British scientists of the 19th century" MacMillan, London.

Cressy E. "Discoveries & inventions of the twentieth century" George Routledge, New York. E.P.Dutton (1914)

Darrow F.L. "Masters of science and invention" Harcourt Brace, New York (1923)

Dibner Bern "Heralds of science as represented by 200 epochal books and pamphlets selected from the Bundy Library" Bundy Library, Norwalk, Conn. (1955)

Fahie J.J. "History of electric telegraphy to the year 1837" F.N. Skoon, London (1884)

Fahie J.J. "History of wireless telegraphy" Wm. Blackwood, Edinburgh & London (1899)

Fleming J.A. "Fifty years of electricity" Wireless Press (1921)

Goldstine H.H. "The computer from Pascal to von Neuman" Princeton Univ. Press (1972)

Hawks E. "Pioneers of wireless" Methuen & Co. London (1927)

Jaffe Bernard "Men of science in America" Simon & Schuster, New York (1944)

Jewkes J., Sawers D. & Stillerman R. "The sources of invention" MacMillan, London (1958)

Johnson P.S. "The economics of invention and innovation" Martin Robinson (1975)

Larsen Egon "A history of invention" J.M. Dent & Sons, London, and Roy Publishers, New York (1971)

MacLaurin W.R. "Invention & Innovation in the radio industry" MacMillan, New York (1949)

Moore C.K. & Spencer K.J. "Electronics - a bibliographical guide" MacDonald, London (1965)

Motteley "Bibliographical history of electricity & magnetism"Griffin, London. (1922)

Pierce J.R. "The beginnings of satellite communications" San Francisco Press(1968)

Pledge H.T. "Science since 1500" H.M.S.O. (1946)

Proceedings of the Royal Society, London.

Rhodes F.L. "The beginnings of telephony" Harper, New York (1929)

Rider K.J. "The history of science and technology" Library Assoc. of London (1967)

Routledge R. "Discoveries & inventions of the nineteenth century" G. Routledge, London (1891)

Shiers G. "Bibliography of the history of electronics" The Scarecrow Press, Metuchen, N.J. (1972)

Singer C.J. "A short history of science to the nineteenth century" Clarendon Press (1941)

Timetable of Technology, Michael Joseph, London (1982)

Tricker R.A.R. "Early electrodynamics" Pergamon Press, Oxford (1965)

Whetham W. "A history of science" Cambridge Univ. Press (1929)

Books on Inventors

(N.B. Some of the older books are now out of print and difficult to obtain)

"A biographical dictionary of scientists"
by T.I. Williams. Adam & Charles Black.
London (1969)

AMPERE "Andre-Marie Ampere and his
English acquaintances" by K.R. and D.L.
Gardiner Brit. Journal for the History of
Science. Vol.2. (July 1965) p.235.

BAIRD "Baird of television - the life story
of John Logie Baird" by R.F. Tiltman.
Selley Service, London (1933)

BELL "Alexander Graham Bell;the man who
contracted space" by Catherine D. Mackenzie
Houghton Mifflin (1928)

   "Bell, Alexander Graham Bell and the
conquest of solitude" by R.V. Bruce.
Victor Gollanz (1973)

BERLINER "Grevile Berliner, maker of the
microphone" by F.W. Wile. Bobbs-Merrill,
Indianapolis (1926)

"Biographical memoirs of Fellows of the
Royal Society" Royal Society, London(Annual)

BRAUN "Ferdinand Braun:Leben und wirken
des Erfinders der Brauchen Roehre, Nobel-
preistraeger" by F.Kurylo. Heinz Moos
Verlag (1965)

CROOKES "The life of Sir Williams Crookes"
by E.E. Fournier D'Albe. Fisher Unwin
London (1923)

EDISON "My friend Edison" by H.Ford.
Ernest Benn Limited.

   "Edison" by M. Josephson. McGraw-
Hill, New York. (1959)

FAHIE "The life and work of John Joseph
Fahie" by E.S. Whitehead. University Press
of Liverpool (1939)

FARADAY "Faraday" by R. &. R. Clark.
Brit. Elec. & Allied Mnfrs. Assoc. (1931)
   "Michael Faraday - his life and work"
by S.P. Thompson. Cassel (1901)

   "Faraday, Maxwell & Kelvin" by
D.K. MacDonald. Doubleday N.Y. (1964)

FESSENDEN "Fessenden - builder of
tomorrow" by H.M. Fessenden. Coward-
McCann (1940)

FITZGERALD "The scientific writings of
the late George Francis Fitzgerald" by
J. Lamor (Ed) Dublin Univ. Press (1902)

FLEMING "Memories of a scientific life"
by Alexander Fleming. Marshall.

HEAVISIDE "Oliver Heaviside" by G. Lee.
Longmans Green, London (1947)

HENRY "Joseph Henry - his life and work"
by T.Coulson. Princeton Univ. Press U.S.A.
                                  (1950)

HERTZ "Gesammelte Werke" by P.E.A.
Lenard Ambrosius Borth, Leipzig (1895)
(Papers in three volumes) English trans-
lations - MacMillan, London.

LODGE "Oliver Lodge - past years, an
autobiography" Scribner, New York (1932)

MARCONI "Marconi, the man and his wire-
less" by O.E. Dumlap, MacMillan (1937)
   "Marconi, master of space" by
B.L. Jacot & D.M.B. Collier, Hutchinson
                                   (1935)

"My father, Marconi" by D. P. Marconi. McGraw Hill, New York (1962)

"Marconi, pioneer of radio" by D. Coe, Julian Messner (1935)

MAXWELL "The life of James Clark Maxwell" by L. Campbell & W. Garnett. MacMillan, London (1882)

"James Clark Maxwell F.R.S. 1831-1879" by R. L. Smith-Rose, Longmans Green, London (1948)

MORSE "The life of Samuel F. B. Morse inventor of the electro-magnetic recording telegraph" by S. A. Prime. D. Appleton New York (1875)

REIS "Philipp Reis" Deutche Bundespost Archiv. fur Deutche Postgeschicht No. 1. (1963)

RONTGEN "Wilhelm Conrad Rontgen and the early history of the Rontgen rays" by O. Glasser. Charles C. Thomas, Springfield, Ill. (1934)

RUTHERFORD "Rutherford - being the life and letters of the Right Honourable Lord Rutherford" by A. S. Eve. Cambridge Univ. Press (1939)

TESLA "The inventions, researches and writings of Nikola Tesla" by T. C. Martin. The Electrical Engineer, New York (1894)

"Ten founding fathers of the electrical science" by B. Dibner. Bundy Library Publications, Norwalk, Conn. (1954) (GILBERT GEURICKE, FRANKLIN, VOLTA, AMPERE OHM, GAUSS, FARADAY, HENRY and MAXWELL)

THOMSON J. J. "The life of Sir J. J. Thomson O. M. sometime Master of Trinity College, Cambridge" by Lord Rayleigh. Cambridge Univ. Press (1942)

"J. J. Thomson and the Cavendish Laboratory of his day" by G. P. Thomson, Nelson, London (1964)

THOMSON W. "The life of William Thomson" by S. P. Thompson. MacMillan, London (1910).

Index

	Page			Page
A.A. Gun radar	23, 115		(nickel-cadium)	72
ACE, computer	129		(nickel-iron)	72
Accumulator	50		(Léclanché)	61
Acoustic Mine	82		(magnetohydrodynamic)	54
APL	139		(Zinc-mercury-oxide)	66
Aerials			(Planté)	60
(Diversity)	100		(secondary)	60
(Ground wave prop.)	85		(solar)	147
(Hertz)	67		(Volta)	49
(Ionospheric prop.)	72		(Weston standard cell)	69
(Maxwell)	60		BARRITT diode	186
(Radio wave prop.)	67		Beam leads	174
(Yagi)	99		Betatron	120
Aeriel 1 Satellite	168		Biosatellite - 1 satellite	180
A.I. radar	22		Boolean Algebra	57
All-electronic clock face	209		Broadcasting	75
Alouette Satellite	168		Bubbles (magnetic)	188, 210
Amateurs (short wave)	91		Bucket-brigade delay circuit	189
Amorphous liquid crystals	211		Cable, fibre optic	215
Amorphous Silicon solar cell	203		Cable insulation	50,57,106
Amorphous semiconductor switch	184		Cable sheathing	56
Amplifier, high quality	130		Cable transatlantic	60, 150
Anisotropic magnet	205		Calculators	170
Angels	25		Capacitors	
Apollo - 7 Satellite	186		(ceramic)	71
Arc lamp (mercury)	59		(glass tubular)	74
Army radars	23		(Leyden jar)	48
Arrays, photodiodes	178		(mica)	61
ASDIC	82		(rolled paper)	62
Assembly Systems			(semiconductor diode)	149
(automatic)	144		(solid electrolyte)	148
(beam lead)	174		Carbon composition resistor	66
(ceramic chip carrier)	194		Carbon film resistor	71
(dip soldering)	136		Carbon filament lamp	64
(dual-in-line)	176		Carbon microphone	63, 64
(flat-pack)	167		Cardiograph	103
(flip-chips)	179		Carrier-domain magnetometer	193
(micromodules)	155		C.A.T.T.	209
(printed circuits)	125		Cathode rays	64
(potted circuits)	129		Cathode ray oscillograph	70, 103
(Sargrove)	131		CCD, A/D converter	204
(thick film)	122, 196		CCD, colour TV camera	211
(thin film)	158, 202		Ceramic capacitors	71
("Tinkertoy")	140		CEEFAX	34
(wire wrapped)	146		Charge coupled devices	191
Astronomy (radio)	107		Chirp radar	131
Atomic change (spontaneous)	74		Circuitry	
Atomic theory	50, 78, 81		(auto volume control)	97
Atomic transmutation	85		(bucket-brigade delay)	189
Atoms, trans-uranian	107		(constant RC stand-off)	110
ATS-1 Satellite	180		(crystal control of frequency)	90
Audio systems	15		(Darlington pairs)	142
Automatic control of crystal growth	197		(Dynatron)	85
Automobile Electronics	36		(digital IC's)	163
Autoradiography	216		(Echo cancelling)	212
A.S.V. radar	22		(energy conserving scanning)	105
A.V.C. circuit	97		(flip-flop)	88
AZUR satellite	187		(hard valve time base)	106
Batteries			(heterodyne & superheterodyne)	80
(Clark cell)	61		(high quality amplifier)	130
(fuel cell)	54		(linear IC's)	161

	Page
(Kirchoff)	56
(long-tailed pair)	112
(Miller integrator)	123
(Miller time base)	89
(multivibrator)	86
(mutator circuit network)	187
(negative feedback)	99
(neutrodyne)	86
(PHANTASTRON)	124
(regenerative)	80
(rotator network)	183
(SANATRON)	124
(saw-tooth time base)	93
(superregenerative)	91
(squegger circuit)	91
(tone control)	142
(transitron oscillator)	98
Circuits (potted)	129
Circuits (printed)	125
Clark standard cell	61
Clock (electronic)	164
Cloud chamber	79
C-MOS integrated circuit	185
Cockroft-Walton accelerator	104
Coherer	68
Cold cathode discharge tube	58
Cold cathode stepping tube	136
Cold cathode trigger tube	112
Collector diffusion isolation	190
COLOSSUS	127
Communication	
(frequency modulation)	106
(information theory)	134
(microwave)	101
(packet switching)	173
(satellite)	129
(shortwave)	91
(single sideband)	83
Components - history	3
Computer Aided Design	32
Computer Aided Manufacture	33
Computers	
(ACE)	129
(APL)	139
(Babbage)	52
(Bell "Complex")	118
(COLOSSUS)	127
(CDC 1604)	160
(CRT storage)	130
(Diff. analyser)	103
(digital ASCC)	118
(EDSAC)	134
(EDVAC)	131
(ENIAC)	126
(History)	29
(Hollerith)	68
(Honeywell 800)	161
(IBM 650)	137
(IBM 701)	138
(IBM 704, 709, & 7090)	144
(Information theory)	116
(Leibniz)	48
(Leprachaun)	150
(Microcomputer)	195
(Microprogramming)	140
(Minicomputer)	165
(one board)	204
(Operation)	29
(Pascal)	48
(RAM 1024 bit)	197

	Page
(RAM 4096 bit)	202
(RAM 16, 384 bit)	203
(RAM 256 k bit)	214
(SAGE)	143
(Scheutz)	58
(SEAC)	132
(theory)	128
(Types)	31
(UNIVAC)	132
(UNIVAC 80/90)	162
(Whirlwind)	160
Condenser microphone	84
Conductor, sodium	74
Constant RC circuit	110
Cores (iron dust)	49
Counter (geiger)	77
Courier 1 Satellite	162
Copper Oxide Rectifier	97
Cracked carbon resistor	95
Cryotron	148
Crystal control of frequency	90
Crystal growth (automatic control)	197
Crystal Detector (Carborundum)	75
Crystal Detector (Perikon)	76
Crystals (liquid)	108, 192
Crystal Microphone	88
Crystal pulling technique	84
Crossbar Telephone Exchange	84
Cyclotron	101
Darlington pair circuit	142
DECCA Navigation	128
Delay circuit (bucket-brigade)	189
Diaphragm microphone	59
Diademe Satellite	182
Diffusion technique	148
Digital IC's	163
Digital voltmeter	142
Diode IMPATT	173
Diode (tunnel)	153
Dip soldering	136
Discharge tube (coldcathode)	58
Discharge tube (low pressure)	59
Discoverer 1 Satellite	158
Diversity reception	100
Dolby noise reduction system	182
Double-beam oscillograph	117
Dual-in-line pack (DIL)	176
Dry etching technique	198
DSCS - 1 Satellite	194
Echo cancelling circuit	212
Echo - 1 Satellite	162
E.C.M.E.	131
Educational Electronics	36
Electron beam lithography	199
Electronic calculator	170
Electronic circuit making equip.	131
Electronic clock	164
Electronic organ	76
Electronic watch	167, 193
Electromagnetic induction	51, 52
Electromagnetism	50
Electron	70
Electron microscope	106
Electrolysis	53
Electrostatic loudspeaker	95
Electrostatics	1
Epitaxy (vapour phase)	162
Epitaxy (liquid phase)	163
ESSA - 1 Satellite	181
Explorer 1 Satellite	154

Facsimile reproduction   56
FAMOS - integrated circuit   194
Femitron   159
Ferreed switch   139
Ferrites   77
Ferroelectricity   90
Field effect transistor   111
Filter (electromagnetic)   83
Fission, nuclear   117
FLAD   206
Flat-pack   167
Flip-chip bonding technique   174
Flip-flop circuit   88
Floating zone refining   147
Floppy-disc recorder   192
FLOTOX   212
Flow soldering (printed circuits)   149
Fluorescent lamp   73
FR - l Satellite   177
Frequency modulation   106
Frequency standard (atomic)   107
Frequency standards (caesium beam) 119
Frequency standards (quartz)   100
Fuel cell   54

Galvanic action   49
Galvanometer (moving coil)   51
Galvanometer (astatic)   51
"GEE" navigation   23, 116
Geiger counter   77
Gemini satellite   173
GGSE satellite   177
GL1 et al radars   23
Glass capacitor (tubular)   74
Glow discharge   59
Goniometer   76
Gramophone   67
Gunn diode oscillator   169
GYROTRON   201

Hall effect   65
HAMMING Code   137
Hard valve time base circuit   106
Heating (induction)   84
Heaviside/Kenelly layer)   72
Heterodune circuit   80
High field superconductivity   102
High quality amplifier   130
H.MOS   206
H2S navigation system   122
History of components   3
Holography   133
Hologram matrix radar   193
Housekeeper seal   88
Hydroplane polishing semiconductors 215
Hydrogenated silicon films   216

Iconoscope   92
Induction (electromagnetic)   31, 52
Induction motor   68
Ignitron   107
IMPATT diode   173
Induction heating   84
Inductor (iron dust cores)   49
Industries from inventions   12
Information Technology (IT)   34
Information theory   116
Infra-red radiation   50
Infra-red emission from GaSb   148
Ink jet printing   169, 213
Intelsat 1 satellite   178
Interdigitated transistor   147
Integrated injection logic   197
Integrated Schottky logic   207

Integrated optoelectronics   210
Ion implantation   136
Ion plating   170, 182
Ionosphere layer   96
IRIS satellite   184
Iron dust cores   49

Johnson noise   96
Josephson effect   167

Kirchoff's laws   56
Klystron   117

Lamp (carbon filament)   64
Lamp (fluorescent)   73
Lamp (neon)   78
LANDSAT satellite   195
Large screen TV projector   119, 213
LASER   157
LASER (semiconductor)   168
Laser annealed polysilicon   206
Laser enhanced plating   212
Laser-deep proton - isolated   198
Laser processing of semiconductors 210
Laser recording system   209
Laser trimming of thick films   181
Leclanche battery   61
LED (light emitting diode)   161, 166
LES - l satellite   177
Leyden jar capacitor   48
Light Bubbles   207
Lightwave powered telephone   208
Linear IC's   161
Liquid crystals   108, 192
LOCMOS integrated circuit   201
Logic IC's   163
Logic-state analyser   199
Logic-timing analyser   199
Long-tailed pair circuit   112
LORAN   124
Low pressure discharge tube   59
Loudspeaker (electrostatic)   151, 95
Loudspeaker (moving coil)   63
Lunar Orbiter satellite   180
LUNIK l satellite   158

Magnet anisotropic   205
Magnetic bubbles   188
Magnetic amplifier   127
Magnetic tape, plastic   89
Magnetic recording   71
Magnetic film recording   135
Magnetism   1
Magnetohydrodynamic battery   54
Magnetostriction   57
Magnetron   119
MAGISTOR   190
Mariner satellite   168
MARISAT satellite   204
MARS satellite   168
MASER   145
Medical Electronics   36, 175
Magnetic avalanche transistor   213
MCZ crystal growth   215
Metal film resistor   81, 89, 98
Memory (plated wire)   152
Memory (RAM)   197, 202, 203, 214
Mercury arc lamp   59
Mercury-Atlas satellite   164, 165
Mercury-zinc oxide battery   66
MET radars   25
Mica capacitor   61
Microcomputer   195
Microscope (electron)   106

	Page
Microelectronics (see also transistors)	
(Aluminium metallisation)	187
(Arrays)	178
(Beam lead)	174
(concept)	141
(C.D.I.)	190
(C-MOS)	185
(digital circuits)	163
(DIL packs)	176
(FAMOS)	194
(flat-pack)	167
(flip-chips)	179
(H-MOS)	206
(integrated optical circuits)	202
(LOCMOS)	201
(linear circuits)	161
(logic circuits)	163
(M.O.S.)	166
(Patent)	158
(Silicon anodisation)	202
(silicon on sapphire)	172
(thick film)	122
(thin film)	158
(versatile arrays)	204
(V-MOS)	196
Microfilming	54
Microphone (carbon)	63
Microphone - carbon granule	64
Microphone (condenser)	84
Microphone - crystal	88
Microphone (diaphragm)	59
Microphone - Reisz	93
Microprocessor (single chip)	200
Micromodule assembly system	155
Microscope (scanning electron)	109
Microwave communication	101
Microwire	135
Molecular beam epitaxy	132
Miller integrator circuit	123
Minicomputer	165
Microcomputer	195
MODEM	139
Morse code	53
Mossbauer effect	157
Motor-electric (Davenport)	53
Motor (induction)	68
Motor (synchronous)	73
MOS/FETs	166
Moving coil loudspeaker	63
Multilayer p.c. boards	160
Multiplier phototubes	111
Multivibrator circuit	86
Mutator circuit	187
NATO 1 satellite	192
Naval radars	24
Navigational aids	25
Negative feedback circuit	99
Negative feedback tone control	142
Negative resistance oscillator	91
Neon lamp	78
Neuristor	159
Neutrodyne circuit	86
Neutron	104
Nickel-cadmium battery	72
Nickel-chromium resistor	152
Nickel - iron battery	72
Nimbus satellite	172
Nitride-over-oxide process	181
Noise (Johnson)	96
Noise (shot effect)	87

	Page
Nuclear fission	117
Numerically controlled machines	35
OAO satellite	179
"Oboe" navigation	23, 120
OFO-1 satellite	192
Ohm's law	51
OMIST	208
One megabit bubble memory	210
Optical fibres	180, 216
ORACLE	34
Organ (electronic)	76
ORIEL-1 satellite	194
Oscar-1 satellite	165
Oscillograph (double beam)	117
Oscillograph (polar co-ordinate)	114
OSO-1 satellite	165
OV1-2 satellite	177
Oxide film resistor	102
Oxide masking process	153
Overlay transistor	174
Packet switching	173
PARCOR speech synthesis	189
Pedestal pulling	153
Paging (radio)	151
Pageos satellite	181
Paper capacitor	62
Pegasus satellite	176
Pentode valve	99
PIN diode	137
Phonograph	63
Photoconductors	158
Photoconductive detector	158, 160
Photodiode arrays	178
Photovoltaic effect	54
Phototube (multiplier)	111
Photosensors	217
Piezo electricity	65
Pioneer 1 satellite	156
Plan position indicator (PPI)	119
Planar process	157
Planté battery	60
Plated wire memory	152
PLUMBICON	152
Pocket TV receiver	205
Polar co-ordinate oscillograph	114
Polythylene insulation	106
Potted circuits	129
Prestel	34, 200
Printed wiring	125
Printed wiring (dip-soldering)	136
Printed wiring (etch-back)	176
Printed wiring ("flow-soldering")	149
Printed wiring (multilayer)	160
Proton 1 satellite	179
Proximity fuse	122
Pulling (crystal)	84
Pulling (pedestal)	153
Pulse Code Modulation	16, 115
Quality control	139
Quality control charts	104
Quantum theory	72
Radio Altimeter	117
Radio astronomy	107
Radio, amateurs	91
Radio broadcasting	75
Radio history	19
Radio paging	151
Radio shortwave communication	96
Radio (single sideband)	83

	Page		Page
Radio wave propagation	60, 79, 85	Mariner-2	168
Radar	93	Mercury-Atlas-4	164
Radar history	22	Mercury-Atlas-6	165
Radar (chirp technique)	131	NATO-1	192
Radiophonic sound and music	102	Nimbus-1	172
RADUGA satellite	203	OAO-1	179
Random access memory(1024 bit)	197	OFO-1	192
Random access memory(4096 bit)	202	Oreol-1	194
Random access memory(16, 384 bit)	203	Oscar-1	165
Random access memory 256 k bit	214	Oso-1	165
RAM Poly R process	203	OVI-2	177
Recording (Magnetic)	135	Pageos-1	181
Recording (sound-on-disc)	98	Pegasus-1	176
Recording (sound-on-film)	99	Pioneer-1	156
Recording (video tape)	154	Proton-1	179
Recrystallised silicon	217	Raduga-1	203
Regenerative circuit	80	Relay-1	166
Relativity theory	75	Salyut-1	194
Relays	53	Samos	199
Relay-1 satellite	166	Score	156
Reliability		Shinsei	192
		Skylab-1	199
(control)	83	Skynet-A	188
(DUANE rel. growth)	166	Snapshot	176
(sampling inspection tables)	127	Soyuz-1	181
(sequential analysis)	125	Sputnik-1	152
(standards)	81	Starlette	203
(quality control charts)	104	Surveyor-1	179
		Surveyor-7	185
Resistors		Syncom-1	169
		Tacsat-1	187
(carbon composition)	66	Telstar-1	165
(carbon film)	71	Tiros-1	162
(cracked carbon)	95	Transit-1B	162
(laser trimming)	181	Tung-Fang-Hung	190
(metal film)	81, 89, 98, 152	Vanguard-1	154
(oxide film)	102	Vela-1	169
(varistor)	154	Venus-1	164
Retarded field oscillator	87	Viking-1	201
Rheotome (waveform plotter)	60	Voskhod-1	172
Robotics	33	Vostok-1	164
Rotator circuit network	183	Westar-1	200
		Saw-tooth time base circuit	93
S.A.G.E. computer	143	SAWS - Surface acoustic wave	
Salyut satellite	194	devices	171
SAMOS satellite	199	Scanning circuit	110
Sampling inspection tables	127	Scanning acoustic microscope	198
Satellites		Scanning electron microscope	109
		Screened-grid valve	97
Alouette-1	168	Score satellite	156
Apollo-7	186	Secondary battery (Plante)	60
Ariel-1	168	Self-induction	52
ATS-1	180	Semiconductor diode capacitor	149
AZUR	187	Semiconductor laser	168
Biosatellite	180	Semiconductor memory system	190
Clarke concept	129	Sequential analysis	125
Courier-1B	162	Shadow mask tube	115
Dideme I	182	Silicon anodisation	202
Discoverer-1	158	Silicon-on-sapphire technology	172
DSCI-1	194	Single crystal fabrication	
Echo-1	162	(germanium)	134
ESSA-1	181	Single crystal fabrication(silicon)	143
Explorer-1	154	Single sideband communication	83
FR-1	177	Short wave (amateurs)	91
Gemini-1	173	Short waves (commercial)	96
GGSE-2	177	Shot effect noise	87
Intelsat-1	178	Skiatron	121
Iris (ESRO I)	184	Skylab satellite	199
Landsat-1	195	Skynet satellite	188
LES-1	177	Snapshot satellite	176
Lunar Orbiter I	180	Sodium conductor	74
Lunik-1	158		
Marisat-1	204		
Mars-1	168		

	Page
Solar battery	147
Solid electrolyte capacitor	148
SONAR	82
Sound reproduction	15
Sound-on-disc recording	98
Sound-on-film recording	99
Soyuz satellite	181
Spontaneous atomic change	74
Sputnik satellite	152
Sputtering process	58
"Squegger" circuit	91
Starlette satellite	203
Stepping tube	136
Stereo reproduction	103
STROWGER Telephone Exchange	68
Submarine cable insulation	60, 150
Superconductivity	78
Superconductivity (Highfield)	102
Superconducting switch	108
Superheterodyne circuit	80, 214
Superregenerative circuit	91
Surface acoustic wave devices	171
Surface barrier transistor	143
Surveyor satellite	179, 185
Switch ferreed	139
Switch (quick break)	66
Switch (Q.M.B.)	67
Switch superconducting	108
Syncom satellite	169
Tacsat satellite	187
Technetron (FET)	153
Tamed frequency modulation	209
Telemedicine	175, 36
Telephone	62
Telephone crossbar exchange	84
Telephone dial	69
Telephone relay exchange	79
Telephone electronic switching	159
Teletext	27
Television (theory)	77
Television (Baird)	96
Television History	27
Television (Farnsworth)	87
Television (large screen)	119
Television (Nipkow)	65
Television (Zworykin)	87
Television Colour	100
Telstar-1 satellite	165
Theory relativity	75
Thermo-compression bonding	138
Thermister	57
Thermoelectricity	51
Thermography	55
Thick film circuits	122
Thin film (direct-bonded copper)	202
Thin film (tantalum)	158
Three electrode valve	76
Thyratron	81
"Tinkertoy"	140
Tiros-1 satellite	162
Tomography	196
Tone control circuit	142
Transatlantic telegraph cable	60
Transatlantic telephone cable	150
Transferred electron device	163
Transferred electron effect	163
Transformer (distribution)	66
Transformer (Faraday)	52
Transformer (power)	66
Transistors (see also microelectronics)	
(crystal pulling technique)	84
(diffusion)	148
(epitaxy - vapour phase)	162
(epitaxy - liquid phase)	163
(field effect)	111
(floating zone melting)	147
(interdigitated)	147
(invention)	133
(ion implantation)	136
(modelling)	174
(MOSFET)	101, 166
(nitride-over-oxide)	181
(oxide masking)	153
(overlay)	174
(pedestal pulling)	153
(planar process)	157
(point contact)	133
(radio set)	147
(single crystal fab.)	134, 143
(surface barrier)	143
(thermo-compression bonding)	138
(U.J.T.)	143
(zone melting)	142
Transit 1 Satellite	162
Transitron oscillator	98
Transuranian atoms	107
TRAPATT diode	181
Travelling wave tube	108, 125
TRIMOS	205
Trigger tube	112
TRINITRON	186
Tubes (see also Valves)	
(cold cathode)	112
(glow discharge)	59
(Klystron)	117
(low pressure discharge)	59
(Multiplier)	111
(Plumbicon)	152
(shadow mask)	115
(stepping)	136
(trigger)	112
(tungar rectifier)	78
Tunnel diode	153
Two-layer resist technique	210
Unijunction transistor	143
Ultra-micrometer	90
Ultrasonics (SONAR)	82
Ultrasonic radar H_2S trainer	126
Ultra-violet radiation	50
Valves (see also Tubes)	
(Housekeeper seal)	88
(magnetron)	119
(microwave - retarded field)	87
(pentode)	99
(screened grid)	97
(two electrode)	94
(three electrode)	76
(vapour cooling)	149
Van de Graaf accelerator	101
Vanguard satellite	154
Varistor (field effect)	154
Vela satellite	169
Velodyne	124
Venus-1 Satellite	164
VIDEO Discs	198
Video games	195
Video tape recorder	154
VIDICON	137

	Page		Page
Viking satellite	201	Wimshurst machine	65
V-MOS	196	WINDOW	23
Vocoder	113	Wire drawing	178
Voltmeter (digital)	142	Wireless (Marconi)	70
Voskshod satellite	172	Wire wrapped joints	146
Vostok-1 satellite	164	Word Processor	36, 176
VT fuse	24		
		Xerography	113
Watch (electronic)	167, 193	X-rays	69
Waveform Plotter (Rheotome)	60	X-rays lithography	191
Waveguides (theory)	69	X-ray scanner	196
Waveguides	113		
Westar satellite	200	Yagi aerial	99
Weston standard cell	69	Yttrium iron garnet (YIG)	150
Wheatstone Bridge	55		
Wiegand wire	197	Zinc-mercuric-oxide battery	66
		Zone melting technique	142

*   *   *